BEHIND THE TUBE

A HISTORY OF BROADCASTING TECHNOLOGY AND BUSINESS

■■ ANDREW F. INGLIS

Ⓕ

Focal Press
Boston London

Focal Press is an imprint of Butterworth Publishers.

Library of Congress Cataloging-in-Publication Data

Inglis, Andrew F.
    Behind the tube: a history of broadcasting
technology and business/Andrew F. Inglis.
        p.   cm.
    Includes bibliographical references.
    ISBN 0-240-80043-5
    1. Broadcasting—History.   I. Title.
HE8689.4.I54   1990
384.54'09—dc20   89-38356
                                                          CIP

British Library Cataloguing in Publication Data

Inglis, Andrew F.
    Behind the tube: a history of broadcasting
technology and business.
    1. Broadcasting services, history
    I. Title
    384.54'09

    ISBN 0-240-80043-5

Butterworth Publishers
80 Montvale Avenue
Stoneham, MA 02180

10 9 8 7 6 5 4 3 2 1

Printed in the United States of America

■■ CONTENTS

List of Illustrations   ix

Preface   xv

Acknowledgments   xvii

1 THE ORIGINS OF RADIO AND TELEVISION TECHNOLOGY   1

Historical Background   1
Electricity and Electromagnetism   2
The Electron and the Electron Tube   9
Sound and Hearing   15
Mechanical Phonographs   18
Communication by Wire   28
Wireless Communications   32
The Founding of RCA and Its Aftermath   53
Summary   55

2 AM RADIO BROADCASTING   57

An Overview   57
Pioneering, 1920–1926   61
The Heyday of the Networks, 1927–1950   74
The Postnetwork Era, 1950–Present   100
A Look Ahead   111

3 FM RADIO BROADCASTING   113

Major Edward Howard Armstrong   113
The Basic Technologies of FM Broadcasting   117
The Early Development of FM   121
The FCC Acts   125
The World War II Hiatus   129

The Great Spectrum Battle   130
Technical Regulations   136
The Postwar FM Broadcasting Industry   140
Postwar Technical Developments   145
The RCA–Armstrong Patent Battle   150
A Look Ahead   154

4 MONOCHROME TELEVISION   155

The Basic Monochrome Television Technologies   158
Mechanical Television Systems   161
Early All-Electronic System Development   168
All-Electronic Television Broadcasting Begins   174
The Great Standards Battle   177
The Wartime Hiatus   185
The Beginning of Commercial Television   188
Television Broadcasting Burgeons   193
Allocation and Assignment Policies   195
The Growth of VHF Television   202
The Tribulations of UHF   204
Intercity Video Circuits   211
Monochrome Broadcasting Systems and Equipment   214
Monochrome Receivers   231
The Demise of Monochrome Television   236

5 COLOR TELEVISION   237

Color Television Is Conceived   237
The Field Sequential System   239
The Dot Sequential System   244
The Color Hearing   245
CBS Color Is Stillborn   266
The NTSC System   267
The Growth of Color Television   272
International Standards   275
From Systems to Products   277
The Tricolor Tube and Color Receivers   277
Color Cameras   283
Again the Japanese   294

6 BROADCAST VIDEO RECORDING   299

Recording on Film   300
The General's Birthday Present   304
The Ampex Breakthrough   310
The Ampex–RCA Competition   328
Quadruplex Recorders: Climax and Denouement   342
The Birth of Helical Scan   344
Helical Scan: Analog Recording   349
Helical Scan: Digital Recording   354
Helical Scan Format Summary   357
The Camcorder   358
Epilogue   358

7 CABLE TELEVISION   360

Cable Television Technology   361
The First Phase: Mom-and-Pop Community Systems   365
The Second Phase: Distant Station Importation   366
The Third Phase: Satellite Program Distribution   374
Cable Television Becomes Big Business   385
Extensions of Cable Service by Radio   386
Fiber Optics and Cable Television   388
The Future of Cable Television   389

8 SATELLITE PROGRAM DISTRIBUTION   392

Geosynchronous Satellites   392
The Technology of Communications Satellites   394
Spectrum Allocation and Use   404
The Orbital Arc and Orbital Spacing   406
The Regulation of Satellite Communications   407
International Satellite Communications Service   412
U.S. Domestic Satellites   416
Satellites and Cable Television   423
Direct Broadcast by Satellite Service   426
Satellites and Broadcasting   433
Summary   437

9 HOME VIDEO RECORDERS AND PLAYERS   439

Home Recording Mediums and Formats   440
The Initial Format Eliminations   440
Disks versus VCRs   444
Videodisks   447
Videocassette Recorders   455
Camcorders   470
A Look Ahead   472

10 THE NEW TECHNOLOGIES   473

High-Definition Television   473
Digital Systems   487
Solid-State Technology   491
Lasers   493
Fiber Optics   494
A Look Ahead   496

Glossary   497

Bibliography   507

Index   511

■■ LIST OF ILLUSTRATIONS

■ FIGURES

1–1   The Royal Institution of Great Britain.
1–2   An electromagnetic wave.
1–3   The electromagnetic spectrum.
1–4   The diode and triode.
1–5   Frequency ranges of sound sources and reproducing systems.
1–6   The evolution of the phonograph, as depicted by Victor in 1901.
1–7   His Master's Voice trade mark.
1–8   1901 Victor sales flier.
1–9   Victor Talking Machine Company phonograph sales, 1901–1929.
1–10   David Sarnoff and Marchese Guglielmo Marconi.
1–11   Lee De Forest and an early radiotelephone (c. 1910).
2–1   The growth of on-air AM radio stations, 1920–1988.
2–2   Radio time sales, 1935–1985.
2–3   Early radio receivers.
2–4   David Sarnoff.
2–5   Radio station coverage by station class, day and night.
2–6   Directional antenna utilization use.
2–7   (*Top*) A 1930 RCA transmitter. (*Bottom*) A 1981 RCA transmitter.
2–8   A radio station studio of the World War II era.
2–9   Examples of microphones.
2–10   Peter Goldmark.
2–11   A radio control console.
2–12   A reel-to-reel recorder.
3–1   Major Edwin Howard Armstrong.
3–2   Armstrong's Columbia University laboratory.
3–3   AM–FM share of the radio listening audience.
3–4   Number of commercial and public FM stations on the air, 1946–1986.

3–5 FM station revenues, 1965–1985.

3–6 Examples of FM transmitters.

3–7 (*Left*) A pylon antenna. (*Right*) A circularly polarized antenna.

3–8 CD playback system.

4–1 The scanning principle.

4–2 Nipkow disk television system.

4–3 Vladimir Zworykin and the iconoscope.

4–4 David Sarnoff demonstrating television at the 1939 World's Fair.

4–5 VHF stations on the air.

4–6 U.S. television homes.

4–7 VHF station revenues, 1946–1987.

4–8 VHF station profits, 1946–1987.

4–9 UHF stations on the air.

4–10 UHF station revenues and profits.

4–11 Television broadcast station system diagram.

4–12 The RCA TK-11 television camera.

4–13 The RCA TK-60 camera.

4–14 A television transmitter system.

4–15 The RCA TT-5A television transmitter.

4–16 The RCA 8D21 power amplifier tube.

4–17 A superturnstile antenna.

4–18 Multiple-antenna arrays on the Empire State Building (*top*) and in Baltimore (*bottom*).

4–19 The RCA TTU-30A UHF television transmitter.

4–20 Beam tilt and null fill.

4–21 (*Above*) A 1939 RCA monochrome receiver. (*Facing page*) The postwar Dumont monochrome receiver.

4–22 Industry sales and average unit prices of monochrome receivers.

5–1 Elmer Engstrom.

5–2 George H. Brown.

5–3 Color receiver sales.

5–4 The shadow mask principle.

5–5 An early RCA color receiver (*top*). A modern Sony color receiver (*bottom*).

5–6 The evolution of color cameras, 1953–1970.

5–7 The RCA TK-76 ENG camera.

5–8 A modern studio camera made by Sony.

6–1 Alexander M. Poniatoff.

6–2 The Ampex recorder development team.

6–3 Photographs of early videotape recordings.

6–4 The Ampex VR-1000 videotape recorder.

6–5 A photograph of a VR-1000 recording.

6–6 Ampex Corporation sales (A) and profits (B), 1956– 1960.

6–7 The first working model of the RCA quadruplex recorder.

6–8 The TRT-1A television tape recorder.

6–9 The RCA TR-22 videotape recorder.

6–10 The sales (A) and profit (B) saga of the Ampex Corporation, 1956–1980.

6–11 A Sony Type C helical scan recorder.

6–12 A Sony broadcast camcorder.

7–1 Typical cable television system layout.

7–2 Typical cable television head-end antenna system.

7–3 Cable television subscribers during the mom-and-pop phase, 1952–1957.

7–4 Martin F. Malarkey, Jr.

7–5 Irving B. Kahn.

7–6 Cable television subscribers during the distant signal importation phase, 1957–1977.

7–7 Ceremony (*top*) and plaque (*bottom*) marking the fifth anniversary of the first cable television satellite transmission.

7–8 Gerald M. Levin.

7–9 Ted Turner.

7–10 Cable television subscribers during the satellite distribution phase, 1977–1988.

8–1 A satellite in geosynchronous orbit.

8–2 The Delta 3914 rocket.

8–3 A typical Delta rocket launch sequence.

8–4 The space shuttle shortly after lift-off.

8–5 The principal components of an RCA communications satellite.

8–6 A satellite signal footprint.

8–7 Channel allocations in a C-band satellite.

8–8 The prime orbital arcs.

8–9 The orbital locations of communications satellites in 1988.

9–1 VCR videodisk unit sales.

9–2 Akio Morita.

9–3   A half-wrap tape configuration.
9–4   Beta's market share, 1976–1987.
10–1   The aperture response of film and television systems.
10–2   Digitizing an analog waveform.
10–3   Single-mode and multimode fiber.

■ FEATURES

Chapter 1

Maxwell's Equations
The Decibel Unit of Measurement
Radio Wave Propagation
The Evolution of Radio Frequency Transmitters
Amplitude Modulation
Pre–Electron Tube Transmitter Modulators
Regenerative and Superheterodyne Receiver Circuits

Chapter 2

Overlapping Sidebands from Adjacent Channels
Station Service Areas
AM Transmitter Modulation Methods
The Principle of Magnetic Tape Recording
Electrical Bias for Magnetic Recording

Chapter 3

FM Waveform and Spectrum
Stereo Transmission Spectrum

Chapter 4

Bandwidth Requirements and Interlaced Scanning
The Iconoscope Tube
The Video Waveform
Television Broadcasting Channel
The Image Orthicon

Operation of the Vidicon Tube
Antenna Gain
Operation of the Klystron Tube

Chapter 5

The Primary Colors
The Field Sequential Color System
The Dot Sequential Color System
Burst and Color Subcarrier
Operation of a Charge-Coupled Device

Chapter 6

The RCA Video Recording System
The Three Critical Quadruplex Recording Technologies
Distribution of Signal and Sideband Energy
Helical Scan Recording
Tape-Wrap Configurations
Analog and Digital Recorder Performance

Chapter 9

The RCA Videodisk Recording System
VCR Technical Parameters

■ ■ PREFACE

The *front* of the television tube, displaying its vast cornucopia of program material, is viewed by almost everyone in the civilized world. It has something for everybody. For mass audiences there are news, entertainment, and sporting events. For more specialized audiences there are programs intended to inspire, educate, or persuade. One is not guilty of hyperbole in comparing the recent and future impact of television as a communications medium with that of the printing press.

The performers and public figures who appear on the front of the tube are well known to viewers, and many are household names. By contrast, the names and achievements of the scientists and engineers responsible for the development of the complex electronic systems *behind the tube,* which make television and its older sibling, radio broadcasting, possible, are largely unknown outside of a small circle within their profession. As this book's title suggests, the history of these achievements and a description of the current status of radio and television technology form its subject matter.

This history is a rich mixture of astounding technical progress, strong-willed and colorful individuals, big and little business, violent controversies, and rapid economic growth. Leading scientists and engineers have exceptional minds, but they are subject to the same motives and emotions as the rest of humankind—ambition, desire for achievement, desire for recognition, desire for financial reward, pride, and jealousy. They have often been at odds, not only with each other but also with the organizations that finance and regulate their efforts. The chronicle recorded in this book is replete with these conflicts, and it is an exciting adventure story.

The development of radio and television technology cannot be adequately presented or understood as an isolated phenomenon since it occurred in the larger context of the business and regulatory environment. Technical progress has been inextricably intertwined with business growth and governmental regulation. Each depends on the others in a major way. The narrative includes, therefore, a history of the major trends in the business and regulation of both radio and television.

The subject matter of this volume will be intensely interesting and professionally rewarding for those established in or preparing for a career in television broadcasting or one of its related industries. It will also be of interest to lay persons who want to learn more about modern technology. Most of all, it is my wish that this book will convey to its readers some of the excitement that was experienced by the participants during the growth of these industries.

■ ■ ACKNOWLEDGMENTS

My first acknowledgment is to Henry Kissinger (whom I have never met). In the Foreword to the first volume of his monumental memoirs, *White House Years* (Little, Brown, 1979), he stated a rationale for histories written by participants, a practice not universally admired by professional historians. He conceded that a participant's view of history may be less than totally objective: "Obviously, his perspective will be affected by his own involvement; the impulse to explain merges with the impulse to defend." He pointed out, however, that a participant can make a unique contribution to the writing of history; more than any nonparticipant, he *knows* what really happened.

I was a participant in many of the events described in this volume, and I have endeavored to enrich the historical account of radio and television technology with my own experiences, while retaining a reasonable level of objectivity. The reader will judge my success in achieving this aim.

I must also acknowledge my debt to the institutions where I received my formal education and to the individuals who were instrumental in guiding me into a broadcasting career. I had planned to be a physicist, and I received an excellent undergraduate education at Haverford College. This was followed by a year of graduate work at the University of Chicago, which had one of the world's most distinguished physics faculties. Although I had not anticipated it, my education in physics was the best possible foundation for a career in broadcasting's rapidly evolving technology.

World War II ended my formal education, and I spent the war years in the U.S. Naval Reserve as an instructor at a radar training school at Bowdoin College. It was here that my interest in a broadcasting career was aroused, particularly by the school's commanding officer, Dr. Noel C. Little, chairman of the Bowdoin physics department in peacetime, and a fellow instructor, Martin V. Kiebert.

After the war, I was introduced to broadcasting as an engineering consultant by Frank H. McIntosh, my partner and the inventor of the McIntosh high-fidelity audio amplifier. After eight years with McIntosh, I joined RCA where I was involved in its broadcast, television, and satellite activities for thirty years. My experience with McIntosh and RCA was a never-ending education, and I am grateful to the many individuals who contributed to it. I was indeed fortunate to have had the opportunity to participate in a major way in such an exciting industry.

This brings me to the acknowledgment of the contributions of the many individuals who have helped immeasurably with the writing of this volume.

Albert Warren, president of Warren Publishing, Inc., and publisher of *Television Digest*, and several members of his staff reviewed the entire manuscript and made invaluable suggestions and corrections based on their decades of experience in the industry.

David Dietz, manager of the Book Division of *Broadcasting-Cablecasting*, was of great assistance in my negotiations with prospective publishers and was the catalyst for my relationship with Focal Press.

Many of my former colleagues from RCA offered information and advice. Three of them now hold top positions in Sony—Neil Vander Dussen, president and chief operating officer of the Sony Corporation of America, and Peter Dare and Laurence Thorpe who hold senior executive positions in the Sony Broadcast Products Company. They supplied a wealth of information on broadcast and home video recorders, HDTV, CCDs, and camcorders, and they reviewed drafts of Chapter 6 (Broadcast Video Recording), Chapter 9 (Home Video Recorders and Players), and Chapter 10 (The New Technologies).

Dr. James J. Tietjen, my successor as president of RCA Americom and now president and chief operating officer of the David Sarnoff Laboratories, together with Dr. Kerns Powers (now retired) and Jack S. Fuhrer, director of television research, provided copious information on high definition television (HDTV) systems. Dr. Powers also reviewed Chapter 10.

Walter Braun, now senior vice president, government and technical operations, for GE Americom (formerly RCA Americom), brought me up to date on developments in satellite communications since my retirement, and reviewed Chapter 8 (Satellite Program Distribution).

Finally, I must thank many retired RCA colleagues for advice and assistance. These include Sidney Bendell, Kenneth Bilby, Dr. George Brown, J.E. Hill, Loren Jones, A.H. Lind, Dana Pratt, Walter Varnum, and Ray Warren.

One of the most pleasant aspects of writing this history has been renewing acquaintances with industry figures and making new ones. They have been very generous with their time by providing background information and reviewing parts of the manuscript for emphasis and factual accuracy.

Gerald M. Levin, a pioneer in the use of satellites for the distribution of pay-TV serivces to cable systems, the founder of HBO, and now vice chairman of Time Inc., reviewed Chapter 7 (Cable Television) and made a number of important suggestions.

Irving B. Kahn, the founder of TelePrompTer Corp. and a pioneer multiple-system cable TV operator, also reviewed Chapter 7 and provided a wealth of historical information concerning cable TV and fiber optic systems.

Donald E. Tykeson, a pioneer broadcaster and cable TV operator and now owner of Bend [Oregon] Cable Communications, provided important insights into the telephone company–cable TV controversy. Mary Chase, general manager of the Bend cable system, provided photographs of its head-end for use in Chapter 7.

Martin F. Malarkey, cable TV pioneer and founder of the National Cable Television Association (NCTA), and the inimitable Ted Turner provided photographs and biographical information.

Joseph A. Flaherty, Jr., vice president and general manager, engineering and development, for the CBS Operations and Engineering Division, supplied copious background material concerning worldwide efforts to establish standards for HDTV.

Charles P. Ginsburg, now retired from the Ampex Corporation and who led the famed team that developed the first practical video tape recorder and revolutionized the broadcast industry, provided detailed background information concerning the dramatic events that occurred during the execution of this epic engineering program.

Elizabeth Dahlberg, my colleague in the consulting business and now a member of the consulting firm of Lohnes & Culver in Washington, D.C., provided a critical review of Chapter 2 (AM Radio Broadcasting) and Chapter 3 (FM Radio Broadcasting), particularly with regard to Federal Communications Commission technical regulations during the past three decades.

Paul Smith of the Motorola Corporation provided background and technical information concerning the Motorola system for AM stereo.

I am especially grateful to Mrs. Everett Dillard for the gift of her late husband's extensive collection of pre-World War II radio engineering textbooks. They give a unique insight into early radio technology, and I drew on them extensively for Chapter 1 (The Origins of Radio and Television Technology) and Chapter 2.

The staffs of the two major industry trade associations, the National Association of Broadcasters (NAB) and the National Cable Television Association (NCTA) were very cooperative. Mark Fratrik of the NAB supplied important industry revenue and profit statistics not available from other sources, while Jodi I. Hooper of the NCTA and Tonia Ballard of the NAB supplied historical information concerning their respective organizations.

I am especially grateful to the members of the academic community who reviewed all or parts of the manuscript. They have given me invaluable insights, pertaining to both content and factual accuracy. They include professors J. Robert Craig of Central Michigan College, Robert Hilliard of Emerson College, Donald R. Mott of Butler University, and David Pine of Haverford College.

The photographs and artwork are an essential part of this volume, and a number of individuals have contributed to collecting and producing them.

Frieda Schubert and Lynn Detweiler supplied many of the photographs from RCA's archives.

James Gimbel graciously gave me his back issues of *Broadcast News,* an RCA house publication that was the source of a number of photographs as well as historical information on the development of broadcasting equipment.

Mrs. I.M. McCabe, information officer for the British Institution, provided historical information concerning that unique institution which played such an important role in the development of radio's scientific foundation.

Elliot Sivowitsch of the Division of Electricity and Modern Physics of the Smithsonian Institute was most helpful in locating photographs of historical events and individuals from the Institute's vast files.

The excellent artwork was prepared by Thom Porterfield of Porterfield Design in Bend, Oregon.

Last, but by no means least, I want to thank the Focal Press editor, Philip Sutherland, who has patiently given me the benefit of his professional skill and knowledge in full measure.

Finally, I want to add a disclaimer on behalf of all of those who gave me assistance. Any remaining errors are mine, not theirs!

1

■ ■ THE ORIGINS OF RADIO AND TELEVISION TECHNOLOGY

■ HISTORICAL BACKGROUND

Radio and television technology had its origins in an extraordinary blossoming of research, engineering, and invention during the nineteenth and early twentieth centuries. The convergence of these technical achievements with the vision of farsighted entrepreneurs in the 1920s led to the initiation of the vast systems of radio and television that have developed in the United States and throughout the world.

This technology began with the scientists. They established the physical laws describing the nature and behavior of electricity, electromagnetism, and sound, and they discovered the electron.

In the meantime, inventors and engineers were discovering a useful and exciting application for electricity in long-distance point-to-point communications by wire. For the first time in recorded history, people could communicate nearly instantaneously over long distances. Wire communications were initially accomplished by *telegraphy*, the transmission of coded signals. Later, *telephony*, the transmission of the human voice by wire, became possible as the result of major technical breakthroughs.

An even more exciting development was the use of the newly discovered electromagnetic waves for long-distance *wireless* communications. As with wire communications, wireless began with *radiotelegraphy*, which then evolved into *radiotelephony*. Radiotelephony, a point-to-point service, then evolved into point-to-multipoint broadcasting. All these evolutionary steps in the prehistory of broadcasting required the solution of difficult technical, regulatory, and business problems.

Research, invention, and engineering were carried out in an atmosphere of optimism and intense excitement. Scientists were confident that they were on the verge of unlocking the ultimate secrets of the physical universe. Engineers and inventors were undeterred by any doubts that technical progress was anything but an unalloyed benefit. Some of these views may seem naive today, but the technical leaders of the time shared the conviction that they were advancing both the knowledge and the well-being of humankind.

■ ELECTRICITY AND ELECTROMAGNETISM

The discovery of the nature of electricity and electromagnetism by nineteenth-century European scientists was the bedrock of broadcasting technology. Their names are immortalized in our nomenclature for electrical and magnetic units:

Unit	Scientist	Nationality
Volt (electromotive force)	Volta (1745–1827)	Italian
Ampere (electric current)	Ampère (1775–1836)	French
Coulomb (electric charge)	Coulomb (1736–1806)	French
Ohm (electrical resistance)	Ohm (1787–1854)	German
Watt (electric power)	Watt (1736–1819)	Scottish
Farad (electrical capacitance)	Faraday (1791–1867)	English
Maxwell (magnetic flux)	Maxwell (1831–1879)	Scottish
Oersted (magnetic flux intensity)	Oersted (1777–1851)	Danish
Gauss (magnetic flux density)	Gauss (1777–1855)	German
Henry (inductance)	Henry (1797–1878)	American
Hertz (frequency)	Hertz (1857–1894)	German

There was an irony in the fruits of the research of these scientists. Reflecting the European academic culture of their time, most of them had little interest in the practical consequences of their discoveries. In fact, there was a feeling that such an interest was unscholarly and even ungentlemanly. It was expressed in a humorous toast of the Cambridge University mathematics department: "Here's to mathematics; may it never be of any use to anybody." Their motive, rather, was to expand human knowledge of the physical universe—to acquire knowledge for its own sake.

The results of their labors were totally different from their goals and quite unexpected. As science, their discoveries were later shown to be but special cases of much broader scientific truths. But their practical consequences were enormous.

Static Electricity

The ancient Greeks knew of the existence of an electric charge, or static electricity, which is produced on the surface of amber and some other non-conducting materials when they are rubbed. The charge causes the material to attract small scraps of paper or other light objects. Queen Elizabeth's physician, William Gilbert, studied the phenomenon and named it *electrica*, the Latin translation of the Greek word for amber. This was later translated to the English word *electricity*.

Serious scientific studies of static electricity began in the eighteenth century. They included the famous experiments of Benjamin Franklin, among which was his demonstration that lightning was an electrical phenomenon. He showed this by flying a kite with a key attached to it during a thunderstorm. (He was lucky he was not killed.) Franklin was a leading proponent of the "one fluid" theory of electricity—that is, that electric charges all have the same sign—as opposed to the "two fluid" theory, which postulates positive and negative charges. His papers on this subject brought him international renown.

Important contributions to the knowledge of static electricity also were made by Aepinus, Cavendish, Volta, Coulomb, Poisson, and Faraday. By the middle of the nineteenth century, the science of *electrostatics* was well developed.

Current Electricity

Static electricity is an interesting natural phenomenon, but its practical importance is comparatively minor. It was the discovery of current electricity, the motion of electric charges along a conductor, that led to the development of the electrical, electronics, and broadcasting industries.

Current electricity was first discovered and studied by the Italian scientists Luigi Galvani (1737–1798) and Alessandro Volta. Galvani was a physiologist, not a physicist, and his work was an early example of interdisciplinary research. In 1780, while studying the nervous system of frogs, he noted that a frog's legs contracted suddenly when its nervous system was subjected to a spark from an electrostatic generator. He attributed this to the flow of electricity from the muscles to the nerves. (Galvani's name also lives on in our language. A *galvanic* pile is a type of battery, and to *galvanize* can mean either to coat a metal by electrolysis or to stimulate to action.)

Volta investigated this phenomenon further and concluded that the role of the nerves was simply that of an electrical conductor. He did extensive research in the flow of electric charges and in the process developed the voltaic cell—a battery consisting of alternate layers of copper, zinc, and paper moistened with a salt solution—which could generate a continuous flow of electric current. He published the results of these experiments in 1800.

Volta's paper stimulated a burst of activity by European physicists who recognized the significance of his research. One of the most important investigations was carried out by George Ohm. He published a paper in 1826 suggesting that the flow of electricity in a conductor is analogous to the transfer of heat along a rod that is heated at one end; the rate of heat transfer is then proportional to the difference in temperature between the two ends. This led to the formulation of *Ohm's law:* "The magnitude of the current flow in a conducting path is equal to the voltage between its two ends divided by its electrical resistance."

Many other contemporary scientists engaged in research that contributed to our understanding of electric currents. All of this was a necessary prelude to the discovery of *electromagnetism* and *electromagnetic waves*, the most basic technical ingredients of broadcasting.

Magnetism and Electromagnetism

The relationship between electricity and magnetism was first noted by the Danish physicist, Hans Christian Oersted, who showed that a magnetic compass needle could be deflected by placing it in the vicinity of a current-carrying conductor. This relationship was studied further by André-Marie Ampère. But the most definitive research in this subject was carried out by the English physicist and chemist Michael Faraday.

Michael Faraday and Electromagnetism Michael Faraday was unique among European scientists of his time in that his family did not have money, social position, or academic credentials. His father, in fact, was a blacksmith. His formal education was limited, and it came to an end at the age of fourteen when he was apprenticed to a bookbinder. He developed an intense interest in science by reading books on the subject that came through the bindery. His interest became so great that he screwed up his courage at the age of twenty-one and wrote a letter to Sir Humphry Davy, a distinguished scientist in the Royal Institution of Great Britain (Figure 1–1), asking for a position in which he could begin a career in science. In spite of Faraday's lack of formal education, Davy granted him an interview and was so impressed that he offered him a position as laboratory assistant.

Faraday's appointment was dated March 1, 1813, and he continued his association with the Institution for more than fifty years until his retirement in 1865, two years before his death. He occupied a variety of posts at the Institution and was appointed director of its laboratory in 1825. He was active outside the Institution as well, serving as consultant and lecturer at the Royal Military Academy. He was held in the highest regard by the contemporary scientific community, and he was the recipient of numerous honors. He was living proof that an exceptionally talented, hardworking individual—helped perhaps by a bit of luck—could rise from humble beginnings to the top, even in England's class-conscious society.

Shortly after his appointment, Faraday accompanied Davy on an eighteen-month tour of Europe. He met nearly all of Europe's most distinguished chemists and physicists, discussed their work with them, and established lasting friendships. He stated that this trip was his university.

After his return to London, he began his career in research. He started with experiments in chemistry but soon switched to physics and the study of electricity. During his lifetime, he made numerous major contributions to scientific knowledge in both chemistry and physics, but from the

Courtesy The Royal Institution of Great Britain, with permission.

■ **Figure 1–1**   The Royal Institution of Great Britain. This institution is uniquely British. It was founded in 1799 by Count Rumford and is supported entirely by private funds—donations, bequests, and membership dues. It was granted a Royal Charter in 1800 but otherwise has no official relationship to the government. It is not affiliated with any university, but members of its staff have maintained close relationships with university faculties. Its purpose is to further the advance of science, both by sponsoring the research of distinguished scientists and by disseminating scientific information to the public through lectures and papers.

Its staff has included many extraordinarily gifted scientists and engineers. Michael Faraday was probably the most famous, but many other scientists and engineers of distinction have served there through the years. Among them was Lord Rayleigh, who carried out basic research in the nature of sound. Still others have used its facilities to announce their discoveries, notably Sir J.J. Thomson for his discovery of the electron.

The building in which it is housed, The House, is on the original site on Albemarle Street in London. It has been repeatedly enlarged and modified. The present front facade shown in the photograph dates from 1838. [Description of building courtesy A.D.R. Caroe, *The House of the Royal Institution* (London: The Royal Institution of Great Britain, 1963).]

standpoint of radio and television technology, the most important was the expansion of our understanding of the relationship between electricity and magnetism.

He found that a force was exerted on a current-carrying wire when placed in a magnetic field. This led him to construct the first *electric motor*. Later, he wrapped two coils of wire around a doughnut-shaped iron ring and found that when current was started or stopped in one of the coils, current was induced in the other. This was the first *transformer*. Finally, he determined that an electromotive force (voltage) is induced in a conductor moving through a magnetic field. This permitted him to construct the first *dynamo*.

Perhaps his most significant discovery was the rotation of the plane of polarization of polarized light when passing through a magnetic field, a phenomenon now known as the *Faraday effect*. This suggested that light was electromagnetic in character. It also led to intuitive speculations that magnetic and electric disturbances could be transmitted through space. These speculations were published in an article titled "Thoughts on Ray Vibrations" in the *Philosophical Magazine*. This article aroused some criticism, since his views had no firm scientific basis. But after Faraday's death, the Scottish physicist James Clerk Maxwell subjected these speculations to mathematical rigor and confirmed, in theory, the existence of *electromagnetic radiation*. Still later, this theory was confirmed experimentally by the German physicist Heinrich Hertz.

James Clerk Maxwell and Electromagnetic Radiation   Unlike Faraday, Maxwell came from distinguished parents, and he had the benefit of a fine education. He graduated from the University of Edinburgh and continued his studies at Cambridge University. He held professorships at the University of Aberdeen and King's College in London before being called to become a professor of experimental physics at Cambridge in 1871. There he supervised the construction of the famous Cavendish Laboratories.

Throughout his professional career, he exhibited an extraordinary talent for mathematics that bordered on genius. He made contributions to scientific knowledge in a number of fields, including astronomy and thermodynamics, but he is best known for the development of *Maxwell's equations*, which describe electromagnetic radiation in mathematical terms as a wave motion (see feature on next page and Figure 1–2).

The development of these equations with no direct experimental evidence that the phenomenon they described actually existed was a truly remarkable feat of mathematical deduction. Their underlying theory was first published in 1873, and it was immediately recognized as a landmark contribution to physical knowledge.

Maxwell died in 1879 at the age of forty-eight. He did not live to see the existence of electromagnetic radiation as described by his equations verified experimentally by Hertz in 1887.

The Experiments of Hertz   Heinrich Hertz, a distinguished German physicist, is best known for his experimental confirmation of Maxwell's equations. He accomplished this in 1887 while a professor at Karlsruhe

MAXWELL'S EQUATIONS

$$\nabla \cdot D = 0$$
$$\nabla \cdot B = 0$$
$$\nabla \times H = \frac{\partial D}{\partial t}$$
$$\nabla \times E = \frac{\partial B}{\partial t}$$

where: D = electric flux density
B = magnetic flux density
H = magnetic field
E = electric field

These equations describe the fundamental wave properties of electromagnetic radiation in free space—radio waves, infrared light, visible light, ultraviolet light, and X rays—in the arcane language of vector analysis.

The velocity of the waves, about 186,000 miles/second, or 3×10^8 meters/second, is a fundamental physical constant of our universe. Einstein's relativity theory states that its measured value is independent of the relative motion of the source and the observer. For example, an astronaut moving toward the moon at a very high rate of speed would measure the same velocity for the moonbeams as another astronaut moving away from it. This fact is not intuitively obvious, to say the least.

Another difficult intuitive concept is that electromagnetic radiation exhibits the properties of particles or small bundles of energy as well as of waves. In the ranges of wavelengths used for broadcasting, the energy of each bundle is so small that they are not individually detectable. But they are readily measurable at the shorter wavelengths of visible light and X rays.

Polytechnic. He found that waves emitted by the sparking of an induction coil could be detected at a distance. (Today these emissions would be called static.)

His experiments went far beyond the mere detection of these waves. He showed that they had many of the same properties as light—measurable velocity and wavelength, refraction, reflection, and polarization—and thus confirmed the growing realization that visible light was a form of electromagnetic radiation.

The Electromagnetic Spectrum During the twentieth century, our knowledge and usage of the electromagnetic spectrum (Figure 1–3), the array of electromagnetic radiations arranged in the order of their wavelength or frequency, has expanded enormously. The location of radiant

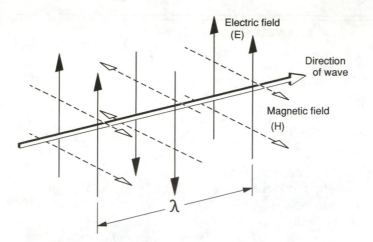

■ **Figure 1–2** An electromagnetic wave. This drawing illustrates the features of an electromagnetic wave traveling through space. It consists of an alternating electric field (that is, a voltage difference in space) and an alternating magnetic field, which are at right angles to each other and to the direction of travel. The direction of these fields reverses every half wavelength so that an observer passed by the wave will see the field direction reverse at twice the wave's frequency. The wavelength, λ, the frequency, f, and the velocity, c, are related by the equation $c = f\lambda = 3 \times 10^8$ meters/sec.

By convention, the polarization of the wave is denoted by the direction of the electric field. AM broadcasting uses vertical polarization. In the United States, FM and television broadcasting use horizontal polarization.

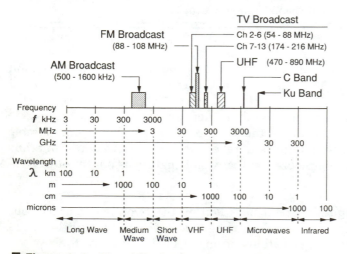

■ **Figure 1–3** The electromagnetic spectrum.

energy in the spectrum can be indicated either by its frequency or its wavelength. It is now customary to use frequency in the portion of the spectrum used for radio and wavelength for infrared, visible light, ultraviolet, and X rays.

The portions of the spectrum used for broadcasting are medium wave for AM radio, very high frequency (VHF) for FM radio and television, ultrahigh frequency (UHF) for television, microwave for intercity terrestrial and satellite network connections, and infrared on fiber-optic cables. The characteristics of electromagnetic radiation at different frequencies vary enormously, and these differences have had a profound effect on the development of the broadcast services.

■ THE ELECTRON AND THE ELECTRON TUBE

Discovering and Measuring the Electron

The scientific investigations described in the previous section were directed at determining the behavior of electricity. They were paralleled by equally active efforts aimed at ascertaining its basic nature. They involved finding answers to two questions: Is electricity an infinitely divisible fluid, or is it discrete individual particles? If the latter, what is the nature of the particles?

The particle view steadily gained support during the latter half of the nineteenth century. This view was reinforced by simultaneous investigations of the nature of the atom, but final confirmation of it required the discovery and measurement of the elemental particle.

The process of discovery and measurement began with experiments in the passage of electricity through a partial vacuum in a *cathode ray tube* (CRT), a glass bulb containing an electrode charged to a high negative voltage (the cathode). During the 1860s, two physicists, Heinrich Geissler in Germany and Sir William Crookes in England, noted strange effects as the gas was progressively removed from the bulb. At first it was filled with a pinkish glow. As the pressure was reduced further, a dark space (now known as the Crookes space) appeared in front of the cathode. With further reduction, a bluish beam extended from the cathode. This beam was called a *cathode ray* and the CRT is sometimes called a *Crookes tube*.

Crookes had many other scientific interests, and neither he nor Geissler followed up on these observations to the point of determining the nature of the blue ray. This was left to the famed British scientist Sir J.J. Thomson (1856–1940).

Sir J.J. Thomson Discovers the Electron   Thomson was born in Manchester to a middle-class family. His father was a publisher and bookseller.

His schooling began at Owens College in Manchester, but at the age of twenty, he transferred to Cambridge, where he remained for the rest of his life.

He joined the Cambridge faculty and in 1884 succeeded Lord Rayleigh (see the section on sound later in this chapter) as the Cavendish Professor of Physics, one of the most prestigious posts in the European scientific community. In 1918, he became master of Trinity College (part of Cambridge University).

He engaged in research in many areas of electrical science, but his fame is principally based on his confirming the existence of the electron and measuring the ratio of its electrical charge, e, to its mass, m, or e/m. This research was accomplished in the 1890s, and Thomson first announced the results publicly at a lecture in the Royal Institution in 1897.

Thomson was noted not only for his research but also for his skills as an author and lecturer. He wrote a number of distinguished scientific works, the best known of which is *The Conduction of Electricity through Gases* published in 1903. He frequently lectured in the United States, and he was highly respected by American scientists. He was knighted in recognition of his extraordinary accomplishments, and in his later years he enjoyed great eminence and affection, both for his scientific achievements and his personal characteristics.

Thomson's experiments showed that electric charges are composed of discrete, negatively charged particles, and he developed a method for calculating their e/m by measuring the magnitude of the magnetic field required to exactly offset the deflection of a beam of electrons by an electric field. He also determined that the particles were much lighter than hydrogen atoms, but he did not accurately measure either their mass or their charge. This was accomplished later by Robert Millikan (1868–1953) in his oil drop experiment.

Millikan Measures the Charge of the Electron   The magnitude of the electron's electric charge was measured precisely by Robert Millikan of the University of Chicago in 1910. He used an imaginative experimental technique involving the measurement of the rate of fall of small oil droplets suspended in an ionized atmosphere between two electrically charged plates. The droplets occasionally lost or gained a charge by contact with the atmosphere, and the amount of the change could be calculated by determining the change in the rate of fall. Millikan noted that the change in the charge was always an integral multiple of the same number, which he took to be the charge of the electron. By averaging the results of thousands of measurements, Millikan calculated a charge of 16.019×10^{-20} coulombs. This figure was verified by later measurements using different techniques.

Shortly after the end of World War I, Millikan left Chicago to become director of the Norman Bridge Laboratory of Physics at the California Institute of Technology. He received a Nobel Prize in 1923 for measuring the charge on the electron.

Inventing the Electron Tube

With the existence of electricity, electromagnetism, and the electron confirmed and the physical laws governing their behavior established, the basic scientific facts needed for broadcasting technology were in place. It was now necessary for the engineers and inventors to take over. They were most eager to do so, and, in fact, some of them jumped the gun and made inventions before the underlying scientific principles had been established. The invention of the electron tube, perhaps the most important single component of broadcasting products and equipment, was a good example.

The simplest types of electron tube are the *diode* and the *triode* (Figure 1–4). The electron source in both is a heated surface on the tube's negative electrode or cathode.

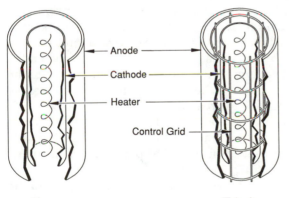

Diode   Triode

■ **Figure 1–4** The diode and triode. The diode has two elements—a cathode, which is coated with a material that emits electrons when heated, and an anode (or plate), which collects the electrons when it is operated at a positive voltage, thus causing current to flow through the tube. (Unlike the cold cathode in the Crookes tube, which requires a high anode voltage for electron emission, the hot cathode of the diode can operate at low voltages.) Since electrons are negatively charged, no current will flow when the anode voltage is negative.

The triode is a diode with a third element, the control grid, added. This is a screenlike structure, mounted between the cathode and anode, through which the electron current can flow. By varying the voltage of the grid, the flow of current through the tube can be controlled: the more positive the grid, the higher the current.

Edison and the Edison Effect   The emission of electrons from a heated surface such as a tube's cathode was probably first observed by the American inventor Thomas Alva Edison (1847–1931) in 1883. Just as Faraday and Maxwell epitomized the nineteenth-century scientists whose interest was the discovery of natural laws, Edison epitomized the Yankee inventors whose interests were the practical application of their discoveries. Edison was somewhat contemptuous of scientists (scientists did not think too highly of him either), whom he said were interested in studying subjects such as the "fuzz on a bee."[1] This was unfortunate because his productivity as an inventor could have been even higher had he had more respect for and knowledge of theory. As it was, he relied largely on intuition combined with meticulous and systematic experimentation.

Fortunately, Edison lived in a time when technology was sufficiently simple that his approach to technical problems could be effective. The output of his laboratory was phenomenal. In addition to his discovery of the emission of electrons from a heated cathode, referred to as the *Edison effect*, he invented the first phonograph and the light bulb, and he made many contributions to the development of electric power generating and distribution equipment. He was granted 1,033 patents during his lifetime.

In his experiments with carbon filament lamps, he noted that carbon was deposited on the inner surface of the bulb when the filament was operated at a high voltage. He suspected that some unknown electrical force was the cause, and he confirmed this by connecting a small metal plate suspended inside the bulb to an external battery. When the plate was connected to the positive pole of the battery, he observed the flow of current with a sensitive current-neasuring instrument. He had created the first diode.

The electron had not yet been discovered, and Edison was not particularly interested in the scientific significance of his discovery. Seeing little practical use for it, he filed a patent application but pursued it no further.

Fleming and the Diode   The discovery of a practical application of the diode was made by the English engineer Sir John Ambrose Fleming (1849–1945) in 1904. He was a university professor who had been retained as a consultant to the Marconi Wireless Telegraph Company to improve its transmitting and receiving apparatus. One of the problems was to develop a more sensitive *detector* for radio signals—a device called a *rectifier*, which converts the alternating current produced by electromagnetic waves to direct current. Fleming had read of Edison's work, and he recognized that the diode, which can pass current in only one direction, might meet this requirement.

Fleming built a number of diodes, and they were modestly successful as detectors of radio signals, although they did not provide an order-of-

---

[1] Wyn Wachhurst, *Thomas Alva Edison: An American Myth.* (MIT Press, 1981), 35.

magnitude improvement over the crystal detectors that had been used previously. Nevertheless, Fleming received recognition for his work when the device was named the *Fleming valve*. (This term for a diode has become obsolete, but the term *valve* for the electron tube continues to be used in England.)

The utility of the diode was limited by its inability to amplify electrical signals. This limitation was removed with the invention of the triode, a diode with a grid structure added. By applying a low-power input signal to the grid, an output signal of much greater power could be produced at the anode. The electron tube could now be used as an amplifier, a key function in communications and broadcasting systems. The inventor of the triode was the American engineer Lee De Forest (1873–1961).

De Forest and the Triode   The Nobel Prize physicist Isidor Rabi characterized Lee De Forest's invention of the triode in 1906 as almost one of the greatest inventions of all time. Since the electron tube was the fundamental building block of the electronics industry, this assessment is a fair one. Had De Forest limited his activities to invention, his reputation would have been unblemished. But he also tried to be a businessman, and his reputation was stained by bankruptcies and charges of fraud.

De Forest was born in Council Bluffs, Iowa, the son of a a Congregational minister. When he was six, the family moved to Talladega, Alabama, where his father became president of Talladega College, a college for blacks. The local whites were not sympathetic to the purposes of the college, and De Forest's family suffered from social ostracism. Nevertheless, they were able to send him to an excellent university, the Sheffield Technical School of Yale University, for both undergraduate and graduate study. He was awarded his Ph.D. there in 1899.

De Forest was a prolific inventor, with three hundred patents to his credit. Two of them, the triode (which he called an *audion*) and the *regenerative detector*, were of significant practical importance. Unfortunately, both were involved in lengthy and acrimonious litigation.

His invention of the triode in 1906 grew out of his desire to improve the performance of radio receivers. Whereas the diode could act only as a *rectifier*—that is, it could pass current only in one direction—the triode had the indispensable ability to act as an *amplifier*. It was not, however, immediately useful or profitable.

Surprisingly, De Forest did not have a clear idea of how the triode worked. He believed that its ability to amplify was due to, or at least aided by, gas in the tube. This faulty explanation would cause him trouble in subsequent patent litigation.

A more serious problem was the need for extensive external circuitry in order to use the triode's unique ability to amplify electrical signals. The development of this circuitry took nearly a decade, and electron tubes did not become a major factor in radio communications equipment until World War I provided an urgent stimulus.

The practical use of the triode also was delayed by the bitter and incredibly complex patent disputes relating both to the triode itself and to its applications. They involved most of the leading inventors and research institutions of the radio industry, including De Forest, Fleming, Edwin Armstrong, Reginald Fessenden, Irvin Langmuir of General Electric (GE), and American Telephone & Telegraph (AT&T).

The courts seemed unable to make their way through the technical subtleties of the conflicting claims, and their decisions, usually reached after years of litigation, often seemed to engineers to be wrong or to put the litigants in a catch-22 situation. For example, the controversy with Fleming reached an impasse when a court ruled that the patent for the diode portion of the triode was Fleming's and the control grid was De Forest's. Its injunction forbade either party to use the other's patent, thus making it impossible for either to proceed.

The litigation over the regenerative detector patent lasted twenty years. The main antagonists were AT&T, which had bought the rights to De Forest's patent, and American Marconi, which had title to Armstrong's (see Chapter 3). General Electric, with a patent by Langmuir, and the German company Telefunken also were in contention. Although the courts finally ruled in De Forest's favor, Armstrong had a large band of enthusiastic supporters in the industry, and this conflict did not help De Forest's already dubious reputation.

In the end, the disputes were settled by negotiations between the parties that resulted in extensive cross-licensing agreements and the formation of the Radio Corporation of America (RCA) patent pool in 1920. These agreements were critical to the growth of the industry, and its development would have been seriously inhibited had resolution of the patent problems been left to the courts.

In spite of the controversies that plagued De Forest's career and his contentious personality, the importance of the triode invention brought him many honors. Among others, he served a term as president of the prestigious Institute of Radio Engineers (IRE).

The Climax and Denouement of the Electron Tube

Stimulated by the growing demands of communications and broadcasting, and later of computers, radar, and a host of other markets, electron tube technology developed rapidly and continuously for fifty years after the invention of the triode. Both World Wars provided an incentive for accelerated engineering development and production. Hundreds of tube types were offered in the marketplace, each filling a particular need in the growing universe of electronic devices. The electron tube industry reached its peak at the end of World War II, when its sales were measured in the billions of dollars and the outlook for the future was exceedingly bright.

The brightness of this future dimmed suddenly in 1947 when William B. Shockley of Bell Laboratories announced the invention of the *transistor*, a device that used a class of materials known as semiconductors. In principle, it could perform most of the functions of an electron tube in a tiny space and with much lower power consumption. It clearly became the wave of the future, as families of transistors duplicating the performance of tubes were developed.

This did not happen overnight. Forty years of engineering development and manufacturing experience were embodied in the electron tube, and the transistor was still a laboratory device. But the incentive was great, and the engineering profession rose to the challenge. Inexorably, transistors were developed to replace tubes in one application after another. During the 1950s, transistors were widely used in consumer products—a portable transistorized radio was popularly called a transistor—and by the 1960s they had replaced tubes in all but a few specialized applications.

The transistor had an enormous effect on the design of broadcasting receivers and station equipment. But progress never ends, and the transistor was in turn replaced in many applications, especially those using digital technology, by the integrated circuit.[2]

■ SOUND AND HEARING

The Science of Sound

The development of the technology of sound was quite different from that of electricity. Sound is a part of our everyday experience, and it is transmitted in the air, a palpable medium. The discovery of its nature required no scientific breakthroughs or extraordinarily creative insights. It was accomplished, rather, by the systematic efforts of many scientists and engineers over several decades.

The British physicist Lord Rayleigh (1842–1919) was perhaps the first and certainly the most distinguished scientist to conduct systematic research in the nature of sound. Unlike many of his contemporaries who were knighted in recognition of their accomplishments, Rayleigh inherited his title. But the excellence of his research in a wide variety of scientific disciplines was so great that he could have earned a knighthood on his own merits.

He was a professor of experimental physics at Cambridge University's

---

[2] For an excellent description of the development of the integrated circuit, see T.R. Reid, *The Chip* (New York: Simon & Schuster, 1984).

Cavendish Laboratory from 1879 to 1884, a member of that extraordinary faculty that contributed so much to the basic science of our electric and electronic industries. He was on the staff of the Royal Institution from 1887 to 1905, and in 1908 he became chancellor of Cambridge University. He did important research not only in sound but also in optics, vision, and hydrodynamics. He discovered the element argon. In 1904, he was awarded the Nobel Prize in physics. The results of his research in sound are described in his massive work, *A Treatise on Sound*. Although this volume appeared nearly one hundred years ago, most of its material is still valid.

The research carried out by Rayleigh and his contemporaries and successors established the following facts:

1. Sound consists of vibrations in a transmission medium that are transmitted by a longitudinal pressure and motion wave.
2. Its velocity (about 1,100 feet per second in air) varies with the medium, being more rapid in water or metals.
3. Sound waves exhibit the properties of all wave phenomena—*reflection, refraction,* and *diffraction.*

Their research was the basis for the science of *acoustics,* the study of the factors that determine the quality of the reception of sound in a radio or television studio, auditorium, or other controlled environment.

Hearing

Hearing is the sense by which we perceive sound. Early scientific investigations of hearing were directed at its physiology and psychology. For example, the German philosopher and physicist Hermann von Helmholtz (1821–1894) showed that the perceived quality of a musical tone is determined by the number and amplitude of its harmonics—that is, integral multiples of its fundamental frequency.

The invention of the telephone stimulated extensive experimental work in hearing by the Bell System and other organizations. Its purpose was to quantify the human perceptions of sound and the intelligibility of speech in order to provide guidance for the designers of telephone systems. The most important parameters were determined to be the following:

1. Intensity, or loudness
2. Frequency, or pitch
3. The ratio of the sound level to unwanted noise (hiss, crackle, and hum), or the signal-to-noise ratio
4. Distortion

Weber's Law and the Decibel   In 1846, the German scientist Ernst Weber (1795–1878) determined that the perception of the magnitudes of all physical sensations is based on their *ratios* rather than their numerical differences. This characteristic of human perception is known as Weber's law.[3] This law applies to all the hearing parameters previously listed. For example, the difference in loudness between two sounds with an energy ratio of 2 to 1 will be perceived to be the same whether they are two very loud sounds or two very soft ones.

The discovery of this characteristic of human perception led to the establishment of the *bel* (named for Alexander Graham Bell) as the unit of sound energy. A bel represents an energy ratio of 10 to 1 (see below). Since it is inconveniently large for many engineering purposes, it is common practice to use the *decibel* (dB), which is one-tenth of a bel and represents an energy ratio of 1.25 to 1. Very approximately, one decibel is the smallest difference in sound volume that can be noted by the human ear. The decibel is such a

---

THE DECIBEL UNIT OF MEASUREMENT

The number of bels difference, N_b, in power level between two sources, P_1 and P_2, is:

$$N_b = \log 10\left(\frac{P_1}{P_2}\right)$$

The difference in decibels, N_{dB}, is:

$$N_{dB} = 10\log 10\left(\frac{P_1}{P_2}\right)$$

Power ratios corresponding to selected decibel levels are:

Decibels	$\dfrac{P_1}{P_2}$
1	1.25
3	1.95
5	3.16
10	10.0
20	100.0
30	1000.0

---

[3] Weber's law in scientific jargon states that the increase in stimulus necessary to produce an increase in sensation in any of our senses is not an absolute quantity but depends on the proportion that the increase bears to the immediately preceding stimulus.

useful unit that it is now used almost universally in the electronics industry for the specification and measurement of electrical power ratios of all kinds.

Sound Wave Frequency, or Pitch　Some of the most important research in hearing was related to the perception of sound wave frequency, or pitch. This research was necessary to establish one of the basic specifications of a communications or broadcasting system—its responses to different frequencies in the transmitted signal.

The frequency range of audible sound for individuals with exceptional hearing was found to be 20 Hz to 20,000 Hz. The sensitivity of the ear at the extremes of the range is low, and very little is lost by limiting it to 30 Hz to 15,000 Hz, even in a high-fidelity sound system. Reasonable fidelity for voice and music can be obtained with an even smaller range. If intelligibility is the primary criterion, as in telephone circuits, a range of 300 Hz to 3,400 Hz is satisfactory. Figure 1–5 shows the frequency ranges of sound sources and reproducing systems.

It was subsequently found that improving the fidelity of sound reproduction requires a balanced improvement in all of the basic parameters—dynamic range (the ability to reproduce very loud and very soft sounds), frequency range, noise, and distortion. If, for example, there is noise in the signal, as from a scratchy phonograph record, more pleasing results are obtained by limiting the frequency and dynamic ranges.

The Stereophonic Effect　The *stereophonic effect*, the ability of the ear to perceive the sources of sound in three dimensions, is an amazing hearing attribute. It enables one to determine the direction of the source and to pick out a single voice in a babble of sounds. It adds realism to the reproduction of music, and nearly all musical recordings today are made in stereo.

Stereophonic perception requires *binaural* hearing—that is, with both ears. At lower frequencies, it depends on small phase differences in the sound as it arrives at the two ears. At higher frequencies, it depends on differences in intensity as the head shadows the ear from sources on the opposite side. The stereo effect can be simulated in sound reproduction systems by recording two channels from microphones separated in space and playing back the channels through separated loudspeakers.

■ MECHANICAL PHONOGRAPHS

The mechanical phonograph was the precursor of radio broadcasting. By making the reproduction of music in the home possible, it established the home entertainment industry. Its basic recording medium, an undulating groove on the surface of a disk, continued to be used in the electronic recording and playback systems used in radio and television, and there was

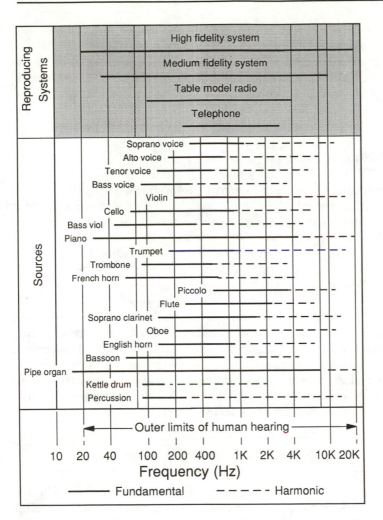

■ **Figure 1–5** The frequency ranges of sound sources and reproducing systems.

no need to invent a new medium for recording sound. Mechanically recorded disks could be played back on electronic phonographs, and the extensive library of mechanically recorded records became available for radio programs.

The effect of mechanical phonographs on the market for radios was equally important. Their popularity had demonstrated a large demand for an instrument that could reproduce music in the home. The large segment of the public that had bought phonographs provided a ready market for a radio broadcasting service and gave radio broadcasters a preconditioned audience. The network of retail stores established to sell phonographs and records was available to perform the same function for radios.

At first, radio broadcasting was considered to be a competitor of the record industry. In time, however, the industries developed a symbiotic relationship. Radio needed records as a source of programming, and records needed radio for sales promotion.

The Edison Phonograph

The first working model of a practical phonograph was produced by Edison in 1877. The recording medium was a spiral groove on a cylinder covered with tinfoil. The undulations at the bottom of the groove were vertical; this is sometimes called *hill-and-dale* recording. Edison intended it to be used for voice recording as in a dictating machine rather than for music or entertainment. He stated, "I don't want the phonograph sold for amusement purposes. It is not a toy."[4]

The first spoken words to be recorded by Edison, "Mary had a little lamb," were barely intelligible when played back, but with time major improvements were made in sound quality. The use of a layer of wax instead of tinfoil for the recording grooves was an important advancement developed by Chicester Bell and Charles Tainter.[5] But the cylinder had a fatal weakness: There was no practical way to produce large numbers of copies cheaply for the mass market. As a result, it lost the competitive battle with disk recordings, which could be stamped out by the thousands.

The Berliner Gramophone

Emile Berliner (1851–1929) was born in Germany and emigrated to the United States in 1870 at the age of nineteen. Like Edison, he was a prolific inventor. He invented two types of microphones and an aircraft engine that was widely used in military planes during World War I. He completed his fruitful career of invention and innovation by introducing the use of acoustic tile in sound recording studios during the 1920s. He was best known for his 1887 invention of a phonograph that used a disk rather than a cylinder as the recording medium.

Berliner's first experiment with phonographs was based on a cylinder like Edison's, but the groove undulations were lateral. The groove was traced on a lampblack coating on the surface of the cylinder and then transferred to copper by an etching process. This device had the same

---

[4] "The 50th Birthday of Victor Red Seal," *Huber News*, April 1953.
[5] Ibid.

weakness as Edison's—the difficulty of making copies—but Berliner soon found a way to solve the problem by using a spiral groove on a disk.

He constructed the first working model of a disk recorder, which he called a Gramophone, in 1887 and began selling the recorders in small quantities in 1888. They were crude devices—the disk had to be rotated with a hand crank—and the market for them would never have been large. But by a fortuitous accident of history, Berliner met Eldridge Johnson, a skilled mechanic and an astute businessman. Johnson supplied the necessary ingredients of mechanical ingenuity, marketing flair, and executive ability to make the phonograph an enormously successful consumer product. The result was the formation of the Victor Talking Machine Company and an exciting new business.

The Technology of Mechanical Phonographs

It is remarkable that mechanical phonographs worked at all. In the absence of electrical amplification, all the energy for operating the recording stylus had to come from the sound wave. (The energy in a voice wave at the level of ordinary conversation is only about 70 microwatts.) Similarly, on playback, all the sound energy had to be generated by the motion of the needle in the groove. A high level of mechanical and acoustical design skill was required to produce equipment that would work satisfactorily. It is equally remarkable that few of the inventors or other technical personnel had much formal scientific or engineering training. They were clever mechanics who achieved results by experimentation, trial and error, and innovative thinking. They continued to make technical improvements during the twenty-five years when mechanical phonographs were popular, roughly from 1900 to 1925, and it is doubtful that better results could be obtained today in an all-mechanical system.

Continual improvements have been made in the techniques for manufacturing disks, and the composition of disks has changed markedly through the years. The disks used for mechanical recording were brittle and contained an abrasive material that would cause intolerable record scratch in a high-fidelity system. With the smaller needles, lower needle pressure, and lower lateral needle force used for electrical playback, a smooth and flexible disk material such as vinyl can be used.

By modern standards, the tone quality of mechanical phonographs was not very good. The low-frequency response was limited by restrictions on the lateral excursion of the recording groove. The high-frequency response was limited by the inability of the recording and playback styluses to record or follow rapid groove undulations. As a result, the frequency range was little better than that of a telephone, nominally 200 Hz to 3,000 Hz, and recordings had a tinny sound.

The dynamic (loudness) range of mechanical recordings likewise was small. Restrictions on the lateral excursion of the groove, together with the absence of amplification, limited the maximum loudness. The need to override record scratch established a minimum loudness level.

Since the technical requirements for recording the human voice are less demanding than those for recording instrumental music, the sound quality of vocal selections was considerably better than that of orchestral pieces. Recordings by famous singers constituted a major part of the record companies' catalogs. A number of these recordings, such as Enrico Caruso's, have been rerecorded electronically with a new orchestral accompaniment with reasonably satisfactory results.

Eldridge Johnson and the Victor Talking Machine Company

Eldridge R. Johnson (1867–1945) was the first giant among the business leaders of the home entertainment industry. He was born in Wilmington, Delaware, and was trained as a machinist in the Spring Garden Institute in Philadelphia. In 1888, he was employed by the Scull Machine Shop in Philadelphia as foreman and manager. This small firm had been started by Andrew Scull, a sea captain, to provide a business for his son John, who was a graduate mechanical engineer from Lehigh University. John died suddenly, his father had no particular liking for the business, and he gave Johnson considerable authority. After two years, Johnson resigned to seek his fortune in the West. An expert machinist, he found no difficulty obtaining employment in Washington State at good wages, but he saw no opportunities for advancement. In 1891, he returned to Philadelphia to rejoin Scull as a partner in the firm of Scull & Johnson.

The next years involved hard work and meager profits. As Johnson put it in his autobiography, "Being the proprietor of a repair machine shop twenty years ago was well calculated to either break a man's spirit or fit him for better opportunities." Scull became discouraged and in 1894 sold his interest to Johnson. Scull & Johnson became Eldridge R. Johnson.

The opportunity that Johnson was seeking came quite by chance in 1896 when Berliner came to his shop to have one of his phonographs repaired. Here is how Johnson described the meeting:

> During the model-making days of the business one of the very early types of talking machines was brought to the shop for alterations. The little instrument was badly designed. It sounded much like a partially-educated parrot with a sore throat and a cold in the head, but the little wheazy instrument caught my attention and held it fast and hard. I became interested in it as I had never been in anything. It was exactly what I was looking for. It was a great opportunity . . . .

My years of hard experience in model making and repair work had well qualified me to cope with intricate designs and processes. I immediately undertook a course of experimenting with talking machines and made discovery after discovery until a talking machine of the Gramophone type, capable of . . . reproducing the tone true to the original sound, stood in my laboratory.

It cost me $50,000 and two and one-half years of desperately hard work, but the Victor Company's factory is a standing testimonial that justifies the cost.[6]

The two and one-half years of "desperately hard work" that followed Berliner's initial visit were filled with complex technical, business, and patent problems.

On the technical side, Johnson made major improvements in the design of the machine and record manufacturing processes. He added a spring-driven motor so that it was not necessary to crank the machine continually. He developed a process for making metal masters from the original wax recording. In addition, he made countless smaller improvements that enabled the phonograph to reproduce a "tone true to the original sound." Figure 1–6 shows how the phonograph evolved.

On the business side, he began manufacturing phonographs under the Berliner patents. In Johnson's words, "The Trade could not get enough of them from the start." Sales were not a problem in starting the business.

The patent problems were more difficult to solve. He and Berliner resolved a dispute over their patent agreement in 1901 by forming the Victor Talking Machine Company, which combined the Johnson and Berliner interests. This did not solve the problem completely, however, because third parties challenged the validity of the Berliner patents as well as his use of the Gramophone trade name. Johnson's legal fees and out-of-court settlements were very costly to the new company.

Licenses for the Berliner patents were granted to the Gramophone Company, Ltd., in England and to the Columbia Company, later to become Columbia Records. This did not significantly increase competition because the Gramophone Company did not attempt to operate in the United States (Victor bought a half interest in the company in 1920) and Columbia was ineffective against Victor's excellent management.

From Gramophone, Johnson acquired the U.S. rights to the trademark "His Master's Voice," showing a fox terrier, Nipper, listening to a phonograph (Figure 1–7). Gramophone had bought the original painting from Francis Barraud, and it soon became the world's most famous trademark.

The years of hard work that preceded the formation of the Victor Talking Machine Company in 1901 were reflected in the extensive product line it was able to offer at the outset (Figure 1–8). The new phonographs were

---

[6] B.L. Aldridge, *The Victor Talking Machine Company* (Camden, NJ: RCA Sales Corporation, 1964).

ANCESTORS

of today's phonograph

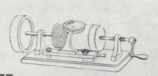

1877...THE EDISON

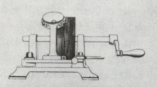

1881...THE BELL AND TAINTER

1888...THE BERLINER

1895...THE BERLINER (II)

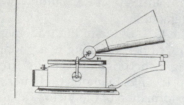

1898...THE ELDRIDGE JOHNSON

Courtesy RCA Corporation.

■ **Figure 1–6** The evolution of the phonograph, as depicted by Victor in 1901.

Courtesy RCA Corporation.

■ **Figure 1–7** His Master's Voice trademark.

popular in the marketplace, sales increased rapidly, and Johnson demonstrated the extraordinary versatility of his talents. The plant in Camden, New Jersey, where Johnson had established the Victor Company was continuously expanded and at its peak encompassed more than two million square feet. The number of employees eventually exceeded twenty thousand.

Like Henry Ford, Johnson believed in vertical integration. Only raw materials came into the plant. It had its own cabinet and parts fabrication shops. It produced the record mastic according to a secret formula. It had its own power station and water purification and sewage disposal facilities. Its employees included some of the world's finest machinists and cabinetmakers. While extreme vertical integration is not the most economical way to run a factory, it was an important aid in achieving the rigid quality control Johnson demanded.

Johnson was equally adept at handling the temperamental stars of the musical world. The services of leading opera singers were eagerly sought,

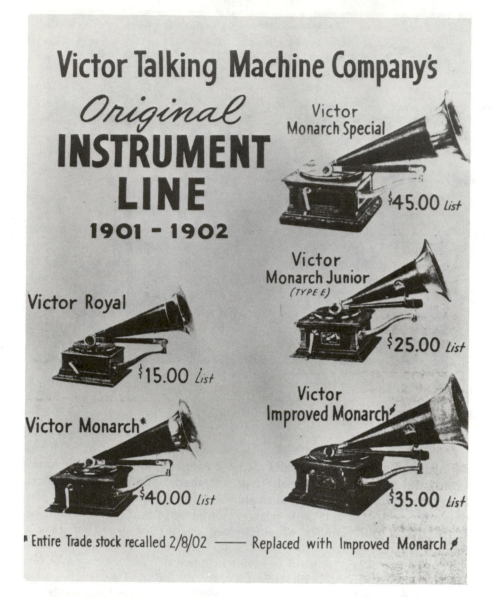

Courtesy RCA Corporation.

■ **Figure 1–8** 1901 Victor sales flier.

not only because of the popularity of their records but also because of the prestige they brought to a new industry.

Victor's greatest success came with the famed Italian tenor Enrico Caruso. Gramophone Ltd. had placed him under contract before his reputation was fully established but later balked at his high recording fees. Johnson, however, was willing to pay, and Victor bought the rights to his services from Gramophone. Caruso's fee was one hundred pounds (then about $500) for ten records—an unheard of sum in those days—but it paid off handsomely for both parties. Caruso received his first contract from the Metropolitan Opera without a live audition solely on the basis of his records. He went on to be a world-famous tenor, his public reputation enhanced by his recordings. His fees and royalties from Victor ultimately exceeded $3 million ($60 million in 1989 currency).

Caruso's records were among Victor's best-sellers—with reissues more than one million were sold—and his stature enhanced its reputation. (Victor's all-time best-seller was a record of the "Wreck of the old 97" and the "Prisoner's Song"; it sold more than six million copies.) Most of the famous musical artists of the time, including Ernestine Schumann-Heink, signed exclusive contracts with Victor. To provide additional prestige for its top stars, Johnson created an elite product line of Red Seal records.

Victor's marketing programs were first-rate. Johnson established a strong network of distributors and dealers, and the popularity of his product made it possible to establish a markup structure that was profitable for everyone. He made heavy use of advertising, and his record catalogs were classics.

As a result of the superb management of Johnson and his staff, the Victor Talking Machine Company enjoyed great prosperity for the twenty-five-year period from 1902 to 1927. Annual phonograph sales rose from 42,000 units in 1902 to a high of 570,000 in 1917 and an average of 350,000 during the early 1920s (Figure 1–9). During the same period, nearly 1 billion records were sold.

The bottom line of Victor's business was as healthy as the top. Annual dividends increased from 6 percent of stock par value in 1902 to a high of 80 percent in 1916. During the days when $1 million was considered a large sum of money, the company was estimated to have created more than thirty millionaires among its top executives and stockholders.

These halcyon days began to come to an end in the early 1920s as mechanical phonographs felt the competition of radio. It became clear that the superior quality of electronic reproduction would doom the mechanical phonograph. Unit sales fell alarmingly from 348,000 in 1921 to 214,000 in 1925. Victor responded in 1925 by entering into a joint venture with RCA and offered lines of radios and Orthophonic electronic phonographs.

These products were modestly successful, but Johnson, whose talents were mechanical, had little interest in them, and he decided to sell his controlling interest in Victor. In 1926, he negotiated with Speyer & Co. and J & W Seligman, New York banking firms, to buy his interest for about $27 million. It turned out to be a bargain, as they resold the stock to the public for $53 million. Three years later, RCA acquired the company, and the

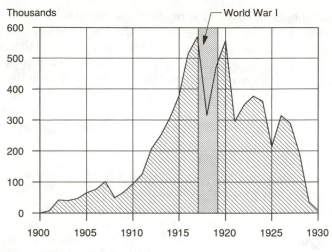

Source: Aldridge, B.L. loc cit.

■ **Figure 1–9** Victor Talking Machine Company phonograph sales, 1901–1929.

Victor plant in Camden became the site of major developments in radio and television.

Johnson devoted the remaining twenty years of his life to philanthropy. He also lived well. Among his amenities was a railroad siding in back of his mansion in Moorestown, New Jersey, where he could board a private car without the hassle of going to the station. He also had the satisfaction of knowing that he had created a major industry.

■ COMMUNICATION BY WIRE

The desire to communicate rapidly with distant points was so great that wire communication was one of the first practical applications of the newly discovered phenomenon of electricity. The Greek word *tele*, meaning far, soon found itself in the vocabulary of electricity—telegraph, telegraphy, telephone, telephony, and finally television.

Telegraphy

After current electricity was discovered near the end of the eighteenth century, a number of scientists experimented with methods of using it for communication. Ampère, for example, constructed a receiving device consisting of twenty-six electromagnets, one for each letter. A compass needle placed near each magnet was deflected when the magnet's winding was energized by an electric current. This apparatus required twenty-six wires to connect it to the sending station.

The full development of telegraphy depended on two inventions by an American artist, Samuel S.B. Morse. These were the *relay repeater* and the transmission of letters and numbers by a series of electrical pulses in a dot/dash code that still bears his name, the *Morse code*.

The relay repeater is an electromagnetic amplifier. Without this device, the range of telegraphy was limited to twenty miles or less because of the electrical resistance of the line. By inserting relays in the line at intervals, the range can be extended almost indefinitely. Use of the Morse code with a make/break key at the sending end and a sounder at the receiver makes it possible to transmit numbers and letters of the alphabet, *alphanumeric* characters, over a single circuit.

In 1838, Morse demonstrated his system to President Martin Van Buren and his cabinet. In 1844, Morse persuaded Congress to appropriate $30,000 for a test circuit from Baltimore to Washington. His successful transmission of the biblical phrase "What hath God wrought?" is one of the legends of the industry.

Although the test was successful, the postmaster general declared that it would never be practical, and the government withdrew its support. Future developments in the United States were largely financed by private funds.

The first transcontinental telegraph circuit was completed in 1861. An even greater technical achievement was the completion of the first transatlantic cable in 1866 after a number of failed attempts. This undertaking required the solution of many problems, including raising money to finance the risky enterprise, mechanical difficulties, and unique electrical problems. The money was raised by American businessman Cyrus Field, and the electrical problems were solved by engineers working under the direction of Lord Kelvin, who was knighted for his work on the transatlantic cable.

One of the electrical problems was the inability to overcome line losses by inserting relay repeaters. Kelvin solved this by using stranded copper wire of very high conductivity and developing extremely sensitive detection devices. The other problem was the high electrical capacitance between conductors. Kelvin calculated that, owing to the capacitance, the transmission speed was inversely proportional to the square of the length of the circuit. This problem was solved by a technique known as *line loading*.

The technology of telegraphy has drastically changed as the result of the availability of more sophisticated and higher speed transmission and switching systems. Modern wideband transmission media and terminal devices—such as coaxial cable, microwave, satellites, fiber optics, facsimile, electronic mail, and computers—have made enormous advances in the capability to transmit alphanumeric information. Nevertheless, old-fashioned telegraphy consisting of a telegrapher transmitting Morse code by key and sounder has had a long and distinguished history and is still used, particularly in wireless communications.

Telephony

The success of wire telegraphy led naturally to a desire to communicate over long distances by voice. Early investigators, including Charles Bousel in France and Philip Reis in Germany, attempted to transmit voice using the make/break technique of telegraphy. It was a crude attempt to transmit voice by *digital* means (on/off pulses). Their experiments were unsuccessful, and it was up to Alexander Graham Bell (1847–1922), an American, to comprehend and enunciate the principle of transmitting voice by continuous, or *analog*, electrical signals. This transmission mode is basic to both radio and television systems, and its conception was a major technological advance.

Telephony made three additional contributions:

1. The carbon telephone transmitter (mouthpiece), which was the basis for microphones widely used in the early years of broadcasting
2. The magnetic telephone receiver (earpiece), which was the basis for early loudspeakers
3. The electron tube amplifier

The carbon microphone and the telephone-type loudspeaker have long been superseded in broadcasting by more modern devices, and electron tubes have been largely (but not entirely) replaced by solid-state (transistorized) components in amplifiers, but analog transmission for voice and other audio (and video) signals continues.

Alexander Graham Bell and Analog Transmission In 1874, Bell stated that he could transmit speech "telegraphically" (sic) if he could make an electrical current vary in intensity precisely as the air varies in density during the production of sound. This was the first enunciation of the principle of analog telephone transmission. This principle seems obvious today, but it was not at all obvious at that time, and it was a major engineering breakthrough. It is of interest to note that the analog format was considered to be a more advanced technology than digital until the 1930s. (See Chapter 10 for a description of the analog and digital formats.)

Bell was born in Edinburgh, Scotland, and was educated at the Universities of Edinburgh and London. He came to Canada in 1870 with his father, a widely known authority on phonetics and defective speech. He moved to the United States in 1871 and became a professor of vocal physiology at Boston University. In 1872, he started a school in Boston for training teachers of the deaf. It was during this period that he became interested in the electrical transmission of voice. In addition to his pioneering work in telephony, he developed a rudimentary phonograph that resembled Edison's and the photophone, a method of recording sound on film that was eventually adapted for talking motion pictures.

Conceiving the principle of voice transmission was but the first step in constructing a telephone system, and it was necessary to develop the apparatus as well. By 1875, Bell had constructed a crude telephone transmitter and receiver, both based on an electromagnetic principle. The first telephone conversation of record was with his assistant, Thomas A. Watson. The message was less biblical than Morse's. He said, "Mr. Watson, come here, I want you."

Bell patented his device, but his patents were soon challenged by Elisha Gray, who had filed conflicting ones only a few hours later. Lengthy litigation followed, but Bell's patents were ultimately upheld. With the aid of financial backers, Bell founded a company to exploit his patents by building commercial telephone systems. He obtained distinguished backers (one of them Gardiner Hubbard, became his father-in-law), and he wisely left the administration of the company to others. He achieved great financial success, and he was able to devote the rest of his life to educational and technical pursuits.

The telephone company founded to use Bell's patents evolved into the American Telephone & Telegraph Company (AT&T) through a series of complex business transactions. AT&T became the dominant telephone company in the United States, and its research arm, Bell Laboratories, has made enormous contributions to communications and broadcasting technology.

Telephone Transmitters and Microphones   Emile Berliner invented the carbon telephone transmitter in 1877. A cavity in the transmitter, or microphone, is filled with particles of carbon in loose contact with each other. The compression of the carbon particles varies as a diaphragm on one side of the cavity vibrates in response to the sound wave. An external battery causes a current to flow through the carbon, and this varies as the electrical resistance of the carbon changes in response to the changes in compression.

Carbon microphones have a much higher output signal than the electromagnetic type Bell had first used, and this discovery was a major step forward in telephone technology. Its performance was satisfactory for telephone systems, and transmitters of this type are still in use today. An adaptation of this device was used as the microphone in early radio stations. Its quality was not adequate for the more demanding requirements of broadcasting, however, and it was replaced by other types of microphones.

Telephone Receivers and Loudspeakers The electromagnetic loudspeakers used in early radios operated on the same principle as telephone receivers. A flexible diaphragm of magnetic material is placed in front of the poles of a permanent magnet equipped with a magnetizing coil. The diaphragm vibrates in accordance with the signal current passing through the magnetizing coil to produce the sound wave. For acoustic amplification, the diaphragm is mounted at the throat of a horn similar to those used in mechanical phonographs. Like the carbon microphone, this type of electromagnetic loudspeaker has largely become obsolete.

Electron Tube Amplifiers The losses in long-distance telegraph lines carrying simple on/off pulses could be overcome by the use of relay repeaters, but these were of no value for continuously varying voice signals, which required electron tube amplifiers.

As noted earlier, De Forest had been slow to develop the potential of the electron tube triode amplifier after its invention in 1906 partly because of his patent disputes and partly because he did not understand the mechanism of the tube's ability to amplify signals. By 1912, however, he had constructed a crude amplifier for voice circuits that he was able to demonstrate to AT&T engineers.

His timing was fortuitous, for AT&T was desperately searching for an amplifier that would make long-distance telephony by wire possible. AT&T's engineering staff accelerated its work on electron tube amplifiers, and its patent department commenced negotiations with De Forest for patent rights.

The AT&T engineers were successful. They understood the mechanism of amplification, and in 1915 they demonstrated a transcontinental telphone circuit using electron tube amplifiers, or repeaters. They also developed much of the electron tube apparatus required for wireless telephony and broadcasting transmitters.

The patent negotiations between De Forest and AT&T were exceedingly complex, with accusations of bad faith on both sides. In an initial agreement signed in 1913, AT&T agreed to pay De Forest $50,000 for the right to use his patents in telephone amplifiers. Later agreements extended the coverage but excluded amateur radio equipment. The amateur exclusion became a big loophole after commercial broadcasting began.

■ WIRELESS COMMUNICATIONS

For citizens of the late nineteenth century, communication by wireless with no tangible connection between the terminals must have seemed even more miraculous than communication by wire. It would have seemed still more miraculous had it been known that this new medium would evolve into broadcasting to the general public. The inventor of this startling new medium, or more accurately of the equipment and techniques that made it possible, was Marchese Guglielmo Marconi (1874–1937).

Marconi Invents Wireless Communications

Marconi (Figure 1–10) was born in Bologna, Italy, of an Italian father and a Scotch-Irish mother. Both parents were affluent: His father had a landed estate, and his mother, Annie Jameson, was an heiress in a family of brewers.

His education at the University of Bologna was informal but very effective. He was intensely interested in telegraphy, and this led to studies of the work of Helmholtz, Hertz, and other electrical pioneers. He audited the classes of Auguste Righi, a professor of physics at Bologna and himself an early experimenter in electromagnetic waves. Encouraged by his mother, he set up a laboratory in his home, where he conducted his own experiments. His mother also helped him become fluent in English, a skill that would be of great importance in his future career.

Hertz's demonstration of the existence of electromagnetic waves had been accomplished in the laboratory with a short distance separating the transmitter and receiver. Marconi was determined to extend the range to a distance that would make wireless communications practical. To accomplish this, he experimented with all three elements of a radio communications system—the transmitter, the receiver, and the antenna for coupling these devices to the "ether."

Marconi developed a more powerful arc to serve as the transmitter, and he developed an improved *coherer* to detect the signals. Perhaps his most important discovery was that the range of transmissions could be increased dramatically by connecting one of the arc terminals to an antenna, in his design a large metal plate elevated above the ground and connected to a spark terminal with a wire.

By 1895, he was able to communicate for a distance of a mile on his father's estate. In 1898, he established a link across the English Channel. And in 1901, he performed the remarkable feat of wireless telegraphic communications across the Atlantic from Cornwall, England, to Cape Cod, Massachusetts.

His mother, having come from a business family, saw the commercial potential of wireless communications, and she attempted to gain the support of the Italian Ministry of Posts and Telegraphs. Finding no interest there, she turned to her native country, Great Britain. The British Post Office, which was responsible for wired communications, was mildly interested and even offered to buy Marconi's patents for a modest sum. But in 1897, his mother's family decided the Post Office's offer was inadequate and

Courtesy Smithsonian Institution, with permission.

■ **Figure 1–10**   David Sarnoff and Marchese Guglielmo Marconi.

decided to form a private company, the Wireless Telegraph and Signal Company, Ltd., with its own funds. In 1900, its name was changed to the Marconi Wireless Telegraph Company, Ltd. Marconi sold his patents to the company and became its chief engineer. With good management, the

Marconi company prospered and became the most successful of the several wireless companies that sprang up throughout Europe and the United States.

The Technical History of Wireless

Technical progress came slowly but steadily. Inspired by Marconi, a host of scientists, engineers, innovative inventors, technicians, and amateurs (hams) made contributions to wireless technology from 1901 to 1920. The relationships among these groups were not always friendly. Scientists and engineers with lengthy professional training tended to be contemptuous of the empirical and undisciplined methods of those with less formal education. The latter, like Edison, tended to regard scientists as living in ivory towers and having no interest in or understanding of the practical uses of technology. In retrospect, both groups were wrong. The great technical progress in wireless communications during this twenty-year period required the contributions of everyone from Ph.D. scientists to hams.

Three basic technical problems had to be solved:

1. Establishing the physical laws that govern the propagation of electromagnetic or radio waves in the proximity of the earth's surface (Maxwell's equations applied to waves in free space)
2. Inventing or developing transmitters for generating radio waves of substantial power and, for radiotelephony, *modulating* the waves with the voice or music signals to be transmitted
3. Inventing or developing equipment for the reception, detection, and amplification of radio signals

In addition to these purely technical matters, the negotiation of international agreements for the regulation of the transmission of radio signals was closely related to technical progress.

The Propagation of Radio Waves

With our present knowledge of radio wave propagation, its laws seem very logical and straightforward. To the radio pioneers, however, they must have seemed enormously complex and confusing. Our knowledge has not come from a single brilliant insight. Rather, it is the result of years of patient measurements, analysis, and theoretical studies.

Prior to 1920, only the frequencies below 3,000 kHz (100 meters) were used extensively, and it was found that their propagation resulted from two type of waves, *ground waves* and *sky waves*.

Ground Waves Ground waves travel along the surface of the earth and can follow the earth's curvature. Their range depends on their frequency, the soil's conductivity, and transmitter power: the lower the frequency and the higher the conductivity, the greater the range. Higher power, of course, also increases the range, but its effect is relatively small compared to the effects of frequency and conductivity. Marconi's first transatlantic transmission was possible with modest transmitter power and a crude receiver because it used a low-frequency wave propagated over high-conductivity seawater. At frequencies above about 3,000 kHz, the range of ground waves is so limited that they have little practical value.

Sky Waves Sky waves result from the reflection of radio signals by layers of ions (electrically charged molecules) in the upper atmosphere. Ionization is caused by the sun's rays; hence the degree of ionization varies from day to night and with the latitude, the season, and the sunspot cycle.

The existence of an ionized layer was first postulated almost simultaneously in 1902 by two physicists—an American, A.E. Kennelly, and an Englishman, Oliver Heaviside—on the basis of their studies of Marconi's transatlantic transmissions. For many years, it was known as the Kennelly–Heaviside layer. Later it was found that there are three and sometimes four distinct layers ranging from the D layer at 50 to 90 kilometers above the earth to the F_2 layer at 200 to 400 kilometers. Because of the complexity and variability of these layers, hundreds of thousands of measurements were required to obtain quantitative information concerning their reflectivity at different times of day, during different seasons, and at different latitudes and to establish a reasonably reliable statistical data base for making predictions. A major breakthrough occurred in 1925 when a radarlike pulse method for measuring the height and reflectivity of the layers was developed.

Long Waves, Medium Waves, and Shortwaves Long waves (10 to 300 kHz), which are propagated by a combination of ground and sky waves (see illustration on next page), were universally used in the early days of wireless communications. They provide a steady, reliable signal night and day, and their low attenuation gives them an extended range. Also, they were the easiest to generate with the equipment available at the time. They have the disadvantages of a high level of static (static generated by electrical storms also is transmitted over long distances) and a limited spectrum space. They are used primarily for telegraphy rather than telephony.

In the medium-wave AM broadcast band, 550 to 1,600 kHz, ground waves are the primary transmission mode. Sky waves are attenuated during the daytime, but at night they provide long-distance service from *clear*

RADIO WAVE PROPAGATION

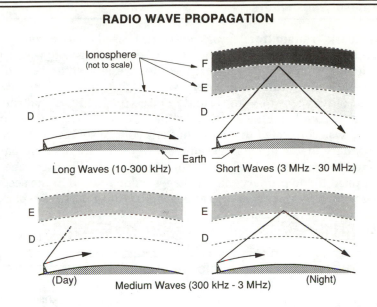

The ionosphere has three distinct layers called D, E, and F. The F layer, in turn, sometimes splits into two layers called F_1 and F_2. Their heights are as follows: D layer, 50 to 90 kilometers; E layer, 90 to 140 kilometers; F_1 layer, 170 to 200 kilometers; F_2 layer, 200 to 400 kilometers. The D layer nearly disappears at night, and the other layers become weaker.

Long waves are transmitted, as though in a waveguide, in the region between the earth's surface and the D and E layers. Medium waves are reflected back to earth by the E layer at night, but in the daytime they are absorbed by the D layer, which forms in the denser regions of the atmosphere. Shortwaves penetrate the weak F layers at night and continue into space. In the daytime, the heavily ionized F layers become very efficient reflectors of shortwaves, and long-distance communication is possible with low-power transmitters.

channel stations, which are the sole occupants of their channels (see Chapter 2).

The sky-wave propagation of shortwaves (3 to 30 MHz) provides an effective long-distance transmission medium during the daytime, and shortwaves are still used for international communications and broadcasting. Sky-wave propagation is too weak to be of practical use for communications at frequencies above 30 MHz. Under abnormal conditions, however, it can be strong enough to cause objectionable interference at distant stations. This was the reason for relocating the FM band upward to its present position in the spectrum (see Chapter 3).

Wireless Transmitters

The Spark Transmitter *Spark transmitters* based on an extension of the method used by Hertz for the generation of electromagnetic waves in 1887 (see below and on next page) were used for wireless communications for more than twenty years. Their basic element was an *induction coil,* which consisted of two coils of wire wound in close proximity so that they were magnetically coupled. One of the coils, the primary, had a small number of turns and was connected to an intermittent voltage source. The other coil, the secondary, had a much larger number of turns, and its terminals were connected to small metal balls to form a spark gap. When voltage pulses were applied to the primary, high voltages were induced in the secondary, causing radio frequency sparks to appear across the gap. These sparks were the source of the radio waves.

THE EVOLUTION OF RADIO FREQUENCY TRANSMITTERS:

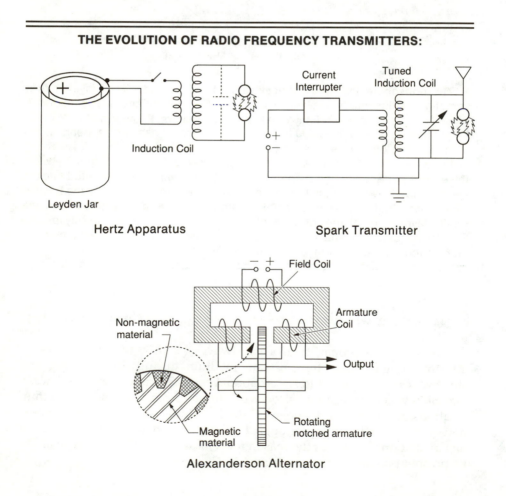

Leyden Jar

Induction Coil

Hertz Apparatus

Current Interrupter

Tuned Induction Coil

Spark Transmitter

Field Coil

Non-magnetic material

Magnetic material

Armature Coil

Output

Rotating notched armature

Alexanderson Alternator

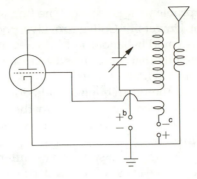

Electron Tube Transmitter

The radio frequency generator used by Hertz was an induction coil that had its primary connected to a Leyden jar through a switch and its secondary connected to a spark gap. The Leyden jar was charged, and when the switch was closed, the sudden surge of current induced a high voltage in the secondary, causing a spark. The current was oscillatory—that is, it surged back and forth at the resonant frequency of the secondary as determined by its inductance and distributed capacity. The amplitude of the oscillations diminished or damped rapidly owing to ohmic and radiation losses.

The operation of the spark transmitter used the same principle, but a continuous stream of electrical impulses was fed to the primary. In the original version, it emitted a damped waveform that was rich in unwanted sidebands and harmonics. By adding highly resonant tuned circuits, the amplitude of the wave was smoothed out, thus producing continuous waves (CWs) and greatly reducing the amplitude of the sidebands and harmonics. To make the signal more audible through interference and static, the CW signal was interrupted at a rate of about 1 kHz.

In principle, the Alexanderson alternator was a conventional rotating electric generator, but its operation at frequencies up to 100 kHz required new and difficult design techniques. It was capable of very high power up to 200 kilowatts.

The electron tube transmitter produced continuous waves, and its use soon became universal, particularly for radiotelephony, because it could be easily modulated.

Spark transmitters had many disadvantages. The bursts of radio energy emitted by early models had a wide band of frequency components, and very few stations could have shared the limited electromagnetic spectrum had a solution for this problem not been found. It was never possible to develop a satisfactory method of modulating the amplitude of the waves for use with radiotelephony. Their efficiency was low, and the high-power spark produced a terrible racket. In recognition of this phenomenon, the radio operator on board ships was traditionally called sparks.

Only one of these problems was solved. The English physicist Sir Oliver Lodge (1851–1940) of University College in Liverpool discovered in 1897 that the energy bursts could be smoothed into a continuous wave of constant amplitude by circuits that were tuned to resonate at the transmitter frequency (he called them syntonic). This technique increased the transmission efficiency, reduced the off-frequency radiation from the transmitter, and made it possible to tune the receiver to the frequency of a single transmitter. The current system of frequency allocations and usage depends on this fundamental invention.

In spite of its disadvantages, the spark transmitter did good service for many years. It saved many lives at sea and provided the basis for the wireless communications industry.

The Alexanderson Alternator The Alexanderson alternator was a rotating electrical generator that operated at radio frequencies rather than the 60-Hz power frequency. Its development began with a letter written in 1901 by Reginald Fessenden, an early radio pioneer, to Charles Steinmetz, the renowned GE engineer. Fessenden asked GE to build an electrical generator having a frequency up to 100 kHz for use by his communications company.

Nearly everyone thought a rotating generator at this frequency was impossible, but Steinmetz and GE accepted the challenge. Steinmetz first designed a generator to operate at 10 kHz, then gave the assignment for designing a 100-kHz model to a young Swedish engineer, Ernst Alexanderson, who had just joined the company. Alexanderson delivered the first working 100-kHz model in 1906, and he continued to improve its design over the next ten years. The alternator could generate much higher powers than spark transmitters, and power levels as high as 200 kilowatts were achieved. This made it ideal for long-distance point-to-point and shore-to-ship communications.

GE's sales were limited at first, however, because Marconi, whose company had the bulk of the world's wireless business, steadfastly stuck with spark transmitters. But by 1915, even Marconi had to recognize the alternator's advantages, and the Marconi Wireless Telegraph Company ordered a quantity of them from GE. The delivery of the alternators to Marconi was postponed by World War I, and Marconi's efforts to take delivery after the war precipitated a major reorganization of the U.S. communications industry.

The alternator had serious limitations. Its maximum frequency was 100 kHz, and like the spark transmitter, it was not well adapted for telephony because it was difficult to modulate. These limitations were removed by the emergence of the electron tube.

Electron Tube Transmitters The electron tube eventually solved most of the problems that had limited the usefulness of earlier tramsmitters. Electron tube technology developed rapidly during World War I, and electron tube transmitters were universally used for radio broadcasting after the war.

Modulation Information cannot be transmitted by a continuous radio signal that does not vary in frequency or amplitude. The transmission mode used exclusively for radiotelephony and radio broadcasting until the late 1930s was *amplitude modulation* (see below), or AM. The amplitude of the radio frequency transmission, the *carrier*, is varied in accordance with the signal waveform at the transmitter. At the receiver, the signal is recovered by passing the modulated carrier through a detector, which passes current in only one direction.

Attempts to amplitude modulate the spark transmitter and the Alexanderson alternator were only marginally successful (see illustration on next page). In one of their outstanding achievements, engineers at AT&T developed a successful method of amplitude modulating a carrier with electron tube circuits. This enabled AT&T to demonstrate simultaneous transatlantic and transpacific radiotelephony in a spectacular "first" in 1915. The development of a satisfactory modulation method was one of the key links in the chain of broadcasting technology.

Wireless Receivers

The basic functions of a wireless, or radio, receiver are tuning, amplification, and detection—that is, the conversion of the radio frequency signal into an audible sound. The tuned circuits developed by Lodge and his

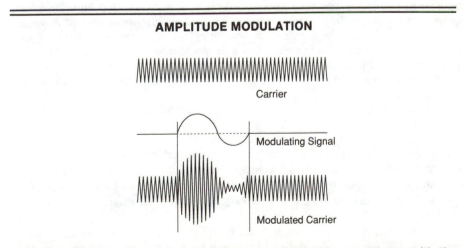

AMPLITUDE MODULATION

Carrier

Modulating Signal

Modulated Carrier

The amplitude of the carrier is varied or modulated in accordance with the signal waveform. The signal is recovered at the detector by rectification.

PRE–ELECTRON TUBE TRANSMITTER MODULATORS

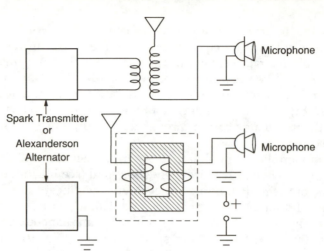

A number of schemes were devised in attempts to amplitude modulate spark transmitters and Alexanderson alternators. None of them worked very well. The lack of a suitable means of amplifying voice signals and the difficulty of varying the output of these generators over a wide power range were almost insuperable barriers.

The most common methods are shown here. The impedance of a circuit element in series with the antenna circuit was varied in accordance with the modulating signal, thus modulating the antenna current. Microphones were sometimes used as the variable element. The power-handling capability of a carbon microphone was only about 5 watts, and considerable ingenuity was exercised in developing microphones that could handle more power. A better solution was the use of a magnetic amplifier in which the reluctance of the magnetic path and hence the inductance of the antenna coil were made to vary with the modulation. Neither of these devices was capable of producing a high level of modulation. These problems were solved by the electron tube modulator (see Chapter 2).

successors for transmitters were readily adapted for receivers, thus providing the tuning function. Developing amplifiers was more difficult—in fact, it was impossible prior to the availability of electron tubes—and early receivers had to operate on the energy extracted from the wave by the antenna.

The invention of the triode in 1906 made the development of amplifiers possible. The amplifying ability of the triode was greatly enhanced by the invention of the *regenerative* amplifier by Armstrong, De Forest, and others in 1913. The regenerative amplifier had a very high gain, but if misadjusted it would oscillate, or howl.

Armstrong superseded his own invention by developing the *superheterodyne* receiver circuit while in the Army during World War I. The performance of this circuit is so superior that it continues to be widely used today in receivers of all types. Like electron tube generators and modulators, it is one of the landmark inventions in the history of radio technology.

The process of detection requires a *rectifier* or some other circuit element to convert the alternating radio frequency signal to a direct current that is proportional to the amplitude of the signal. Various detectors were used in the early days of radio.

One of the first was the *coherer*, a container of loosely packed iron particles. The conductivity of the mass of particles increases when subjected to a radio frequency current, and this change can be used to develop an output signal.

The *electrolytic detector* invented by Fessenden also worked on the principle of conductivity change. A fine platinum wire was immersed in a

---

REGENERATIVE AND SUPERHETERODYNE RECEIVER CIRCUITS

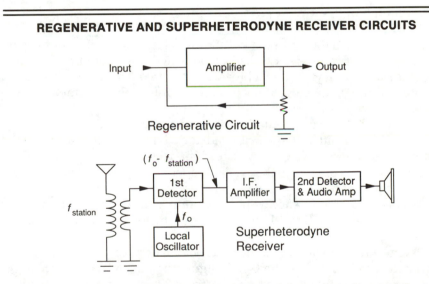

In the regenerative circuit, a little of the output signal is fed back to the input and reamplified. The total gain can be very high, but the amount and phase of the feedback must be carefully controlled or the circuit will oscillate.

In the superheterodyne circuit, an internal, or local, oscillator is tuned so that its frequency remains at a fixed increment (the intermediate frequency, typcially 456 kHz in a radio receiver) above that of the station being received. Most of the receiver's amplification occurs at this fixed frequency, and the amplifier's design is enormously simplified.

container of nitric acid. When a current was passed through the wire to an electrode at the bottom of the container, small bubbles formed on the wire, insulating it from the acid. The bubbles could be removed and the conductivity restored by passing the radio frequency signal through the wire. Thus the current through the device depended on the strength of the radio frequency signal.

The *crystal detector* was used by generations of radio amateurs as well as by commercial communication companies. It consists of a small galena crystal that is just touched by a fine wire, or "cat's whisker." If the point of contact is chosen carefully, it will act as a rectifier.

As with all the elements of wireless systems, the best solution to the problem of detection was the electron tube. Fleming's invention of the diode, an electron tube rectifier, was in response to Marconi's need for a better detector. De Forest improved on it by inventing the triode, which can act as both amplifier and detector.

Regulation of the Airways

Chaos reigned in the airways soon after wireless communications began. The frequency of early spark transmitters drifted, they could not be accurately tuned, and they emitted a broad band of frequencies. Anyone could operate on any frequency, thus producing intolerable—sometimes deliberate—interference.

The technical improvements in spark transmitters reduced the bandwidth of the transmitted frequencies and added the capability of tuning the transmitter to a desired frequency. These changes increased the number of stations that could operate in the available spectrum space, but government regulation was necessary to assign channels to both commercial and military users so that interference would be minimized. And since radio waves do not respect international boundaries, international agreement was required.

The first international Conference was held in Berlin in 1903, having been convened at the request of the German government. A major objective of the German delegation was to break Marconi's near monopoly of wireless communications, which Marconi enforced by refusing to allow Marconi-equipped stations to communicate with stations having equipment made by other manufacturers. Marconi, supported by the British and Italian governments, refused to yield on this point. Although a Conference resolution stated that coastal stations were obliged to communicate with all ships, regardless of their equipment, it was meaningless because the British government refused to enforce it. Little else was accomplished at the Conference, but it did give the participants an opportunity to discuss the many legal and regulatory problems caused by the new industry.

A second Conference held in Berlin in 1906 was more fruitful. The issue of Marconi's obligation to communicate with all stations was finally resolved against Marconi. Specific regions of the spectrum were allocated for commercial and military use, and many rules of a technical nature were adopted. The United States refused to ratify the Treaty that emerged from the Conference partly because there was strong opposition to international regulation of "American ether." A more reasonable objection was the lack of adequate knowledge of radio wave propagation for setting international rules.

As the use of wireless continued to grow, the need for increased regulation became pressing. A third Conference was convened in London in 1912. There, detailed allocations and rules were worked out for low and medium frequencies, and the Treaty resulting from the Conference was ratified by all the major countries.

Underlying the 1912 agreement were the principles that the electromagnetic spectrum is public property, that regulation of its usage is a legitimate governmental function, and that international agreement is required. These principles are taken for granted today, but they were often strongly contested in the early days of wireless.

After the London Treaty of 1912 was ratified, the U.S. Congress passed the Radio Act of 1912. This was still in place when commercial broadcasting began nine years later. Authority for regulation was vested in the Department of Commerce, which received little guidance and few restrictions on its licensing authority. From the standpoint of broadcasting, the most significant provisions were as follows:

- Intrastate wireless communication was not regulated.
- Frequencies available for licensing under the Act were those below 187.5 kHz and above 500 kHz.
- Five hundred kHz and 1,000 kHz were reserved as distress frequencies for ships.
- Amateur stations were required to operate above 1,500 kHz. (The utility of shortwaves for long-distance communications had not yet been discovered, and these frequencies were judged to have little commercial value.)
- There were requirements for "pure waves" and "sharp waves"— that is, limitations on off-frequency radiation and the rapidity of damping from spark transmitters. If strictly enforced, these rules might have placed minor limits on the use of amplitude modulation (which clearly was not contemplated by Congress).
- The licensing of stations for experimental use was specifically permitted.
- The Act did not specifically contemplate broadcasting, but in 1919 the Commerce Department permitted experimental radiotelephone stations to operate as "limited commercial stations."

Commercial Wireless Communications

The Entrepreneurs   Marconi's inventions and demonstrations of wireless communications stimulated the formation of several entrepreneurial companies in Europe and the United States to exploit this exciting new medium. Their evolution followed a strikingly similar pattern. An inventor developed a new product or system, conceived a practical application for it, found financial support from private investors or by selling stock to the public, and formed a business. Thus he played a dual role as inventor and entrepreneur. The major participants in this new industry were Marconi in England, Slaby-Arco in Germany, and De Forest and Fessenden in the United States.

Marconi   Marconi had pioneered the technology of wireless, and his family had pioneered the business. They were good businesspeople, and they managed the company (the Marconi Wireless Company) wisely. It concentrated on telegraphy and was not diverted by attempts to establish a telephone service. After a few difficult start-up years, the company became profitable.

Guglielmo Marconi's family recognized that his unique talents lay not in administration but in technology and in the conception of profitable uses of wireless for industry, government, and the general public. They employed competent administrators to manage the day-to-day affairs of the business, thus allowing Marconi to devote his time to the functions for which he was so eminently qualified.

Marconi realized from the outset that over-water transmission was the best market for wireless communications. There was no competition from wire circuits in ship-to-shore or ship-to-ship traffic. There was competition from cable for transoceanic circuits, but cable was expensive and its availability limited. Further, by a happy chance of nature, radio waves travel farther over water than they do over land owing to the high conductivity of seawater.

To serve this market, Marconi offered leased systems that included a communications service from shore stations and equipment for ship stations. In part this was a strategy born of necessity because Marconi found it difficult to persuade shipowners and other customers to make the large capital investment required for transmitting and receiving stations. But it also had a side effect that benefited the company for a few years.

By controlling the complete system, the company was able to establish a near monopoly. Its operators were forbidden to communicate with stations equipped with non-Marconi equipment. And since it had installed the most extensive network of shore stations throughout the British Empire and in the United States, ships and shore stations with other equipment found it difficult to communicate. This created tremendous resentment in other countries. The Germans were infuriated in 1902 when Marconi stations refused to communicate with the *Deutschland*, a German warship carrying the kaiser's brother on a goodwill trip to the United States. And, of course, there was the issue of safety-of-life at sea. It would have been inhumane to refuse to communicate with a ship in distress. As described earlier, the issue came to a head at the Berlin Conference of 1906 where Marconi's policy was forbidden by international agreement.

By 1906, however, Marconi had a large head start on its competitors, and it remained the dominant international wireless company until the end of World War I. Its start-up costs were high, and it did not turn a profit until 1910, but thereafter it was highly successful. Its most famous feat was its communication with the liner *Titanic* in 1912 when the ship struck an iceberg and sank, resulting on the loss of 1,517 lives. For many years, it was widely reported that the messages were received by David Sarnoff, then an employee of Marconi, at its station at the Wanamaker store in New York. Subsequent research[7] has shown that Sarnoff's role was exaggerated and that the primary traffic from the disaster scene was through the Marconi station at Cape Race, Newfoundland.

Marconi eventually became conservative in his approach to new technologies. He was reluctant to abandon the spark transmitter with its damped oscillations in favor of continuous waves. He was equally reluctant to adopt the Alexanderson alternator, and he contributed very little to radiotelephony. But his role as the original pioneer of wireless communications is unchallenged. In recognition of this, he received the Nobel Prize in physics in 1909.

De Forest Lee De Forest (Figure 1–11) was a gifted engineer, and he had visions for the uses of wireless that went beyond point-to-point communications. His entrepreneurial ventures in communications repeatedly failed, however, and he made and lost three modest fortunes. He was not a good administrator, his ethical and moral standards were elastic and those of his associates even worse, and he made the mistake of attempting radio broadcasting too early.

He founded the American De Forest Wireless Telegraph Company in 1901 and eventually found a backer, Abraham White. White was more interested in selling stock to the public, often fraudulently, than in managing the business, and De Forest, happy to share in the proceeds, did not object. A 1902 entry in his diary reads: "Soon, we believe, the suckers will begin to bite. Fine fishing weather, now that the oil fields have played out. 'Wireless' is the bait to use at present."[8] For a time, White was successful, and De Forest was temporarily well off.

---

[7] Kenneth Bilby, *The General* (New York: Harper & Row, 1986).

[8] De Forest, Diary, February 9, 1902.

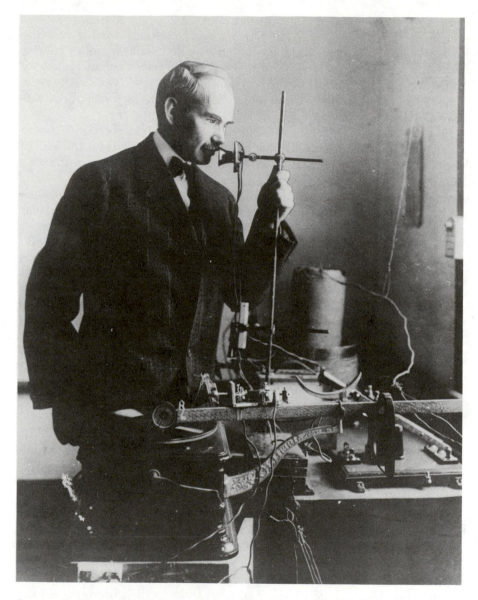

Courtesy Smithsonian Institution, with permission.

■ **Figure 1–11**   Lee De Forest and an early radiotelephone (c. 1910).

The company offered both communications services and equipment. It promoted its services heavily, and it achieved some successes, including the use of its equipment by the press in the Russo-Japanese War. Soon, however, its sins and weaknesses caught up with it. The company was not profitable, and, when Fessenden won a major judgment against it for patent infringement in 1906, White resigned, leaving De Forest with a company that was virtually bankrupt. Meanwhile, De Forest had continued with his inventions, and it was during that low point in his fortunes that he invented the triode.

De Forest soon tried to go into business again, but this time he had a new idea—radio broadcasting. He was probably the first to consider seriously the use of wireless for transmitting programs directly to the general public. In 1907, he formed the Radio-Telephone Company to engage in an elementary form of radio broadcasting. Through the French connections of his wife's family, he was able to lease the Eiffel Tower to mount an antenna for a broadcast of phonograph music in 1908. Later, he carried out a series of broadcasts from the Metropolitan Life Insurance Tower in New York.

The Radio-Telephone Company went bankrupt in 1909, but De Forest soon formed North American Wireless. In 1910, he attempted a live broadcast from the stage of the Metropolitan Opera House in New York, a performance of *Pagliacci* with Caruso in the leading role, in an effort to sell stock in North American Wireless. The stock sale was unsuccessful, and De Forest was forced to become a salaried employee of the Federal Telegraph Company. In 1915, he tried broadcasting once again, this time to an audience of enthusiastic radio amateurs. This effort came to an end when the United States entered World War I in 1917 and amateurs were required to cease operations.

From a commercial standpoint, all De Forest's attempts as a broadcaster failed and for the same reasons. He was not a good businessman, and he did not have a clear idea of how to derive revenue from the business. He was often more interested in promoting the sale of stock to the public than in creating a sound business—he very nearly went to prison in 1912 for fraudulent practices in selling stock of the Radio-Telephone Company. But the most important reason was that neither the spark transmitter nor the Alexanderson alternator was capable of producing a signal that even approached broadcast quality. De Forest was simply ahead of his time.

Fessenden   Fessenden's life was comparatively short (1866–1932), but it was bewilderingly complex. He was born in Canada and taught at Bishop's College and in Bermuda. In 1886, he came to the United States to work in Edison's laboratory but was laid off in a reorganization in 1890. For the next ten years, he taught electrical engineering at Purdue and the University of Pittsburgh, was a consultant to Westinghouse, and worked for the Weather Bureau to establish a system of wireless communications for weather reporting.

While working for the Weather Bureau, he made his most important contribution to wireless, and ultimately to broadcast, technology. He discovered the superiority of continuous waves (CW) to the damped oscillations that were the natural output of spark transmitters.

In 1902, he resigned from the Weather Bureau to form his own company, the National Electric Signalling Company (NESCO) with the backing of two Pittsburgh businessmen. The Company survived for ten years, but it ultimately failed as a result of mismanagement and the inability of Fessenden and his backers to agree on a consistent business strategy.

NESCO was primarily a radiotelegraph company, but Fessenden experimented with radiotelephony and broadcasting. In 1906, using an Alexanderson alternator, he transmitted voice and music from his station in Brant Rock, Massachusetts, to nearby ships—perhaps the first radio broadcast. But like De Forest, he was unable to develop a transmitter that would produce a signal even approaching broadcast quality. NESCO was eventually forced to declare bankruptcy, and Fessenden lost control of the patents he had assigned to it.

Fessenden's final years were sad. He became bitter over the loss of his patents and distrustful of others almost to the point of paranoia. He died believing that he had been cheated of both money and recognition.

AT&T and the Radiotelephone   AT&T's attitude toward wireless communications was initially ambivalent. Telegraphy was competitive with its telephone system, and wireless was competitive with its wired system. In addition, there were sincere doubts as to the quality of radiotelephony. But AT&T, with its desire to monopolize voice communications, could not risk being left out if radiotelephony became practical. It would break AT&T's telephone monopoly, and it might force it to break its policy of noninterconnection with other systems, a policy it followed as zealously as did Marconi (and with more success).

Accordingly, AT&T followed developments in radiotelephony closely and engaged in research in its own laboratory. After its successful development of a radiotelephone apparatus in 1915, it added a radiotelephone service to its system, mainly for communication with ships at sea and with remote areas that were not yet wired.

Naval Wireless Communications

Early History   With the unique ability of wireless to communicate with ships at sea, one would have expected the U.S. Navy to embrace it immediately and enthusiastically. In fact, the Navy did embrace it but only slowly and reluctantly. The officer corps was conservative and reluctant to try a new and untested system. Ship captains resisted the weakening of the absolute authority they had enjoyed while at sea and out of touch with their superiors. And the performance of wireless systems was marginal at first.

There also were problems in procuring complex technical systems (which continue to plague the military). Ideally, each component of the system is procured by competitive bidding on the basis of detailed specifications. In theory, each bidder offers the same product, and the only variables are price and delivery. But what if a supplier offers a superior system that is protected by patents? And what if individual components of other systems are incompatible? How, then, does the buyer obtain a competitive price for the superior system or even components of it when only one supplier can submit a responsive bid?

In spite of these difficulties, the advantages of wireless systems were so obvious that the Navy proceeded to investigate them seriously. Marconi demonstrated his equipment to the Navy in 1899. The demonstration was so successful that the Navy wished to buy an initial test system, but Marconi's terms were so onerous that it could not accept them. It was sufficiently impressed, however, that it investigated wireless equipment manufactured by other European manufacturers and by Fessenden and purchased some experimental equipment from them.

In 1904, President Theodore Roosevelt established the Interdepartmental Board of Wireless Telegraphy to study the problems that had arisen in connection with the operation of wireless systems by the government. Among these were the relationship of the government to commercial communications companies; the responsibilities of the different government departments; the rights to the ether, or the electromagnetic spectrum; and the rights of inventors.

The Board's report was a clear victory for the Navy. It was given responsibility for all the government's coastal stations. Further, the claim of Marconi and other commercial companies to special rights to spectrum was denied, and the government was given preemptive rights. Most of these recommendations were later codified in the international conferences in Berlin in 1906 and London in 1912.

With these policy issues settled, the construction of shore stations and ship installations proceeded more rapidly, although still haphazardly. Communications was not looked upon as the road to high rank by the professional naval officer corps, and it did not attract its most ambitious members. The ship's radio officer was usually an ensign. Thus, at the outbreak of World War I, the vital importance of wireless communications was just beginning to be realized.

World War I　World War I was a proving ground for the military use of radio, and it provided a portent of the explosion in electronic technology that would occur in World War II twenty-five years later. Senior military officers, both Army and Navy, came to realize that radio communications would play an indispensable role in future warfare. The war also accelerated technical development, particularly of the electron tube and its associated circuitry.

The Amateurs

A history of broadcasting technology must include a recognition of the major contributions of radio amateurs. The number of licensed amateurs in the prebroadcast years was not great—about six thousand in 1920—but there were many more unlicensed ones, and no license was required to operate a receiver. They formed a ready-made and enthusiastic audience for early broadcasters; it was estimated that there were more than three hundred thousand listeners to the broadcast of the Dempsey–Carpentier fight in 1921, many of them amateurs.

Amateurs also constituted a pool of experienced engineers and technicians who enthusiastically served the armed forces during World War I. After the war, they gave invaluable support to the growth of broadcasting, which owes much to their knowledge and dedication.

Many felt that the amateurs had been shortchanged when they were allocated only the frequencies above 1,500 kHz in the Radio act of 1912. These frequencies were thought to be worthless because of their limited ground-wave range.

The attitude of the professionals toward shortwaves was illustrated by an anecdote in Paul Wright's best-selling account of the British intelligence services.[9] In 1920, Wright's father was employed as an engineer by Marconi Wireless, and he persuaded Marconi to authorize the construction of a shortwave communication link from England to Australia. During a trip to New York, the elder Wright described his plan to David Sarnoff, then the general manager of American Marconi. Sarnoff expressed deep skepticism that it would work, and he told Wright, "You can kick my ass all the way down Broadway if it does." It did work, and the younger Wright reported, "My father's only regret was that he never took the opportunity to kick Sarnoff's ass all the way down Broadway!"

The amateurs made a virtue of necessity and discovered the potential of sky-wave transmission for long-distance communications in the shortwave portion of the spectrum, a capability that is still used for international broadcasting.

The Professionals

While the amateurs were making important contributions to radio technology, it was being brought to full fruition by a group of able and dedicated professional radio engineers. Radio engineering was initially regarded somewhat contemptuously by the members of electrical engineering facul-

---

[9] Paul Wright, *Spy Catcher* (New York: Dell Publishing, 1987).

ties who believed it to be a more appropriate subject for technicians. But radio engineers of international renown, such as John Morecroft and Michael Pupin at Columbia and G.W. Pierce at Harvard, soon gave the profession solid respectability.

The Institute of Radio Engineers (IRE), a professional engineering society, was founded in 1912, and its *Proceedings* provided a means of disseminating information about new developments and discoveries to its members. Its pages became an archive of technical progress.[10]

■ THE FOUNDING OF RCA AND ITS AFTERMATH

The Founding of RCA

During World War I, the U.S. Navy became concerned because Marconi and the British had a virtual monopoly of the United States' foreign wireless traffic. At the outbreak of the war, all the radio transmitting stations in the United States with sufficient power for transatlantic communications were owned by foreign interests—three by the American Marconi Company and two by subsidiaries of German companies, Telefunken and HOMAG. To maintain its neutrality, the United States assumed a degree of control over these stations to prevent them from being used for espionage.

When the United States entered the war, the German stations were seized as enemy property, and Marconi's stations were placed under the control of the Navy. But the Navy still chafed at having to depend on a foreign-owned company, even though a friendly one, for its international radio communications. In addition, Marconi had not made many friends in the government with its arrogance and intransigence in exploiting its near monopoly.

The issue came to a head at the war's end when GE advised the government that it was preparing to fill the order for Alexanderson alternators that American Marconi had placed before the war. The secretary of the Navy, Josephus Daniels (remembered in naval circles for forbidding the consumption of alcoholic beverages on board ship), was furious that this result of American technology was going to a foreign company. He communicated with President Wilson, who was in Europe having his own troubles with the British at the peace negotiations. Wilson immediately decided that the order for the alternators must be canceled. Franklin D. Roosevelt, Assistant

---

[10] The first issue of the *Proceedings of the IRE* was published on January 1, 1913. The following articles appeared in this issue; M. Pupin, "Experimental Tests of the Radiation Laws of Antennas"; S.M. Hills, "High Tension Insulators for Radio Communications"; L. De Forest, "Recent Developments at Federal Telegraph"; A.E. Kennelly, "Daylight Effects in Radio Telegraphy."

Secretary of the Navy, then convened a historic meeting of Rear Admiral Bullard, the director of Naval Commuications; Commander Hooper, one of the Navy's pioneers in the use of radio; and Owen D. Young, GE's general counsel. From this meeting, a radical plan emerged.

It was proposed that a new company be formed as a subsidiary of GE. It would buy out the stock of American Marconi and operate an international communications company under American ownership. The plan was approved by the U.S. government and the GE board of directors with amazing speed, and it remained only to negotiate the terms with Marconi.

Marconi, probably sensing the inevitable, negotiated in a cooperative fashion. Within three months, the deal was struck, and GE's subsidiary, named the Radio Corporation of America (RCA), bought American Marconi for $3.5 million. Most of its staff, including its president, Edward J. Nally, and its commercial manager, David Sarnoff, came with it.

RCA's Charter

Soon after the sale was consummated, an internal dispute arose over the new company's fundamental strategy. Nally's interest was international radiotelegraphy. He believed that this was RCA's charter, and he proposed that RCA remain strictly a radiotelegraph company. With his broad experience in dealing with foreign companies (including his former employer), he was successful in the critical task of arranging contracts with "correspondents," companies in foreign countries that handled the overseas terminals of communications circuits. The business showed promise of prospering, and he saw a bright future for it.

There was, however, another faction led by Sarnoff who felt that this was far too limited a vision for the new company. Sarnoff proposed, in his famous "radio music box" memorandum in 1915, that its horizons be broadened to include manufacturing, radiotelephony, and broadcasting. Sarnoff found an eager ally in Young, who sold Sarnoff's proposal to the GE board.

The RCA Patent Pool

RCA's broader charter soon caused serious patent problems. Certain key patents necessary for the manufacture of broadcasting equipment were owned by Westinghouse, AT&T, and to a lesser extent United Fruit. This created an impasse because none of the companies owned sufficient patents to manufacture radio receivers or equipment without infringing on others.

Owen D. Young, general counsel of GE and chairman of RCA and a powerful figure at GE who would soon become its chairman, believed "If you can't lick 'em, join 'em.'' He commissioned Sarnoff to make AT&T and Westinghouse an offer they couldn't refuse—RCA stock in return for their patents, which would be placed in a common pool. The negotiations proceeded swiftly, and agreements were reached with AT&T in July 1920 and with Westinghouse and United Fruit in March 1921. At the conclusion of the negotiations, the ownership of RCA was as follows:

Company	Percent	Key Patents
General Electric	30.1	Marconi's wireless patents
Westinghouse	20.6	Armstrong's regenerative detector and oscillator
AT&T	10.3	De Forest's audion
United Fruit	4.1	Loop antenna, crystal detector
Public	34.9	

The "Radio Trust"

The agreement among the companies broke the patent impasse, but it went far beyond the patent pool—it was in fact a trust. It was agreed that AT&T would manufacture radio transmitters. GE and Westinghouse would manufacture receivers but with predetermined market shares—60 percent for GE and 40 percent for Westinghouse. RCA would continue to operate the international telegraph business inherited from Marconi and would act as sales agent for GE and Westinghouse receivers. With the trust's lock on all the essential patents, no one else could legally manufacture radio receivers or equipment.

The manufacturing agreement could not last. It was a per se violation of the antitrust laws, and it was soon attacked on those grounds. But it was also a bad business arrangement. A successful business requires the unified direction of a single organization. The trust attempted to manage its affairs with a clumsy committee structure, but it was doomed to failure, although it did survive for a few years.

■ SUMMARY

As the world entered the 1920s, the chain of technologies required for radio broadcasting was complete.

- The basic laws of electricity and electromagnetism were known and understood.
- The propagation characteristics of long-wave and medium-wave radiation had been determined.

- The means for generating AM radio signals for radiotelephone systems had been developed, as had microphones and loudspeakers.
- The basic laws of sound and hearing had been determined, and primitive but workable transmitting and receiving equipment had been developed.
- The patent impasse had been solved, albeit with an agreement that was later judged to be illegal.
- It now remained for men of vision to apply these technologies to practical broadcasting systems.

2

■ ■ AM RADIO BROADCASTING

■ AN OVERVIEW

Radio broadcasting ranks with electricity, the telephone, the automobile, and indoor plumbing when measured by its impact on American life. It captured the imagination of the American public and created an atmosphere of excitement that touched everyone.

For technicians and engineers, there was the excitement of a remarkable new technology. For entrepreneurs, there was the excitement of a profitable new business opportunity. For educators, there was the excitement of a new medium for the wider dissemination of culture and learning. For performers, there was the excitement of a glamorous outlet for their talents. For advertisers, there was the excitement of a new medium for reaching the buying public. And for the public, there was the excitement of an endless source of news and entertainment brought directly into the home. Radio receivers, with their glowing tubes and odor of hot Bakelite, seemed almost alive as they brought in distant voices and music. For everyone who was involved in this new industry in the 1920s and 1930s, it was a never-to-be-forgotten experience.

AM radio broadcasting has prospered for nearly seven decades from its beginning in 1920 until the present time. With the help of advances in technology, it has been an exceedingly resilient medium, and it has continued to prosper in spite of wars, depressions, and competition from television.

The Growth of the AM Broadcasting Station Population

One measure of the size of the AM broadcasting industry is the number of stations on the air (Figure 2–1). The attraction of radio broadcasting as a business and communication medium was so great that the number of on-air stations rose to more than seven hundred shortly after the first broadcasts in 1920. Since there was no effective instrument for technical regulation at the time, anarchy soon prevailed. Interfaces between stations was so severe that the number of stations remained essentially constant for a number of years.

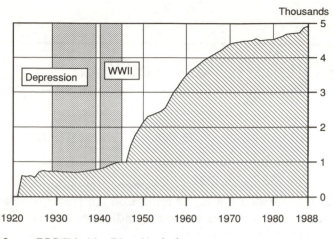

Source: FCC/Television Digest Yearbook.

■ **Figure 2-1** The growth of on-air AM radio stations, 1920–1988.

In 1927, the Federal Radio Commission (FRC) was established with the authority to assign powers and frequencies in a manner that would minimize interference. In the short term, this caused a small drop in the number of stations as the FRC thinned their ranks, but the establishment of a policy for frequency assignments laid the groundwork for major increases in future years.

In 1934, the Federal Communications Commission (FCC) was established to replace the FRC. Its attention was focused primarily on nontechnical issues such as programming, but its engineering department expanded and improved the rules for frequency assignment that had been established by its predecessor.

The meteoric growth in the number of AM stations at the end of World War II, inspired by the profitability of the older stations, was made possible by the introduction of the directional antenna in the late 1930s. This key technical development permitted many more stations to coexist on the same channel with a level of interference judged to be tolerable.

The Growth in Radio Time Sales

Advertising time sales are another measure of the radio broadcasting industry's size (Figure 2–2). Even after adjustment for inflation, the growth rate shown in Figure 2–2 is impressive. Although radio has been oversha-

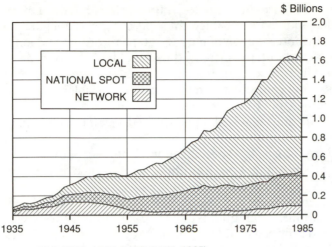

$ Billions

LOCAL
NATIONAL SPOT
NETWORK

Source: FCC (1935–1979); NAB (1979–1985).

■ **Figure 2–2**   Radio time sales, 1935–1985. (These data were compiled by the FCC, which terminated the service at the end of 1979. FM station sales (minor) are included through 1960. After 1960, the sales of AM/FM stations are included.) National spot sales are made by broadcast stations to companies that advertise on a national basis—for example, soap manufacturers. Rates for national advertisers are higher than for companies that advertise only locally.

dowed in recent years by television, it continues to be one of the most cost-effective means of reaching the buying public.

The Heyday and Decline of the Networks

The shift in the mix from network sales to local sales is as impressive as their overall growth. Prior to and during World War II, the networks dominated broadcasting. Network time sales constituted more than 40 percent of the total, while revenues from local advertising constituted less than 30 percent. With the advent of television in 1947, network advertisers began moving to the new medium, and radio network sales declined rapidly after 1951. By 1956, network time sales were less than 10 percent of the total, while local sales had risen to 60 percent.

Radio broadcasting had become so dependent on the networks that for a time their decline threatened radio's future. Fortunately, the resourcefulness of the free enterprise system prevented this.

In response to the loss of network income from the plush national accounts, radio stations adopted drastic cost-reduction measures, many of them made possible by advances in technology. Broadcasting station equipment became simpler and more reliable, thus permitting reductions in technical staffs. With the introduction of magnetic tape recording and long-playing (LP) records, along with the increased use of older 78-rpm records, stations were able to switch almost entirely from live to recorded programs and announcements. New control equipment made it possible for an announcer to play recordings, put commercial announcements on the air, and adjust signal levels—functions that had once required a staff of highly paid technicians.

Stations also reduced costs by specialization. They are now quasi-officially classified by their primary programming format, such as Country or Classical.

The cost-reduction efforts were successful. Based on reports to the FCC, in 1946, the average annual operating expense of all radio stations was $240,000. By 1956, it had dropped to $145,000 in spite of inflation, which had reduced the value of the dollar by more than one-half.

Advances in receiver technology also were effective. The miniaturization of portable receivers, first by improved electron tubes and then by transistors, created a whole new audience in automobiles and outside the home.

In addition, innovative programming formats were required to appeal to listeners who were away from their television sets. The industry rose to the challenge, and total radio advertising revenues continued to rise, even as the networks' sales were declining.

AM Radio Broadcasting Today

While radio broadcasting was prospering in total, AM broadcasting was losing a major part of its market share to FM because of its technical deficiencies. The coverage areas of most AM stations are severely reduced at night by "monkey chatter," interference from stations on the same frequency. Interference from adjacent stations on adjacent channels also is a serious problem in many locations. The bandwidth of AM channels is not sufficient for high-fidelity service, and there is a wide variation in the daytime coverage of AM stations depending on the power, soil conductivity, and frequency.

Some of these deficiencies can be alleviated by technical advances and improved channel assignment policies, but none of them can be totally corrected. This does not mean that AM will die, however, because it also has significant advantages over FM that are important in some situations. It provides the only coverage of many rural areas, and in densely populated areas there are not enough FM channels to meet the demand. As a result, AM radio continues to play an important though secondary role in radio broadcasting.

■ PIONEERING, 1920–1926

The First Broadcasts

Where and when was the first radio broadcast? And who was the first radio broadcaster? Credit for this accomplishment has customarily gone to Westinghouse station KDKA in Pittsburgh for its broadcast of the Harding–Cox presidential election returns on November 2, 1920.

The idea for the broadcast had been developed by Frank Conrad, a Westinghouse engineer who was also an enthusiastic radio amateur, and Henry P. Davis, a Westinghouse vice president. As a hobby, Conrad had been transmitting opera music and announcing its source over his amateur station. To his surprise, he received a deluge of letters from fellow amateurs acknowledging that they had heard his transmissions. Davis, after learning of the response to Conrad's programs, concluded that regularly scheduled broadcasts would stimulate the sale of Westinghouse amateur receivers. He obtained a license from the Department of Commerce to operate an experimental station on 833 kHz, a frequency below the bands allocated for amateur use, and he gave Conrad corporate financial support to build and operate a station at the Westinghouse plant. The broadcast of the election returns was heavily promoted in the newspapers, and thousands of amateurs were able to receive it.

KDKA's claim to have achieved the first broadcast has been challenged, however. There were many earlier radio transmissions intended for a group of listeners rather than a single receiving terminal. For example, there was Fessenden's broadcast to ships at sea from his Brant Rock station in 1906 and De Forest's broadcast from the Eiffel Tower in 1908. The 1920 election returns also were broadcast from an amateur station, 8MK (soon to become broadcasting station WWJ), installed at the *Detroit News*. And David Sarnoff preceded Davis in conceiving the potential of radio to reach a mass audience for commercial purposes.

Sarnoff's concept was expressed in 1915 when he was commercial manager of American Marconi. He wrote its president, Edward J. Nally, the famed "Music Box" memo:

> I have in mind a plan of development that would make radio a household utility in the same sense as the piano or phonograph. The idea is to bring music into the house by wireless.
>
> While this has been tried in the past by wires, it has been a failure. . . . With radio, however, it would seem to be entirely feasible. . . . The problem of transmitting music has already been solved in principle, and therefore all the receivers attuned to the transmitting wave length should be capable of receiving such music. The receiver can be designed in the form of a "Radio Music Box" and arranged for several different wavelengths, which should be

changeable with the throwing of a single switch or the pressing of a single button.

The "Radio Music Box" can be supplied with amplifying tubes and a loudspeaking telephone, all of which can be neatly mounted in one box. The box can be placed on a table in the parlor or living room. . . . There should be no difficulty in receiving music perfectly when transmitted within a radius of 25 to 50 miles.[1]

Nevertheless, a good case can be made for KDKA's claim of priority. KDKA was one of the first stations to use electron tube technology to generate the transmitted signal and hence to have what could be described as broadcast quality. It was the first broadcast to have a well-defined commercial purpose—it was not a hobby or a publicity stunt. It was the first broadcast station to be licensed on a frequency outside the amateur bands. Most significantly, it was the direct ancestor of modern broadcasting. By contrast, the broadcasts of De Forest and Fessenden used dead-end technology such as the spark transmitter, and they can best be described as collateral ancestors.

The success of the initial KDKA broadcast gave Davis a tremendous boost in the Westinghouse executive suite. He quickly obtained licenses for three more stations, WJZ (now WABC) in New York, WGZ in Springfield, Massachusetts (now in Boston), and KYW in Chicago (now in Philadelphia). A Westinghouse broadcast division was created with Davis as its manager. He brought showmanship to its programming, as the first KYW broadcast was from the Chicago Opera House, featuring the renowned soprano Mary Garden. He began to emphasize the broadcasting of live programs rather than recordings, and the number of listeners grew rapidly.

Westinghouse's success placed Sarnoff in a difficult position. It was his vision that RCA would be the leader in both the broadcasting and manufacturing aspects of the radio industry. He had foreseen the potential of broadcasting earlier than Davis with his "Music Box" memorandum (although Davis was not aware of it). But now Westinghouse, a part owner of RCA, was aggressively moving ahead of it as a broadcaster.

Sarnoff's ability to lead RCA in an attack on its competitors (particularly when a competitor was a part owner) was limited by his position in the corporate hierarchy. At the time, he was only the commercial manager of RCA, and its president, Edward J. Nally, was not enthusiastic about his desire to expand its business beyond radiotelegraphy. Sarnoff, therefore, decided to go over his boss's head and appeal to Owen D. Young, RCA's chairman. He arranged a private dinner meeting with Young in which he

---

[1] Kenneth Bilby, *The General* (New York: Harper & Row, 1986), 39.

reiterated his plans and aspirations for the company. It was a masterful presentation, and Young agreed to support him. He also promoted him to general manager. With Young's backing, Sarnoff had the corporate clout he needed to proceed with his plans.

Sarnoff's first move was a dramatic public relations feat that diverted the public's attention from Westinghouse. He borrowed a transmitter from GE and used it to broadcast the Dempsey–Carpentier world heavyweight championship fight on July 2, 1921. It was an extremely hot day, and both the fight and the transmitter lasted only four rounds. Dempsey knocked out Carpentier, and the transmitter failed simultaneously from the heat. Nevertheless, the broadcast was a huge success, and more than 300,000 people were estimated to have listened, mostly with amateur receivers. It received widespread publicity in the newspapers, and henceforth there could be no doubt that the radio broadcasting industry had been born, with RCA as one of its leaders.

The First Broadcasting Stations

Westinghouse and RCA were not alone in noting the success of the KDKA and Dempsey–Carpentier broadcasts. During the following months, there was a rash of requests for station licenses. In response to these requests, the Department of Commerce recognized broadcasting stations as a distinct class and began licensing them on a single frequency, 833 kHz. By the end of 1921, twenty-eight stations had been granted licenses, and nine of them were on the air. Four of these were Westinghouse stations.

Present Call Letters	City
WABC (formerly WJZ)	New York City
KYW	Chicago (now Philadelphia)
WBZ	Springfield/Boston
KDKA	Pittsburgh
KQV	Pittsburgh
WJAS	Pittsburgh
WWJ	Detroit
KNX	Los Angeles
KCBS	San Francisco

The rush for station licenses accelerated, and 430 more were issued during the first seven months of 1922—all on the single frequency of 833 kHz. The Radio Act of 1912 did not give the Department of Commerce the authority to deny a license to a U.S. citizen, and stations were left to solve the mutual interference problem among themselves.

Chaos in the Airways and Early Attempts at Regulation

Interference became so severe that there was a return of the chaos that had characterized the early years of wireless. The interference problems of point-to-point communications had been largely solved by the Berlin and London treaties and the Radio Act of 1912, but the voracious spectrum requirements of broadcasting had not been anticipated.

At first, stations tried to solve the problem by making cooperative time-sharing agreements. Some followed a practice of doubtful legality and used a channel at 619 kHz that had been allocated for government weather and crop reports. But these provided only temporary relief, and they soon proved to be inadequate as the number of stations continued to increase.

The Department of Commerce and its secretary, Herbert Hoover, did the best they could within the limitations of the Radio Act of 1912 to cope with this situation. The department convened National Radio Conferences in Washington each year from 1922 to 1925. At the 1922 conference, an additional channel, 750 kHz, was assigned for broadcasting. Stations operating on that channel were required to operate with a minimum power of 500 watts and a maximum power of 1,000 watts and to broadcast live—that is, they could not use phonograph records. Stations wishing to operate at a lower power were required to remain at 833 kHz, which became much more crowded.

The number of stations continued to grow. In 1922, the department ran out of three-letter combinations for station calls and began to assign four letters. It soon became clear that the two channels were inadequate to meet the demand. The 1923 conference took the epochal step of allocating the entire band of frequencies from 550 kHz to 1,350 kHz for broadcasting. The 1924 conference extended the band further to 1,500 kHz. These conferences also established classes of stations: *high power* with a maximum of 5,000 watts, *medium power* at 500 watts, and *low power* at 100 watts. The low-power stations all operated on 833 kHz. These three classes were the precursors of the *clear, regional,* and *local* channels later established by the FRC.

With the Department's limited ability to assign the operating frequencies of stations, not even the nearly 1,000 kHz allocated for broadcasting in 1923 and 1924 was sufficient to accommodate the growing demand for licenses. By 1925, more than seven hundred stations were on the air, and the interference problem was becoming steadily worse. At the 1925 conference, Secretary Hoover decreed that no more licenses would be issued.

Unfortunately, all these efforts to solve the interference problem were made ineffective by a court decision in 1925. The secretary of commerce sued the Zenith Radio Corporation to force its station, WJAZ in Chicago, to comply with the frequency and power specified in its license. A federal court, taking a narrow legalistic position, ruled that since the 1912 act did not specify the criteria by which the Department of Commerce should exercise its "discretion" in specifying a station's power and frequency, it could not preempt this power by administrative rulings. Accordingly, the court said, the Department did not have the authority to regulate either the power or the frequency of broadcasting stations. After this ruling, the chaos progressed from serious to intolerable, and necessity forced Congress to pass the Radio Act of 1927. This Act established the FRC and gave it the necessary authority to regulate the technical operation of the stations. The FRC, to be described later, laid the groundwork for the current technical regulation of radio broadcasting.

The Economic Basis of Broadcasting

It is appropriate to ask why neither De Forest nor Fessenden was able to establish a profitable business in broadcasting, a service for which such a strong public demand was demonstrated in 1921 and 1922. The primary reason was technical. Neither the spark transmitter nor the Alexanderson alternator was capable of producing an on-air signal of satisfactory quality. Their broadcasts were of interest to amateurs as a technical feat, but the quality was not good enough to serve as entertainment for the general public.

In addition, neither De Forest nor Fessenden was able to perceive a means of making broadcasting pay. Once electron tube transmitters became available, it was necessary for the broadcasting pioneers to turn their attention to the problem of obtaining financial support. There were four principal possibilities:

1. The broadcasting system could be owned and operated by the government with funds derived from taxation or receiver license fees.
2. Stations could derive their financial support from the sale of advertising.
3. Colleges and universities could operate the stations as part of their overall educational mission.
4. Radio receiver manufacturers could operate stations in order to provide programming availability as an incentive for the purchase of sets.

The first and second possibilities represent opposite extremes of political ideology. The advocates of government ownership argued that the radio spectrum is a valuable resource owned by all the people and that it should not be used by private enterprise for private profit. The advocates of advertising as the basis of financial support argued that the incentive for high ratings in a laissez-faire system is the best assurance that broadcasters will give the public what it wants.

In Europe, public ownership won the day, and most European systems are owned and operated by quasi-governmental organizations such as the British Broadcasting Corporation (BBC) in England. Quasi-governmental ownership also developed in the United States as an extension of educational broadcasting. Today more than thirteen hundred public radio stations, most of them FM, are on the air.

At first, the broadcast of advertisements was not viewed with much enthusiasm. A number of influential radio pioneers, including Sarnoff, objected to it on the basis that it was "unseemly." Secretary of Commerce Hoover stated that broadcasting advertisements was "unthinkable." But advertising rapidly became and continues to be the principal economic basis of radio broadcasting in the United States. Broadcasting does not enjoy the same degree of freedom from governmental regulation as other advertiser-supported media such as newspapers and magazines, but it is essentially a private for-profit institution.

The concept of manufacturer-supported broadcasting did not survive for long, although a number of manufacturers became successful broadcasters. The first commercial message was broadcast over AT&T's New York station, WEAF (later owned by NBC and eventually renamed WNBC), on August 28, 1922. Air time was bought by a local real estate concern to advertise the availability of apartments in a new development in Jackson Heights, Long Island. The fee was $50 for a ten-minute commercial. The advertisement was successful in renting apartments, and from this modest beginning, the use of radio as an advertising medium grew exponentially.

Early Radio Broadcasting Technology

Early radio broadcasting technology was primitive, but with the twin incentives of radio's commercial success and the excitement of new engineering challenges, it developed rapidly. At the end of the pioneering phase, it had progressed sufficiently to provide a solid basis for the meteoric growth of radio that followed.

Transmitters and Antennas The maximum transmitter power permitted during the pioneering phase was 5 kilowatts, and most transmitters were rated at less than 1 kilowatt. They used the modulation method developed by AT&T for radiotelephony (low-level grid modulation). Their efficiency was low—less than 25 percent—which not only increased the power bill but also created a problem in dissipating the wasted power. Most of this appeared as heat on the anode of the final amplifier tube, and water cooling was required for higher power transmitters. The anode was immersed in a jacket through which water was circulated to carry away the heat. Since the anode operated at a high voltage, distilled water (a poor conductor) and meticulously cleaned ceramic plumbing had to be used.

In spite of these difficulties, early broadcast transmitters were comparatively simple devices, and skilled amateurs could and did build them. To do so, however, they had to purchase tubes covered by patents in the RCA pool and licensed for use only in amateur equipment.

Transmitting antennas were simple tower structures or long wires strung betwen poles. The importance of a good grounding system was just being discovered, and many antennas operated without one. Multitower directional antennas were not yet in use.

Radio Receivers Radio receiver design went through a major metamorphosis during the pioneering period. Some early receivers used a "cat's whisker" crystal detector with no electron tube amplification. Earphones were used in place of loudspeakers, and the energy to operate them was extracted from the radio wave. As expected, a strong signal was required, and only nearby stations could be received. These receivers were often sold as kits, and assembling the kits became a favorite occupation of many teenage boys, who used a cylindrical oatmeal box as the form for the tuning coil.

Armstrong's regenerative detector was used in some receivers, but their appeal to consumers was limited by their tendency to howl when improperly adjusted. The tuned radio frequency (TRF) receiver was a more common design. This had a number of amplifier stages in cascade, all tuned to the desired station. To change stations, each stage had to be retuned individually and precisely with an array of knobs on the front panel.

The first loudspeakers were essentially telephone-type earpieces mounted at the throat of a horn that provided acoustic amplification. These were soon replaced by moving coil speakers, which had far better fidelity and greater power-handling capacity and are still almost universally used today.

Early sets were battery operated. An A battery, usually a storage battery, was used to heat the tubes' cathodes, a B battery provided the anode voltage, and a C battery provided a fixed bias voltage for the tubes' grids. In the mid-1920s, these designs began to be superseded by receivers equipped with Armstrong's superheterodyne circuit (see Chapter 1) and 110-volt alternating current (AC) power supplies. The superheterodyne circuit made it possible to tune the set with a single knob and improved its sensitivity. The AC power supply replaced the inconvenient and costly batteries. These advancements revolutionized receiver design and were an important factor in the growth of radio broadcasting during its second phase. Figure 2–3 shows two early radio receivers.

Courtesy RCA Corporation.

■ **Figure 2–3** Early radio receivers. The Radiola II (top), one of RCA's first commercial models, was marketed in the early 1920s. It used batteries and a headset, and the TRF circuitry required three tuning knobs. The Radiola 17 (bottom) was marketed in the mid-1920s. It included three major technical innovations—AC power, a loudspeaker, and superheterodyne circuitry, which reduced the number of tuning knobs to one.

Many of the technical advances were made possible by major improvements in electron tube technology. One of the most important developments was the invention of the screen grid tetrode, an electron tube with a second grid structure mounted between the triode's control grid and the anode. It provided an electrical barrier between the tube's input and output circuits and greatly increased the amplification that could be achieved with a single tube.

The 1929 RCA receiving tube catalog provides a snapshot of the status of electron tube technology as the radio's pioneering phase came to a close. It lists twenty-one tube types—ten triode detector/amplifiers, three tetrode amplifiers, six triode power amplifiers for driving loudspeakers, and two diode rectifiers for converting AC to DC (direct current). A representative sample follows:

Type	Description	Price
UX-240	High-gain triode detector/amplifier; amplification factor = 30	$2.00
UY-224	High-gain tetrode radio frequency amplifier; amplification factor = 400	$4.00
UX-245	Triode power amplifier; power output = 1.6 watts	$3.50
UX-210	Triode power amplifier; power output = 4.0 watts	$9.00
UX-280	Double diode, full-wave rectifier	$3.00

The prices in this list should be multiplied by more than ten to get an idea of what the tubes would cost in 1989 dollars.

Calculating Radio Station Coverage   The development of a quantitative method for calculating the coverage of radio stations and the mutual interference between them was a critical requirement for a sound licensing policy. The basic mechanisms of ground- and sky-wave propagation of radio waves were understood in 1920 when broadcasting began, but there were no methods for making quantitative estimates of coverage or interstation interference. Such methods required the development of a statistical data base from actual measurements. To fill this need, a number of organizations—the National Bureau of Standards, universities, the military, and the broadcasters themselves—made extensive measurements of ground and sky waves. Little by little, a data bank was assembled to provide a rough guide for the deliberations of the FRC in 1927 and 1928 as it tried to bring order out of the chaos.

Trouble in the Radio Trust

The controversy surrounding the FRC's efforts to solve the problem of frequency assignments was matched by conflict in the commercial world of radio. Two conflicts were inherent in the structure of RCA.

The first was a basic conflict of interest among its major stockholders—GE, Westinghouse, and AT&T. Competitive conflicts in the manufacture and sale of equipment had presumably been resolved (probably illegally) by share-of-market agreements. But the agreement was silent as to who had the right to engage in broadcasting. All three stockholders and RCA itself wished to be broadcasters, and this led to serious disputes.

The other problem was the patent pool. If its terms had been rigidly enforced, no one but GE and Westinghouse could have manufactured radio receivers. In practice, these terms could not be enforced. Their legality under the antitrust laws was seriously challenged, and they had a gaping loophole—the exclusion of amateur equipment from the coverage of some of the patents. For example, when De Forest sold the patent rights for the triode to AT&T, he retained its rights for use in amateur equipment. It was not easy to distinguish between receivers built for the public and receivers built for amateurs, and other manufacturers took advantage of this by claiming that their receivers were being built for amateurs.

Sarnoff set out to solve these problems in a manner favorable to RCA. He was determined that RCA should become a broadcaster even though this would put it in competition with its stockholders. He was equally determined to prevent the array of receiver manufactuers that had sprung up—Atwater Kent, Stromberg Carlson, Zenith, and Philco, to name a few—from using the patents on RCA's pool. Being smaller and more agile than the clumsy committee-managed consortium of GE, Westinghouse, and RCA, these manufacturers were dominating the receiver market, which was growing at an incredible speed.

AT&T Enters and Leaves Broadcasting

Although Westinghouse was the first of the trust's members to become a broadcaster, Sarnoff regarded AT&T as his most formidable rival, and AT&T's actions gave him ample reason for concern. AT&T argued, with some justification, that broadcasting was a natural extension of its telephone business. In contrast, Westinghouse and GE were primarily manufacturers, and RCA's role was supposed to be in radiotelegraphy and the marketing of receivers.

AT&T obtained a license for a broadcasting station in New York that went on the air in 1922 as WEAF. It envisaged radio as a toll service similar to the telephone. Anyone wishing to broadcast a message could do so by leasing time, just as one would make a long-distance telephone call. (The first radio commercial was broadcast under this arrangement.) To expand the concept, AT&T established the precursor of the radio networks, which it described as a toll network. It consisted of telephone line interconnections to broadcasting stations in other cities so that a customer could lease time in cities other than New York.

AT&T had a monopoly on telephone service, and for a time it sought to extend its monopoly to broadcasting. In 1923, A.W. Griswold, an AT&T vice president, advised its operating companies:

> We have been very careful, up to the present time, not to state to the public in any way, through the press or in any of our talks, that the Bell System desires

to monopolize broadcasting; but the fact remains that it is a telephone job, that we can do it better than anyone else, and it seems to me the clear, logical conclusion must be reached is that sooner or later in one form or another, we have got to do the job.[2]

AT&T sensed that its relationship with RCA might be a problem in its efforts to become a broadcaster, and in 1922 it sold its stock in the company. Subsequent events have shown that AT&T's attempt to establish a monopoly in broadcasting was bound to fail, as the public would never have tolerated it. But its failure was hastened by AT&T's heavy-handed tactics in dealing with its competitors.

AT&T tried to make use of its right under the RCA patent pool to be the sole manufacturer of transmitters. There was widespread infringement of these patents, and by the end of 1922, only thirty-five of the several hundred stations were using transmitters manufactured by Western Electric, AT&T's manufacturing arm. AT&T demanded that stations using non-Western Electric transmitters pay a license fee or be put off the air. In some cases, it even refused to sell transmitters to broadcasters, suggesting that they buy time on AT&T's stations instead.

AT&T went a step further. When RCA, GE, and Westinghouse announced a plan to compete with AT&T with their own toll network, AT&T refused to lease them the necessary intercity lines. This put AT&T and RCA on a collision course.

AT&T brought the rivalry to a head by its action in an unrelated matter. It announced plans to enter the business of receiver manufacturing, using a Bell Laboratories' design that allegedly avoided all the patents in the RCA pool. This gave an infuriated Sarnoff the opportunity to invoke the arbitration clause in the agreement that had established the patent pool. He challenged not only AT&T's right to manufacture receivers but also its right to engage in broadcasting.

After lengthy hearings, the arbitrator, Ronald W. Boyden, ruled almost completely in favor of RCA. He found that under the original agreement between the companies, AT&T was prohibited not only from manufacturing and selling receivers but also from engaging in broadcasting. AT&T was stunned, and it struck back by obtaining a legal opinion that the entire agreement (of which it was a part) was illegal on antitrust grounds and therefore null and void.

At this point, cooler heads prevailed. Neither RCA nor AT&T could go to court with clean hands concerning an antitrust matter, and AT&T's new president, Walter S. Gifford, was more inclined to compromise. He was convinced that AT&T's gains from broadcasting were not great enough to justify the antitrust risks and the adverse publicity it was receiving. Accord-

---

[2] Ibid., 72.

ingly, Sarnoff and Edgar Bloom, representing the Bell System, engaged in negotiations to settle the matter out of court.

The outcome of their meetings was virtually a complete capitulation by AT&T. It agreed to sell WEAF to RCA for $1 million. It also agreed to discontinue its Washington station, WCAP, and gave up its airtime to the new RCA station, WRC. In return, RCA agreed to lease AT&T's toll network at an annual rate of $1 million. Their agreement, which was signed on July 22, 1926, was a complete victory for Sarnoff. RCA was now able to enter the broadcasting business without fear of competition or harassment from AT&T.

The Founding of NBC

While Sarnoff was negotiating the AT&T–RCA accord with Bloom, he was busily engaged in making plans to take over AT&T's toll network. By this time, his early opposition to broadcasting advertisements had faded, and AT&T's toll network concept evolved in his mind into the network–affiliate relationship that survives to this day. The result was the formation of a broadcast radio network to be owned 50 percent by RCA, 30 percent by GE, and 20 percent by Westinghouse. The new company, named the National Broadcasting Company (NBC), was incorporated on September 9, 1926. The network era of radio broadcasting was about to begin.

Receiver Manufacturing and the Patent Pool

The market for radio receivers grew with incredible speed. In 1922, the second year of broadcasting, the Federal Trade Commission (FTC) reported that industry sales were $60 million (nearly $1 billion in today's dollars). But in spite of its presumed monopoly resulting from its patent position, RCA's sales of receivers manufactured by Westinghouse and GE were only $11 million. The patent protection was not effective, and RCA, dependent on the bureaucratic managements of two large companies for which radio was a sideline, found it difficult to compete with smaller companies that could respond more quickly to new technologies and the demands of the market.

Competing manufacturers attempted to evade charges of patent infringement by a number of ruses. One, already described, was to claim that the receivers were for amateurs. Another was to allege that the tubes they bought from RCA were replacements for tubes in GE or Westinghouse receivers. Another was to sell sets without tubes, leaving it to the retail store or the customer to supply them.

Sarnoff was furious at this turn of events. He believed that inventors were entitled to the fruits of their inventions and that this entitlement passed on to the purchasers of their patents. With the approval of RCA's board, he took drastic actions to stop the infringement. He attempted to block the replacement-tube loophole by requiring distributors to send back burned-out tubes in order to receive a shipment of new ones. He dropped distributors who sold competing brands of receivers, and he began legal proceedings against a number of manufacturers. But unfortunately for Sarnoff, these efforts were not successful.

His problem was that the use of RCA's patents was essential for the survival of the other manufacturers. Faced with the prospect of being forced out of business, RCA's competitors, led by Zenith's Eugene McDonald, fought back savagely. Their main weapon was the antitrust laws, and they used it in Congress, in the press, and in the courts.

For a time, RCA fought back, arguing that its monopoly resulted from a legitimate exercise of its patent rights. But as time passed, the battle became hopeless. The FTC filed a formal complaint against RCA, GE, and Westinghouse in 1923 charging them with a "conspiracy in restraint of trade." The companies were violently attacked in the press and in Congress, and although the courts were slow to act, there was concern that they would hold that the patent pool was illegal.

Faced with these threats, Sarnoff decided that RCA's rigid position was untenable and that he would have to compromise. He proposed that competing manufacturers be permitted to use any or all patents in the RCA pool for the payment of a license fee. The fee was set initially at 7.5 percent of sales but was soon reduced to 5.0 percent. His proposal was approved by his management and presented to competing manufacturers in 1923. Most of them, including Zenith, accepted the plan, albeit with considerable grumbling about "paying tribute," and became RCA licensees.

By taking this action, Sarnoff relieved the antitrust pressure on RCA and established a lucrative source of income. The FTC pressed its complaint with less vigor and eventually dropped it. The relief was only temporary, however, and charges of monopoly were to plague the company for more than forty years. For the moment, however, RCA had solved the problem, albeit at the cost of allowing its licensees to compete in the marketplace on an almost equal basis.

With the patent issue resolved, the production and sales of receivers grew at an even faster pace, even at prices that ranged up to $400 (more than $4,000 in today's dollars). According to the 1939 edition of *Broadcasting Yearbook,* annual receiver sales were as follows:

1922	$ 60 million
1923	136 million

1924	$358 million
1925	430 million
1926	506 million

By the end of 1926, more than five million homes had radios.

■ THE HEYDAY OF THE NETWORKS, 1927–1950

The Networks

The networks, NBC and CBS, and, to a lesser extent, ABC and Mutual, dominated the radio broadcasting industry for a quarter of a century from 1927 to 1950. Their influence was so pervasive that it affected all aspects of broadcasting—programming, regulation, business practices, and technology. When competition from television brought the radio networks' domination to an end in the early 1950s, radio broadcasting was forced to undergo a major revolution in all of these aspects in order to survive.

The Role of the Networks   The relationship between the networks and their affiliates is a textbook example of symbiosis, the living together of two dissimilar organisms in a mutually beneficial relationship. The network provides the affiliates with programming (hopefully with high ratings) that would be too costly for individual stations to produce. The stations derive income directly from the networks and indirectly from the sale of local and national advertising attracted by the network programs. The affiliates provide the network with nationwide coverage of its programming, thus increasing its value to the network advertisers. The relationship has worked so well that it has existed throughout the history of radio broadcasting and was carried over into television as a matter of course.

During the heyday of the networks, their affiliates prospered with them. While a network affiliation was not essential for a radio staion to be profitable, possession of one almost guaranteed it. To own a station with a major network affiliation in a metropolitan area was tantamount to a "license to print money," even in the depression years. With an assured income, network affiliates could afford to construct elaborate and costly technical facilities and to employ large staffs in order to produce quality local programs to complement those produced by the networks.

The radio networks had another effect: They brought to the fore the two individuals, David Sarnoff (Figure 2–4) and William S. Paley, who were destined to become the giants of American radio and television broadcasting for more than forty years.

Courtesy RCA Corporation.

■ **Figure 2–4** David Sarnoff.

David Sarnoff (1891–1971)   David Sarnoff's role as the driving force in the early years of RCA and the radio broadcasting industry has already been described. Largely as the result of his vision and determination, both RCA and the industry became major factors in the nation's economy. Unlike many chief executives, he did not inherit an empire; to a large extent, he created one.

His early years have taken on legendary proportions—how he arrived in New York from Russia in 1900 at the age of nine; how he quickly learned the language and by hard work and ability rose from newsboy to office boy and then to telegrapher for the Marconi Wireless Company of America; how as a Marconi office boy he first met the famous Guglielmo Marconi, who for a time was his mentor and role model; and how he rose in the American Marconi organization to become its commercial manager in 1919. From this position, he eventually rose to become RCA's chairman of the board.

Sarnoff's name will appear repeatedly in this volume. He was a dominant leader in the initiation and growth of radio, television, and color television broadcasting and the technologies upon which they are based. Although he was not an engineer, he had much respect for science and an intuitive understanding of its practical applications. His interests encompassed all aspects of broadcasting, from technology to talent. His influence was enormous, and the history of broadcasting would have been quite different without his leadership.

Like most great leaders, Sarnoff had king-size faults. Although he became chief executive of a company with major manufacturing divisions, he had little expertise in manufacturing. He was intensely interested in research and the marketplace, but he had an imperfect understanding of the intermediate and very difficult step of translating laboratory outputs to manufacturable and salable products. The lack of understanding of product engineering and manufacturing by RCA's top management continued after Sarnoff's retirement and was a major corporate weakness.

Sarnoff also had personal faults, most of them the result of the classic flaw of hubris—arrogance and self-pride. He was not a gracious winner, and his victories often left a residue of bitterness. When his ego or interests were threatened, he could be ruthless. His treatment of Edwin Armstrong in the FM patent dispute (described in Chapter 3) was unnecessarily cruel.

His vanity was boundless. An otherwise excellent biography by his cousin, Eugene Lyons,[3] was nearly ruined by Sarnoff's insistence that every negative comment, no matter how mild or justified, be expunged. As a result, he is portrayed in the book as an unreal, two-dimensional figure—the exact opposite of the intended result. He was especially proud of his rank of brigadier general in the U.S. Army Reserve, and he insisted on being addressed as General. Not satisfied with this rank, he shamelessly— and unsuccessfully—pursued a promotion to major general through political channels (although his efforts were not successful). These weaknesses should not, however, obscure Sarnoff's enormous accomplishments. By comparison, his faults were trivial, and he will be remembered as one of the great industrialists of the twentieth century.

Sarnoff's final years were sad. In 1968, at the height of color television's commercial success, he contracted an incurable neurological disease. His physical and mental faculties slowly ebbed, and he died on December 12, 1971.

William S. Paley (1901–)   Like Sarnoff, William S. Paley was of Russian Jewish ancestry. Both had extraordinary qualities of innovation and leadership, and both were responsible for the growth of major corporations. But

---

[3] Eugene Lyons, *David Sarnoff* (New York: Harper & Row, 1966).

they also differed widely in their personalities and in the particular skills they brought to the business world.

Sarnoff's humble beginnings as an immigrant and the insecurities of his early life led to an enormous ego and a prickly vanity. Paley was one generation removed from the old country, his ancestors were more affluent, his father was a prosperous cigar merchant and manufacturer, and he moved easily and confidently in the best social circles. His second wife was the society beauty, Barbara Cushing, the daughter of the widely known neurosurgeon Dr. Harvey Cushing. Paley excelled in the leisure arts admired in upper-class society; he was a gourmet, a collector of fine art, and a lover of good music. Although, like Sarnoff, he had a well-developed ego, it was more carefully concealed. His memoir contains refreshing touches of modesty and even admissions of occasional mistakes and failures.

Sarnoff's and Paley's approaches to business also were quite different. Paley accurately described these differences in his memoir:

> The general and I had a long, continuing, avuncular relationship down through the years. From the earliest days of radio when he was the "grand old man" and I was that "bright young kid," we were friends, confidants, and fierce competitors all at the same time, and we understood each other and our relative positions. I always had the greatest respect and admiration for him. He had a sharp mind and a keen sense of competition. I always thought his strengths lay in the more technical and physical aspects of radio and television, while mine lay in understanding talent, programming, and what went on the air.[4]

To this, one could add that Sarnoff had a broader vision than Paley. It encompassed the entire broadcasting industry, including the role and promise of technology, whereas Paley's interest was concentrated—with enormous success—on its programming and business aspects. For practical purposes, he was the founder of the Columbia Broadcasting System (CBS), and, coming from behind in both radio and television, he brought it to a position of leadership.

Paley's major business weakness was a lack of understanding of technology, which forced him to depend too heavily on his technical counselors, particularly Peter Goldmark. Although CBS had a superb engineering department at the operational level, Paley frequently received bad advice concerning new technologies at the strategic level. This led to costly mistakes as CBS moved into television broadcasting and manufacturing. In total, they cost CBS hundreds of millions of dollars. But Paley's great talent for programming kept CBS in a position of profitable leadership, both

---

[4] William S. Paley, *As It Happened* (Garden City, N.Y.: Doubleday, 1979).

in radio and television, and in this area he consistently bested Sarnoff and NBC.

As of this writing (1989), Paley at 88 is a highly respected elder statesman of the industry. One of his major contributions in this role was the creation of the Museum of Broadcasting in New York, which preserves the classics of early radio and television programs for posterity.

A Brief History of the Radio Networks   The circumstances surrounding the formation of NBC, the first major radio network, in 1926 were described earlier. NBC's owners, RCA, GE, and Westinghouse, collectively owned two stations in New York City, WEAF, which NBC had bought from AT&T, and Westinghouse's WJZ. To avoid duplicate programming on these stations, NBC was split into two divisions, the Red Network with WEAF as its flagship station and the Blue Network with WJZ as its flagship. Generally, NBC put its most popular programs on the Red Network, and it became the dominant division in the company. NBC was under continuous attack on antitrust grounds, and to reduce the legal risks, both to NBC and themselves, GE and Westinghouse sold their interests in 1930 to RCA, which then became the sole owner.

Paley's founding of CBS resulted from something of an accident. After graduating from the Wharton School of the University of Pennsylvania in 1922, he entered his father's cigar business in Philadelphia, expecting to devote his life to it. His interest in radio began with a highly successful pioneering use of WCAU to advertise the family brand of cigars. He became more and more deeply involved in radio, including a stint as program producer for the La Palina Smoker, a program sponsored by the family cigar business, and this interest culminated in the purchase with his father of a majority interest in a struggling new network, the United Independent Broadcasters (UIB) on September 26, 1928.

It was an unequal contest at first. UIB, which became CBS (it had been identified over the air by its corporate relationship with Columbia records), had only sixteen affiliates and a dozen employees. Its sales in 1929 were about $4 million. In contrast, NBC, with its two networks, had control of most of the intercity transmission circuits and a far greater variety of programs. With a combination of determination, business acumen, an able staff, and Paley's programming skills, CBS steadily narrowed the gap.

In 1948, CBS surged ahead of NBC when Paley's tax attorneys devised a scheme whereby entertainers could incorporate themselves and save hundreds of thousands of dollars in taxes. Paley offered this deal to NBC's top stars, and when NBC refused to match the offer, there was a mass defection from NBC to CBS. The defectors were the elite of NBC's entertainers. They included Amos 'n' Andy, Jack Benny, Red Skelton, Edgar Bergen, Burns and Allen, Ed Wynn, Fred Waring, Al Jolson, Groucho Marx, and Frank Sinatra. CBS's audience soared, and now it was NBC's turn to scramble from behind.

The American Broadcasting Company (ABC) was founded in 1943 as a result of antitrust pressure on the existing networks, which culminated in the FCC's Chain Broadcasting Regulations. These regulations forbade the networks to have more than one affiliate in a single city. This forced NBC to divest itself of one of its networks. NBC sold the Blue Network, the weaker of the two, to Edward Noble, the Life Saver manufacturer, for $9 million. Nobel renamed the network the American Broadcasting Company.

For many years after its founding, ABC struggled to achieve a respectable position as a radio network against its two powerful rivals. It was handicapped by a lack of resources, and this problem became even more severe as it entered the far more costly medium of television in 1947. The resource problem was solved by its merger with Paramount Theaters in 1953, but by this time radio networks were beginning to wane in importance.

In addition to NBC, CBS, and ABC, there was the Mutual Broadcasting System. It found a niche, albeit a marginally profitable one, in providing less prestigious programs to a host of low-power stations.

Technical Regulation during the Network Era

In 1926, the airways were so chaotic that broadcasting probably could not have survived had the Radio Act of 1912 remained in effect. The court decision in the Zenith case (discussed earlier) made the Commerce Department powerless to assign station power or frequency. Each station could go its own way, and the result was a free-for-all. One symptom of the industry's poor health was a drop in receiver sales from $506 million in 1926 to $425 million in 1927. In a belated response, Congress replaced the Radio Act of 1912 with the Radio Act of 1927. The 1927 act was in turn superseded by the Communications Act of 1934, which is still in effect today.

The Radio Act of 1927 and the FRC The Radio Act of 1927 established the Federal Radio Commission (FRC) as an arm of the Department of Commerce. It was supposed to have five members, each representing one of the zones into which the United States was divided. It was originally authorized as a temporary commission with a life of two years, but it was soon made permanent.

The FRC's authority was far broader than that of the Department of Commerce under the Radio Act of 1912. It had the power to do the following:

1. Grant and deny licenses under the general criterion of the "public interest, convenience, and necessity"
2. Establish classes of radio stations
3. Prescribe the nature of service to be rendered by stations

4. Allocate bands of frequencies to the various classes of stations and assign frequencies to stations operating in each class
5. Make regulations applicable to chain (network) broadcasting
6. Regulate the technical operations of stations, including the specifications of the equipment and the qualifications of the operators
7. Conduct administrative hearings
8. Designate stations' call letters

The FRC got off to a rocky start. President Coolidge appointed the five members, but only three were confirmed by the Senate before it adjourned. Two of the three confirmed members died before the Commission could begin to act, and its initial work was carried out by the one confirmed member, Eugene O. Sykes, a former justice of the Mississippi supreme court, and the two unconfirmed members, Henry N. Bellows, a broadcaster, and Orestes H. Caldwell, a writer for electrical technical journals.

Bellows and Caldwell were victims of the dilemma the government so often faces in drafting laws and regulations. The greatest expertise often lies with industry representatives, who are not disinterested parties. The ties of Bellows and Caldwell with the industry they were to regulate made them suspect, and their confirmation was long delayed. But feeling an obligation to the industry, they served as unconfirmed and unpaid members, and they made major contributions to the development of rational assignment policies.

The FRC's Frequency Assignments The FRC faced an enormously difficult task. It was required to establish classes and power levels of stations and to make frequency assignments that would minimize interference while maximizing the total number of operating stations. It had to be done on the basis of radio wave propagation data that were far from complete. Its task was made even more difficult by the political pressure exerted by members of Congress who were seeking choice assignments for their constituents (and sometimes themselves). The pressure was understandable because it was clear that a station license would become an extremely valuable franchise. In view of these problems, the Commission, aided enormously by the expertise of Orestes Caldwell, was remarkably effective in developing policies and administering the assignment of frequencies to stations.

Four classes of stations were established:

Class IA	Clear channel, unduplicated
Class IB	Clear channel, duplicated but nighttime sky-wave coverage protected
Class II	Channel shared with Class IB

| Class III | Regional channel, ground-wave coverage protected day and night |
| Class IV | Local channel, only daytime ground-wave coverage protected |

The maximum power of the clear channels was set at 50,000 watts, regional channels at 5,000 watts, and local channels at 100 watts (later increased to 250 watts, and still later to 1,000 watts). The objective was to provide a measure of service to all areas of the country. The nighttime sky-wave coverage of the clear channels was to serve rural areas beyond the ground-wave coverage of any stations. Small towns and cities were to be served by the local stations. Larger cities were to be served by regional and clear channel stations.

After more than a year of concentrated effort, the FRC decided that it would be impossible to accommodate all 732 stations currently on the air with the spectrum space available and the state of the technology (the directional antenna with its ability to squeeze more stations on the same channel had not yet been developed). Accordingly, it issued "show cause" order to 164 stations requiring them to prove that they should be allowed to continue broadcasting.

These orders became the subject of lengthy and acrimonious hearings in which stations, often supported by their congressional representatives, fought for their right to exist. The Commission received hundreds of thousands of letters inspired by announcements broadcast by the threatened stations urging that they be allowed to continue broadcasting.

One of the Commission's problems was the absence of established standards upon which it could choose between stations when necessary. Programming was considered to be too intangible a factor, and great emphasis was placed on the technical equipment used by the stations. In every case, however, the decision was related to "the public interest, convenience, and necessity," a phrase that the FCC would later make famous.

As a result of the FRC's policy of reducing the number of stations to minimize interference, the number of stations declined between 1927 and 1930. The attrition of educational stations was particularly heavy, with more than fifty of them going off the air. As expected, a number of disappointed stations contested the commission's actions in the courts, but the Commission was upheld in every case.

Under these conditions, no regulatory body could escape severe criticism. Probably the best indications that the FRC's actions were reasonably rational and fair were accusations on the one hand that it was "dictatorial and arbitrary" and on the other that it was a "jellyfish Commission." With the perspective of time, it appears that the FRC did the best it could in solving a problem that had no perfect solution.

The stations that survived the FRC's thinning process and the depression were lucky. They went on to become the fat cats of the industry, rich and envied.

The Communications Act of 1934 and the FCC   The Communications Act of 1934 was passed at the request of President Franklin Roosevelt, who wanted to put the regulation of wired and wireless communications under the control of a single agency. It created the seven-member (reduced to five in 1986) Federal Communications Commission (FCC) with authority to regulate all communications systems except those operated by the federal government. On radio matters, the FCC coordinates with the National Telecommunications and Information Administration (NTIA), which exercises its authority over federal radio systems through the Interdepartmental Radio Advisory Committee (IRAC). It is independent of the Department of Commerce, and its members are appointed by the President and confirmed by the Senate.

The provisions of the Act with respect to radio are not greatly different from those of the Act of 1927, and this provided continuity from the FRC to the FCC. Until recently, however, the FCC took a broader and more aggressive role in the regulation of broadcasting. It involved itself deeply in the programming policies of stations, and it adopted policies and practices that encouraged increased competition among broadcasters.

The FCC's policy of increasing competition was aided by technical advances. The introduction of the directional antenna in the 1930s greatly increased the number of AM radio stations that could be accommodated in the available spectrum space. The introduction of frequency modulation (FM) and the opening of the VHF frequency band doubled the number of radio stations. The development of UHF transmitters and receivers and the allocation of a portion of the UHF spectrum for television broadcasting greatly increased the number of available television channels.

The FCC's Standards of Good Engineering Practice   The publication of the *Standards of Good Engineering Practice for AM Radio Stations* in 1939 was one of the notable achievements of the early FCC. It was prepared under the direction of the FCC's able and irascible assistant chief engineer, Andrew D. Ring, who later became a prominent consulting engineer. The document established minimum performance standards for station equipment and provided detailed rules for AM station frequency assignment. It soon became the bible for engineers planning new stations.

Frequency Assignment Policies and Rules

Spacing between Channels   The *Standards of Good Engineering Practice* formalized the de facto 10-kHz channel spacing standard that had been used in the earliest frequency assignments of the Department of Commerce.

This spacing between channels was adequate for the technology of the time, but it became a serious limitation in later years when high-fidelity programs became commercially important. The sidebands of a high-fidelity signal extend for 15 kHz on either side of the carrier, and the sidebands from stations on adjacent channels overlap (see below). Most AM receivers limit their bandwidth to avoid adjacent channel interference and improve their sensitivity.

To minimize this interference, the *Standards of Good Engineering Practice* established minimum separation for stations operating on adjacent channels, but the restriction is not severe enough to permit high-fidelity transmissions. In the more congested parts of the country, adjacent channel interference is a serious problem even for low fidelity.

The Frequency Assignment Rules For the most part, the frequency assignments of the seven hundred stations that were on the air when the FCC was given jurisdiction over the radio broadcasting industry in 1934 were considered to be a fait accompli. With few exceptions, they were not required to change their frequency or power. The assignment rules were applied mainly to new stations and to power increases for the old ones.

OVERLAPPING SIDEBANDS FROM ADJACENT CHANNELS

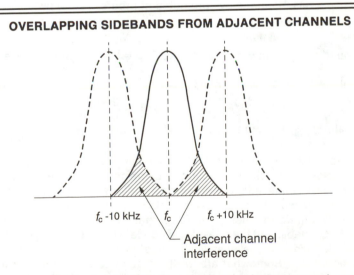

f_c -10 kHz f_c f_c +10 kHz

Adjacent channel interference

Sidebands extend above and below the carrier by an amount equal to the highest frequency component in the audio signal. It would be necessary to limit the audio bandwidth to 5 kHz to keep the bandwidth of the transmitted signal within the 10-kHz channel width. Broadcast stations are allowed to transmit 10-kHz audio signals, and this causes emissions in adjacent channels, as is shown in the drawing.

The most basic policy governing the assignment rules provided existing stations a high degree of protection from interference from new stations. An applicant for a new station was required to demonstrate that it would not cause "objectionable interference" within the "normal service area" of any existing station. The terms *objectionable interference* and *normal service area* were defined very precisely, and they acquired a quasi-legal status. The overall effect of this policy was to give older regional stations a significant competitive advantage over the newer ones. Their coverage was protected, but the newer stations had to accept the interference levels that existed when their applications were filed. The FCC also had a policy of encouraging competition, however, and this caused it to bend the rules to allow more new stations to come on the air—with somewhat unfortunate results. The application of the assignment rules involved a complex set of calculations (see below) that were usually carried out by consulting engineer specialists.

STATION SERVICE AREAS

The service area of a radio station is the area in which its signal strength is sufficient to provide reasonably satisfactory reception. The FCC defined "normal" service areas, bounded by "protected contours," as follows:

Class	Daytime	Nighttime
Clear	0.1 mv/m ground wave	0.1 mv/m ground wave 0.5 mv/m sky wave
Regional	0.5 mv/m ground wave	Existing interference-free ground-wave contour, typically 2.5 to 10 mv/m
Local	0.5 mv/m ground wave	Not defined

To determine the normal day and night service areas of a station, the locations of its protected contours are calculated using the ground-wave nomographs developed by Norton (see text). For regional stations, the level of existing nighttime interference must be calculated to determine the protected contour.

To provide a uniform criterion for judging the severity of cochannel interference, the FCC established the quasi-legal concept of objectionable interference, which was defined as the level at which the aural effect of the interfering signal was not significant. Somewhat arbitrarily, it decreed that an interfering signal was not objectionable if its strength was less than one-twentieth of the desired signal. If one or more interfering signals were already present, any new additional signal was deemed not to be objectionable if its level was less than 70 percent of the strongest existing interfering signal (later modified to less than 50 percent of the root sum square of all the existing signals).

Stations' Service Areas

A combination of the laws of nature and the effect of the FCC's Rules led to enormous inequalities in the size of stations' service areas. The most obvious difference was the wide range of authorized transmitter power—from 250 watts for local stations to 50 kilowatts for clear channels—but differences in ground-wave propagation and sky-wave effects caused equal or greater disparities.

Ground-Wave Service   All classes of stations rely on the ground wave for their primary service. The ground-wave range varies widely with frequency and soil conductivity and, to a lesser extent, the station's power. The calculation of this range requires the solution of a series of complex equations—a difficult task in the precomputer era. This task was accomplished by an able FCC engineer named Kenneth A. Norton.[5]

■ SERVICE RADIUS OF AM RADIO STATIONS

(distance in miles to 0.5 millivolts/meter contour)

Soil Conductivity	5 kw		50 kw	
	600 kHz	1,500 kHz	600 kHz	1,500 kHz
Good	60	40	120	65
Poor	40	20	65	30

Note that a 5-kilowatt station operating at 600 kHz in an area of good soil conductivity has twice the range of a 50-kilowatt station operating at 1,500 kHz where the soil is poor.

Sky-Wave Effects   During nighttime hours, the sky waves provide nationwide secondary service from clear channel stations, which are the sole occupants of their frequencies or have only limited duplication. Although sky-wave transmission is erratic and variable, even at night, it provides usable level of service to rural areas that do not receive primary ground-wave service from other stations. In addition to the positive benefits that clear channel stations derive from sky waves, they do not suffer a nighttime reduction in their primary ground-wave coverage as a result of interference from other stations.

---

[5] K.A. Norton, "The Calculation of Ground Wave Field Intensity Over a Finitely Conducting Spherical Earth," *Proceedings of the Institute of Radio Engineers*, December 1941.

In contrast, sky waves are a serious problem for regional stations and a near-disaster for local stations. In every case, their nighttime ground-wave coverage is significantly reduced by sky-wave interference from other stations on the same channel.

The amount of service area reduction on the regional channels varies widely from station to station because of differing levels of interference. It is generally much less severe for older stations that have had their original service areas protected by FCC policy. Newer stations have usually had to accept much higher levels of interference at their locations. Local stations receive no protection against sky-wave interference, and the result is a cacophony of interfering signals, sometimes at a distance of only three or four miles from the transmitter.

Because of differences in soil conductivity, operating frequency, and interference from other stations, the coverage of AM stations varies widely. Figure 2–5 shows the coverage of the different classes of stations.

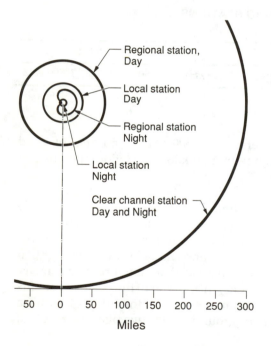

Regional station, Day

Local station Day

Regional station Night

Local station Night

Clear channel station Day and Night

| 50 | 0 | 50 | 100 | 150 | 200 | 250 | 300 |

Miles

■ **Figure 2–5** Radio station coverage by station class, day and night. The clear channel station is operating at 50 kilowatts, the regional at 5 kilowatts with a directional antenna at night, and the local at 250 watts. The same frequency and soil conductivity are assumed for each.

Equipment Technology

The Impact of World War II   AM radio was a relatively mature technology at the outbreak of World War II, and most of the electronics research inspired by the war was directed toward newer technologies such as radar and higher frequency communications. Unlike FM and television, AM broadcasting received very little direct benefit from wartime research.

Instead, the benefit was in the other direction. Engineers and technicians from the broadcasting industry as well as radio amateurs provided a pool of experienced manpower that provided invaluable assistance during the war. The military returned an even larger pool of electronics technicians to civilian life after the war, which made the rapid expansion of radio and television broadcasting possible in the postwar years.

Directional Antennas   The enormous growth in the number of AM stations that began in 1935 and accelerated in the postwar years was made possible by the development of the directional antenna in the mid-1930s by George H. Brown of RCA and others. The directional antenna made an important contribution to the broadcasting industry and brought new radio services to millions of Americans.

A directional antenna consists of an array of two or more towers in which the *amplitude* and *phase* of the currents are adjusted to cause the strength of the radiated signal to vary in different directions. It permitted new stations to operate by limiting their radiation toward existing stations, thus reducing the interference below the level defined as "objectionable" by the FCC. It had a secondary purpose of increasing the signal intensity over the most densely populated regions in the service area, but its most important role was in enabling new regional stations to comply with the FCC frequency assignment rules. (Directional antennas were not needed for clear channel stations and were not permitted for local stations.) Figure 2–6 shows how directional antennas are used.

Most prewar directional antennas were simple two- or three-tower arrays. After the war, they became more complex because of the more stringent requirements imposed by the growing congestion of the AM frequency spectrum and made possible by improved techniques for designing and adjusting arrays. Arrays with six or more towers were sometimes used, and four-tower arrays became commonplace. A paper on directional antennas by G.H. Brown[6] became a standard reference for consultants who specialized in the design and adjustment of antennas.

---

[6] G.H. Brown, "Directional Antennas," *Proceedings of the Institute of Radio Engineers*, January 1937.

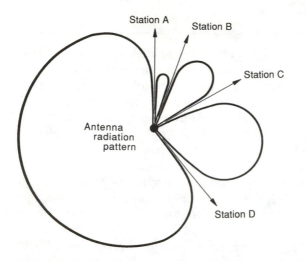

■ **Figure 2–6** Directional antenna use. This directional antenna is used to reduce the radiation in the direction of existing stations A, B, C, and D, operating on the same frequency. The objective is to reduce the interference level at these stations below the level defined as "objectionable" by the FCC.

Prior to the advent of the computer, antenna designers were constrained by the difficulty of calculating the radiation patterns of any but the simplest geometric figures. Most arrays were either in-line or parallelograms. The computer made more complex arrays possible, but by the time it was available, the demand for new directional antennas had fallen off.

Transmitters The development of radio broadcast transmitters proceeded steadily but with no spectacular breakthroughs. At first, the two major manufacturers were Western Electric and RCA. Western Electric, AT&T's manufacturing arm, had been the chosen instrument of the radio trust to manufacture transmitters, but with the breakup of the trust, RCA was free to enter the business. Later, Gates Radio, a small company in Quincy, Illinois, founded by Parker Gates, began the manufacture of transmitters, primarily for lower power stations.

There were many changes in the business of transmitter manufacturing after World War II. Gates was purchased by Harris Intertype, later to become the Harris Corporation, in 1958 and is now a leading transmitter supplier. A number of other companies, including CCA and Continental Electronics, entered the business. Western Electric stopped manufacturing transmitters in 1947 in compliance with an antitrust consent decree, and RCA withdrew completely from the broadcast equipment business in 1986.

The first Western Electric transmitters employed low-level grid modulation (see illustration on next page). They were solidly built in accordance with telephone company standards and were expected to last for decades. The lower efficiency of grid modulation not only increased the power bill, but it necessitated the use of water-cooled tubes to carry away the wasted energy. They were also expensive, and they required extensive routine maintenance.

RCA followed a different technical course, high-level or plate modulation (see illustration on page 90). High-level transmitters were more efficient and used air-cooled tubes. Their disadvantage was that they required an expensive high-powered audio amplifier for the modulator, a formidable technical challenge with available technology.

RCA was winning the competitive battle until a quiet and serious engineer from Bell Laboratories, William Doherty, developed a new circuit, which came to be known as the "Doherty amplifier." It modulated the signal at the final stage without using a high-powered audio amplifier, and its efficiency was nearly as great as with high-level modulation. It was not without its problems; it was difficult to tune, and it did not work well into sharply tuned directional antennas. Nevertheless, it equalized the RCA–Western Electric competition until Western Electric was forced to withdraw.

In the meantime, RCA was being pressed competitively by others, particularly Gates, and its high-level transmitters became overpriced. As a cost-reduction measure, it developed the ampliphase modulation circuit, which had good efficiency but was difficult to tune. In later years, still other modulation methods were developed in an effort to avoid the inefficiency of grid-modulated transmitters, the high cost of plate modulation, and the tuning difficulties of the Doherty and ampliphase transmitters (see page 91).

AM TRANSMITTER MODULATION METHODS

Low-Level Grid Modulation

The carrier signal is modulated at a very early stage in the transmitter, and the modulated signal is amplified in a series of amplifiers that must be adjusted for a very high degree of linearity to avoid distortion of the modulation envelope. The maximum theoretical efficiency of the final amplifier is 25 percent.

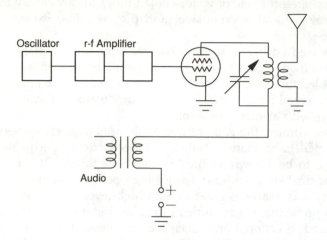

High-Level Plate Modulation

High-level plate modulation is an outgrowth of the Heising, or constant-current, modulated oscillator, named for its inventor, R.A. Heising. The modulating signal is applied to the anode circuit of the final power amplifier. It is capable of efficiencies as high as 80 percent.

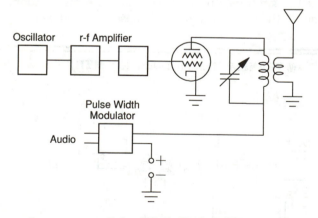

Pulse Width Modulation

This method was developed by Harris and is in common use today. The audio signal is converted into a pulse train with wide pulses at the peaks and narrow pulses in the valleys. The varying duty cycle of the pulse train then directly determines the voltage applied to the power amplifier plate circuit without the use of a modulation transformer or reactor.

Other Modulation Methods

A number of alternative modulation methods have been developed to avoid the low efficiency of grid-modulated transmitters and the need for high audio power with plate modulation.

The Doherty amplifier: This circuit was developed by W.H. Doherty of Bell Laboratories. Through a clever circuit arrangement, it uses high-level grid modulation, which requires very little modulator power. The efficiency is about 60 percent, but it tends to have high distortion, particularly when feeding a sharply tuned directional antenna.

The ampliphase circuit: This circuit was developed by RCA. Two unmodulated carriers are applied to the final stage through phase modulators. In the absence of a modulating signal, the carriers are in quadrature. For 100 percent modulation, their phase relationship is varied from 0 to 180 degrees, thus causing the output signal to vary from zero to twice the carrier level.

Screen grid modulation: This method, developed by Continental Electronics, uses two tetrodes in the power amplifier stage, one of which provides the carrier power and the other the additional power for modulation peaks. The modulation signal is applied to the screen grids, and very little audio power is required.

There has been an increasing use of transistorized circuitry in AM transmitters, first in lower power stages and then as complete low-power transmitters as suitable solid-state devices became available. Figure 2–7 shows two RCA AM radio transmitters, one from 1930 (Figure 2–7, top) and another from 1960 (Figure 2–7, bottom).

The apogee of AM transmitter design was reached with the construction of a huge 500-kilowatt transmitter for station WLW in Cincinnati, which went on the air in May 1934. It had been authorized under the somewhat permissive standards of the FRC at the behest of Powell Crosley, the aggressive founder of the Crosley Radio Corporation, the owner of WLW, and a leading receiver manufacturer.

The transmitter was designed and constructed jointly by RCA, GE, and Westinghouse and was installed under the supervision of WLW's chief engineer, Joseph Chambers, who later became a successful Washington consultant. The construction of the transmitter was an enormously challenging feat, and participating in the program was a rite of passage for the radio engineers of the time. In later years, scores of engineers claimed to have been involved, and it was said that the *Queen Mary* ocean liner was not large enough to accommodate all of them.

The superpower WLW installation also included one of the first AM directional antennas. Its purpose was to reduce radiation in the direction of Canada to the equivalent of a 50-kilowatt station, a requirement imposed by complaints of interference with adjacent channel stations by the Canadian government. It was designed by a team led by George Brown.

Courtesy RCA Corporation.

Courtesy RCA Corporation.

■ **Figure 2–7**   AM radio transmitters. (*Top*) A 1930 RCA 5 kw transmitter.
(*Bottom*) A 1981 RCA 50 kw transmitter.

Unfortunately perhaps, all the expense and effort of WLW's superpower station eventually came to naught. It was discovered that a clause in the North American Regional Broadcasting Agreement (NARBA), a treaty to which the United States was a signatory, forbade the use of transmitter power in excess of 50 kilowatts, and WLW was forced to restore its power to that level.

On a less cosmic scale, the progress in transmitter design has been extraordinary. Consider, for example, the following: In 1940, a typical 5-kilowatt transmitter required 500 square feet of floor space, 25 kilowatts of primary power, and the constant attention of at least one engineer. It cost $35,000 in depression dollars. In 1980, a typical 5-kilowatt transmitter with better performance and reliability required 100 square feet of floor space and 8 kilowatts of primary power. It operated unattended and cost $20,000 in 1980 dollars.

Studios Most radio stations were extremely profitable during the network era, and their affluence was reflected in their studio designs. The studios were often as much showcases as workplaces, designed to impress advertisers and the public with the glamour of the medium. Although affiliates depended on the networks for the bulk of their programming, their studios were lavish. A network affiliate in a medium-size city might have five: a large audience-participation studio with a seating capacity of one hundred or more, two medium-size studios for musical or dramatic groups, a small plush studio for interviewing VIPs, and a small workhorse studio for newscasts and interviews with less exalted individuals. Figure 2–8 shows a typical station studio.

Each studio had its own manned control facilities, which fed into the *master control*. In turn, the master control selected the program source to be aired. These facilities required large and expensive technical staffs.

Great attention was paid to the acoustical design of studios. It was believed that the criteria for designing a studio were different from those for a concert hall because the reverberation of the room in which the receiver was located would be added to the reverberation of the studio. Toscanini was never completely happy with the acoustics of the studio that NBC built for the NBC Symphony Orchestra, which he directed. He called it *sec*, or "dry," meaning that it had a short reverberation time.

A recognized U.S. authority on acoustical design for studios was Harry F. Olson of RCA. His research in sound encompassed acoustics, microphones, loudspeakers, and hearing. He wrote a number of technical papers and definitive textbooks on the subject, and he introduced advanced concepts into the design of microphones and loudspeakers. (Unfortunately, his distinguished reputation as a scientist in acoustics and sound systems was somewhat tarnished by his subsequent failure in video recording technology; see Chapter 6.)

Courtesy RCA Corporation.

■ **Figure 2–8** A radio station studio of the World War II era. Note the elaborate paneling, which was used to control reverberation.

Studio Equipment The design of audio equipment for studios kept pace with the design of studios. Although the individual components of an audio system are relatively simple, the systems themselves are complex, and their design requires specialized engineering talent. The principal components are microphones, amplifiers, recording and playback equipment, and control equipment.

Microphones Microphone technology advanced enormously from the crude carbon devices used in radio's pioneering years. Transducers, the elements that convert sound to electrical energy, were either magnetic, electromagnetic, or piezoelectric. (The last type is based on the use of piezoelectric crystals, which generate a voltage when subjected to pressure.) They also could be classified as *pressure* or *velocity* types. Pressure microphones respond to variations in the pressure of the sound wave. Velocity microphones respond to variations in the velocity of the medium. One of the most popular microphones was the RCA 77-D, an electromagnetic velocity microphone in which the transducer was a thin metallic strip suspended between the poles of a magnet. When it was set in motion by velocity variations of the air transmitting a sound wave, it electromagnetically generated a signal voltage. Its distinctive shape, a cylinder capped by two hemispheres, came to be the standard symbol of a microphone for cartoonists. The small lavaliere microphones that have become so popular in television require the use of transistors, which did not become available until later. Figure 2–9 shows some examples of microphones.

Recording and Playback Equipment   In the earliest days of radio, phonograph records were played over the air by placing a microphone in front of a mechanical playback phonograph. This method was replaced in the mid-1920s by the use of electrical phonographs, which generated signals that could be broadcast directly.

Some stations had facilities for making disk recordings for storage and playback. They used sixteen-inch disks and operated at $33\frac{1}{3}$ rpm. The recording equipment was costly, required considerable skill to operate, and was not well adapted for making copies or for editing. This system was abandoned soon after World War II when high-quality magnetic tape recorders became available.

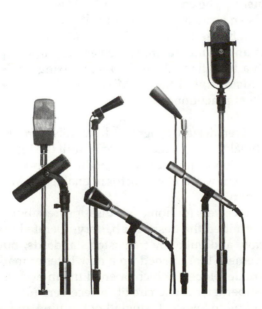

Courtesy RCA Corporation.

■ **Figure 2–9**   Examples of microphones.

There was strong opposition to the use of recordings on the air, both by the musicians' union and by many broadcasters who took pride in "all-live" programming. The networks refused to use any recordings until ABC broke the ice in 1946 by allowing Bing Crosby to use them in order to lure him from NBC. Recordings also had copyright problems. As a result, the use of recordings was limited until the end of the network era made extensive use necessary and the introduction of magnetic tape recording made it possible.

During World War II, Peter Goldmark of CBS began research on the use of $33\frac{1}{3}$ rpm for records. The result was the introduction of long-playing (LP) records by Columbia Records in 1948. They not only operated at a slower speed, but the grooves were finer. As a result, nearly thirty minutes could be recorded on one side of a twelve-inch record, as compared with four minutes on a standard 78-rpm recording (see Figure 2–10). In addition, vinyl was introduced as the disk material, which greatly reduced record scratch.

RCA responded with a 45-rpm system that used a small disk with a large hole in the center to facilitate record changing. About seven minutes could be recorded on each side. A rapidly operating record changer also was developed. In spite of these innovations, 45-rpm records never quite made it in the marketplace, much to Sarnoff's chagrin, and the technology was eventually abandoned.

Control Equipment The console (Figure 2–11) is the basic piece of control equipment. It is an array of volume controls, switches, and sometimes tone controls that can be used to adjust the levels of individual microphone or recorder signals and to mix them. Consoles vary in complexity from a simple device with a few channels to the huge mixing panels used in the production of records or sound tracks. Mixing audio signals is an art that requires highly skilled practitioners.

Receivers Receiver design progressed steadily throughout the network era, but real breakthroughs did not occur until the introduction of the transistor, which coincided with the decline of the networks. The transistorization of receiver design was a major factor in enabling AM radio to adjust to a world of weak networks.

Battery-operated receivers for home use disappeared, but improvements in electron tubes permitted the design of battery-operated portable receivers. They were heavy and bulky by transistor standards, but their design was a major achievement that opened up a new listener market. Automobile radios also were developed, which opened up an even larger market.

The use of the superheterodyne circuit in receivers (see Chapter 1) became universal, leading to low-cost, single-knob tuning and good sensitivity. A few manufacturers, notably Magnavox, produced combination radio–phonograph sets with audio systems that approached high-fidelity performance.

Courtesy CBS.

■ **Figure 2–10** Peter Goldmark. This illustrates the dramatic reduction in the number of records made possible by Goldmark's invention of the 33⅓-rpm LP recording format. The material recorded at the original 78-rpm standard on the albums to the right and left was rerecorded on the small stack of LP records in the middle. Goldmark's inventions in color television and electronic video recording were not as successful (see Chapters 5 and 6).

The reduction in receiver costs, stimulated by vigorous competition, improved designs, and the economies of scale, was impressive. Receiver prices were remarkably stable and in some cases even dropped between 1926 and 1950 in spite of inflation.

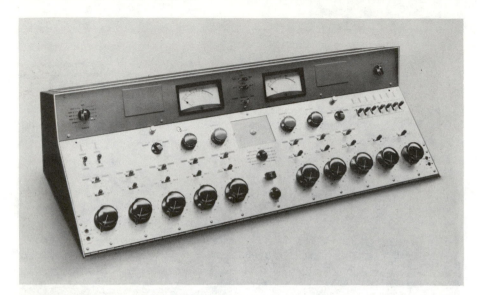

Courtesy RCA Corporation.

■ **Figure 2–11**   A radio control console.

The End of the Radio Trust

The breakup of the GE, Westinghouse, AT&T, and RCA radio trust—which was inevitable from its founding because of its fatal legal and commercial flaws—had already begun in 1923 with the agreement to issue patent licenses to all comers and in 1926 with the withdrawal of AT&T from broadcasting. There still remained the problem of receiver manufacturing, however. The GE, Westinghouse, and RCA consortium, with its clumsy and slow-moving management structure, continued to find it impossible to compete profitably with the more agile companies that specialized in receiver manufacture. All of RCA's radio profits came from patent royalties, and Sarnoff became convinced that the only solution was for RCA to acquire its own manufacturing facilities and operate independently of its corporate parents.

The problems of the Victor Talking Machine Company provided the opportunity. As described in Chapter 1, Eldridge Johnson, Victor's president, had neither the resources nor the will to make the transition from mechanical to electrical phonographs, and he sold his controlling interest to a banking syndicate, which resold it to the public in 1926. Ealier, in 1924, Victor had begun marketing a radio–phonograph combination designed by RCA engineers. This alerted Sarnoff to Victor's problems, which he sensed to be RCA's opportunity.

Sarnoff proposed that RCA purchase the Victor stock, which was now in the hands of the public, and thus acquire Victor's extensive manufacturing facilities in Camden, New Jersey. RCA would use these facilities to manufacture radios and phonographs and would manage the enterprise without interference from its corporate parents.

At first, Westinghouse vehemently opposed the plan, as it did not relish the thought of competition from its offspring. But after a lengthy debate and with the support of Owen D. Young, GE's chairman and Sarnoff's longtime mentor and sponsor, the RCA board—including its Westinghouse members—approved the proposal in 1929. The Victor stock was acquired through exchange for RCA stock, and the RCA Victor Company was incorporated, owned 50 percent by RCA, 30 percent by GE, and 20 percent by Westinghouse.

Sarnoff was still not satisfied with this arrangement. He wanted RCA to own all the stock of NBC and the RCA Victor Company. He proposed this to GE and Westinghouse, and to his surprise they agreed to his plan with little opposition. All the assets of these subsidiaries owned by these companies were transferred to RCA in return for more than six million shares of RCA common stock.

The final step in the dissolution of the trust occurred as the result of an antitrust action brought by the Department of Justice in 1930. The trust was charged with criminal violations, which, had they been proved in court, would have resulted in some executives (possibly including Sarnoff) being sent to prison. RCA also would have lost its licenses for NBC's stations.

Sarnoff decided to make a virtue of necessity and he proposed that all the agreements among the members of the trust be dissolved and that this be proposed to the Justice Department as the basis for a consent decree. Deciding that prudence was the better part of valor, GE and Westinghouse agreed. Discussions with the Justice Department were begun, and after lengthy negotiations, the decree was signed on November 21, 1932.

The decree provided for the complete divorcement of GE and Westinghouse from RCA. They were required to distribute their RCA stock to their shareholders, and all their executives were required to resign from the RCA board of directors. Sarnoff was now free to operate RCA as an independent entity and to lead it in its next momentous step, the development of a television system.

Station Population and Revenues

Except for a slight dip during the depths of the depression, the number of stations on the air increased steadily from 1926 to 1950 (see Figure 2–1). From 1926 until the United States' entry into World War II in 1941, this number increased from about seven hundred to just over one thousand.

Owing to the enormous profitability of stations during the war, there was a rash of applications for licenses immediately following it. As the network era drew to a close in 1950, the number of stations had more than doubled to twenty-two hundred.

The growth in AM broadcasting revenue was equally impressive (see Figure 2–2). In 1935, the first year in which the FCC compiled statistics, revenues totaled $80 million; by 1950 they had grown to $455 million.

■ THE POSTNETWORK ERA, 1950–PRESENT

The Decline of the Radio Networks

In 1950 the radio networks had dominated radio broadcasting for twenty-five years, and they and their affiliates had been enormously profitable. Affiliation with a major network was an eagerly sought after prize. Networks had a near monopoly on the most popular talent, and it was a sellers' market for network time and for national and local spot advertising on the affiliates. It was a time of great prosperity for the radio broadcasting industry.

The advent of television brought this period of prosperity to a close in the early 1950s. The home audience for television exceeded that for radio, especially during the prime evening hours. Television was a more effective medium for advertising, and the choice national accounts switched to it. Lured by higher salaries, the leading personalities in the entertainment world moved to television. Network revenues dropped dramatically from $141 million in 1948 to $48 million in 1956, and their net profit declined until they were all operating at a loss. The economic foundation of radio broadcasting as it had operated in the past crumbled.

The Era of Local Radio

In the face of the disintegration of radio's economic mainstay, the networks, the medium proved to be remarkably adaptive, and the growth in local advertising more than offset the network decline. From 1948 to 1956, local advertising revenues rose from $116 million to $298 million, and by 1979 they had risen to more than $2 billion.

These increases were due in part to inflation and the increase in number of stations, factors that did not increase the profitability of individual stations. But they also reflected the skill of broadcasting station management in adjusting to the new environment. Stations were aided in this by a number of technical and nontechnical developments.

Favorable technical developments included the increase in reliability and stability of transmitters and studio equipment, the introduction of magnetic tape recording, and the advent of small transistorized portable receivers.

Favorable nontechnical developments included the health of the economy in the postwar years, the breakdown of barriers to the use of recorded music, the increase in outside-the-home listeners due to portable and automobile radios, and a reduction in the stringency of FCC programming and operating regulations. All these factors contributed to the reduction in operating costs and the changes in programming formats that were needed to make radio broadcasting profitable.

Radio programming came to be based predominantly on recorded material, either records or tape cassettes. Whereas the use of records by radio stations was once vigorously opposed by the musicians' union and record companies, they later recognized the advertising value of radio and attempted to have their releases featured on the air.

Most stations became specialists in broadcasting programs of a single format. The most common formats were Adult Contemporary, Agriculture and Farm, American Indian, Big Band, Black, Classical, Contemporary Hit, Country and Western, Educational, Ethnic, Gospel and Religious, Jazz, Middle of the Road, News, Oldies, Progressive, Spanish Language, and Variety.

One technical development was favorable to broadcasting in general but unfavorable to AM. This was the development of FM, AM's competitor for the radio audience. FM broadcasting did not become a serious competitor for more than twenty years after the beginning of the postnetwork era, but it eventually prevailed and became the primary radio broadcasting medium.

Technical Regulation during the Postnetwork Era

During the 1960s, a number of regulatory problems came to a head at the FCC. Most of them were related to frequency assignment policies resulting from a demand for spectrum space that exceeded the supply. The FCC had attempted to follow two mutually exclusive policies: On the one hand, it tried to authorize as many stations as possible to provide increased competition, but on the other, it wanted to afford existing stations a high degree of protection from interference. These policies have continued to the present, and the Commission has vacillated between rigid enforcement of its assignment criteria and great permissiveness. For the most part, however, the desire to authorize more stations has taken precedence, and as a consequence most AM stations are suffering from serious interference problems.

Overly lax administration of the frequency assignment rules, exacerbated by malfunctioning directional antennas, has led to serious problems of nighttime interference on regional channels. The problem of adjacent channel interference has become much more serious than was originally anticipated. This has been aggravated by the practice of preemphasis—that is, boosting the higher audio frequencies to improve the signal's fidelity. Yet in spite of the large number of stations that have been authorized, some areas of the country are still not served or are underserved by AM stations, particularly during nighttime hours. A perfect solution to this problem does not exist, but the FCC has continued to attempt simultaneously to increase the number of services and reduce mutual interference.

In 1964, alarmed by the growing complaints of interference, the FCC announced that its assignment rules would be enforced more rigidly. This slowed but did not stem the flow of applications for new stations. In 1968, the FCC announced a freeze during which no applications would be accepted pending an outcome of a study of its policies.

The freeze was lifted a year later with additional restrictions imposed on new stations. One of the criteria the FCC had originally used in making assignment decisions was the extent of the "white area" (the area with no other service) that the new station would cover. *White area* was originally defined as having no other AM service, but after the freeze it was redefined as a much smaller area with neither AM nor FM service. The effects of the new policy were twofold: It significantly reduced the number of authorizations for new AM stations, and it gave a further impetus to the development of FM.

The new policy was not completely effective, however. When it was adopted, there were 4,125 AM stations on the air. Subsequently, 800 additional stations have been added, some of them made possible by the breakdown of the clear channels. In addition, daytime-only stations are now allowed to operate at night. The result is that mutual interference between stations, both cochannel and adjacent channel, is worse than at the end of the freeze in 1969.

Breakdown of the Clear Channels The original FRC/FCC assignments had resulted in two categories of clear channels—Class IA, in which no nighttime duplication of the dominant station was permitted, and Class IB, in which nighttime operation by Class II stations was permitted but with stringent requirements for protection of the sky-wave service of the dominant station.

The Class IA channels were eyed covetously by would-be broadcasters and by the FCC. Hearings were held as early as 1946 for the purpose of investigating the "breakdown of the clears"—that is, allowing nighttime Class II station operation. The licensees of the Class IA stations fought back vigorously and for many years were successful in maintaining their nighttime exclusivity. But the pressure to change the policy eventually became too great; duplication was allowed on Class IA channels, and, in effect, all

Class IA channels became IB. In addition, duplication of clear channels assigned to Mexico and Canada is now permitted, provided only that their borders be protected.

Daytime-Only Stations The AM spectrum can accommodate more daytime-only than full-time stations because sky-wave interference is not significant in the daytime. During radio's network era, the greatest revenue came from nighttime service, and it was assumed that daytime-only stations were economically marginal. One of the results of television competition was to reduce the value of nighttime service for radio stations. At the same time, new radio programming formats had an increasing appeal to daytime listeners in automobiles or out-of-doors. These developments, coupled with reduced operating costs, made daytime-only stations potentially profitable.

As a result of the spectrum availability and the improved profit potential of daytime-only stations, they became increasingly popular. More than half of the new stations authorized during and after the 1950s were daytime-only.

Daytime-only stations were initially licensed to operate from local sunrise to local sunset. This was a reasonable broadcast day in the spring and summer months, but it was a serious handicap in the fall and winter. It was a particular problem in respect to automobile commuters, who were a significant part of the AM radio audience. Daytime-only broadcasters, therefore, put pressure on the FCC to extend the broadcast day.

The FCC's first step was to take advantage of the fact that the transition between nighttime and daytime propagation does not occur sharply at sunrise and sunset. It adopted a complex set of rules for issuing presunrise service authorizations (PSRA) and postsunset service authorizations (PSSA) for stations to begin operations for a period (typically beginning at 6 A.M. local time) before sunrise and to continue for a period afterward. Operation was restricted to low power, 75 to 500 watts, and it was judged that the added service was of greater value to the public than the small loss of service to full-time stations resulting from increased interference.

Later, the restrictions on daytime-only stations were relaxed even further, and full-time operation is now allowed provided the station's power is reduced sufficiently to avoid objectionable interference with full-time stations. Often this requires power levels as low as 10 watts or even 1 watt at night.

Efforts to Improve AM Service Quality

Shaken by the ever-increasing loss of market share to FM, AM broadcasters and the FCC have addressed the problem of improving the technical quality of AM service. The FCC has instituted more stringent control of directional antennas. It has declined to establish standards for AM stereo broadcasting, but the industry is in the process of establishing de facto standards. Looking to the future, the National Radio Systems Committee (NRSC), a joint venture of the National Association of Broadcasters (NAB) and the Electronic Industries Association (EIA), has proposed rule changes that it believes will improve AM reception.

Control of Directional Antennas As the AM band became more crowded, antenna systems became more complex and difficult to maintain in adjustment, and the FCC took steps to regulate the performance of directional antennas more closely. It defined certain antennas as "critical" and required that the current amplitudes and phases be adjusted within tighter tolerances than those required for less critical arrays. These adjustments require the use of special high-precision monitors and more frequent measurement of the adjustment of all directional antennas.

AM Stereo All AM stereo systems suffer from the limited bandwidth of AM radio channels, and AM stereo will never be able to match the fidelity of FM. The ability of AM systems to transmit and reproduce higher audio frequencies would be further limited by attenuation of out-of-band transmissions as recommended by the NRSC. Nevertheless, the use of stereo significantly improves the attractiveness of AM broadcasting and goes partway in matching the technical quality of FM.

The FCC established transmission standards for FM stereo, but it waffled in its attempts to do so for AM. The Reagan administration's policy of deregulation, plus the historical example of serious FCC errors in establishing technical standards (for example, the CBS color system) led the Commission to let the marketplace decide. Left to make the decision, broadcasters had the choice of five stereo systems. Two of these, the Kahn and Motorola systems, very quickly took the lead, and as of this writing, it appears that Motorola's C-Quam® system is headed toward total industry acceptance.

In the C-Quam system, the sum of the left and right signals (L + R) amplitude modulates the carrier, which is then received and detected by standard monaural receivers, thus resulting in compatibility. The difference between the left and right signals (L − R), usually of lower amplitude than L + R, is transmitted by a form of phase modulation that does not affect the amplitude of the carrier. This signal is decoded by special circuitry in a stereo receiver. The left and right signals are recovered by adding and subtracting the L +R and L − R signals:

$$L = \frac{(L + R) + (L - R)}{2}$$

$$R = \frac{(L + R) - (L - R)}{2}$$

The response of the marketplace to the C-Quam system has been good. As of the end of 1988, Motorola reported that it had shipped more than 16.5 million C-Quam decoders to receiver manufacturers.

The Recommendations of the NRSC   In October 1988, the NRSC prepared a report in which it summarized its recommendations to the FCC for improving the quality of AM reception. They included the following:

1. Stricter limits on out-of-band emissions and a standard preemphasis of higher frequencies. This is to reduce adjacent channel interference and to provide a guide to receiver manufacturers.
2. Reduction of the permissible level of sky-wave interference from 50 percent of the root-sum-square (RSS) of existing interfering signals to 25 percent.
3. Reduction of the permissible level of adjacent channel interference from 0 dB to −16 dB (from 1 : 1 to 1 : 40).

The FCC is seriously considering these proposals, some as formal rule changes.

Technology in the Postnetwork Era

Studios   The elaborate radio studios of the network era were among the first casualties of television competition. With the transition from costly live productions to recorded programming, they became an anachronism. All that was needed were announce booths for news broadcasts and disk jockeys and other announcers to put the recorded material on the air. There might be one small studio for the recording of commercials and short program segments, but complex live productions with costly studio facilities became a thing of the past.

Magnetic Tape Recording   The development of a satisfactory method for recording audio signals on magnetic tape was one of the technical breakthroughs that enabled radio broadcasting to adapt to the postnetwork era. Tape recorders made it possible for stations to prerecord commercials and program segments with equipment that was relatively inexpensive and did not require a high degree of technical skill to operate. Magnetic recording was a key element in broadcasting's transition from live to recorded programming and to smaller technical staffs.

Magnetic recording of audio was first proposed by a Danish scientist, V. Poulsen, in 1900.[7] For several decades thereafter, desultory efforts were

---

[7] V. Poulsen, *Ann. Phys.* (Leipzig) 3(1900): 754.

made to develop a satisfactory commercial product using the principle described by Poulsen. Most of them used a thin iron wire pulled past a recording head as the recording medium. None was very satisfactory. The quality of the recorded sound was poor, and the iron wire was easily broken or snarled.

The breakthroughs that led to the excellent performance and versatility of today's recorders were made by German engineers during World War II. They replaced the iron wire with a strip of plastic tape coated with fine grains of a magnetic material (see below). Plastic tape worked far better, mechanically and electrically, than wire. A second breakthrough was the use of high-frequency bias (see illustration on next page), which made low-distortion recordings possible. The exact mechanism of the bias technique is still not completely understood, but it is extremely effective.

The performance of the equipment developed by the Germans was good enough for on-air use by their radio stations. The value of the German technology was immediately recognized by American engineers who entered Germany at the beginning of the occupation, and it was transmitted quickly to the United States.

Tape recorders originally were reel-to-reel (Figure 2–12) and usually used

THE PRINCIPLE OF MAGNETIC TAPE RECORDING

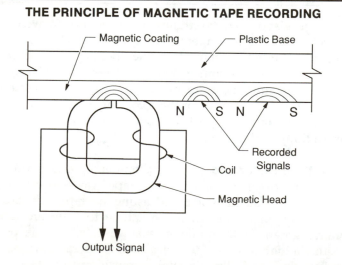

As the tape is pulled past the magnetic head, the magnetic pattern that has been recorded on its surface causes variations in the flux in the magnetic path through the coil. These changes induce a replica of the signal voltage in the coil winding.

ELECTRICAL BIAS FOR AUDIO RECORDING

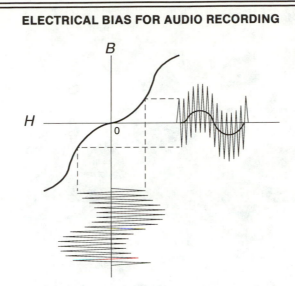

This drawing, the normal BH curve, shows a fundamental problem with magnetic recording. The curve is very nonlinear, and if the audio signal were recorded directly, a high level of distortion would result. (This was one of the problems with early wire recorders.) This difficulty is overcome by adding (not modulating) the audio signal to a high-frequency (50- to 100-kHz) bias signal. This carries the composite recorded signal well out of the nonlinear region of the BH transfer characteristic.

quarter-inch tape, although some special multitrack recorders used wider tape. Standard tape speeds of 15 and 7.5 inches per second (ips) were established, with corresponding audio bandwidths of 15 and 7.5 kHz.

Subsequently, there were major improvements in tape and recording head technology that permitted equal or better performance at lower tape speeds. Standard tape speeds of 3.75 and 7.5 ips have been established for reel-to-reel recorders and 1.875 and 0.938 ips for cassette recorders. Recorders are available in 4-track configurations, and a frequency response up to 20 kHz can be achieved.

The development of the tape cassette greatly increased the utility of tape recording for radio stations. Cassettes are more easily handled, and it is practical to use a separate cassette for each commercial or programming segment. This led to partially automated radio stations. The individual program segments are prerecorded on individual cassettes, and automatic players play back the segments in sequence with no human intervention.

Courtesy RCA Corporation.

■ **Figure 2–12**   A reel-to-reel recorder.

Transistors   Probably no single technical innovation has had as profound an effect on broadcasting, both radio and television, as the transistor. Its use became ubiquitous, both in professional broadcasting equipment and in receivers and recorders used by the public. In its total impact, it was even more important than the tape recorder in radio's adjustment to television.

Transistors began to be used in station equipment in the early 1950s, and little by little they replaced electron tubes as a greater variety of transistor types became available. The resulting equipment designs were smaller, required less power and maintenance, were more reliable, and ultimately were cheaper. These designs permitted major reductions in technical personnel and were an important factor in the cost-reduction policies stations were forced to adopt.

The impact of transistors on receiver design was even greater. As television preempted the home audience in the prime evening hours, radio programming was directed more to the outdoor and automobile audiences. Transistors made truly portable radios for outdoor use possible, and *transistor* became a generic term for a small hand-held radio. With their lower heat generation, greater reliability, and smaller size, transistor radios were superior for automobiles, and their use became universal.

The influence of the transistor is illustrated dramatically by the growth in the number of portable radios sold in the United States:

■ RADIO SALES IN THE UNITED STATES

(thousands of units)

	Table/Clock	*Automobile*	*Portables*
1955	5,245	6,863	2,082
1960	8,291	5,501	9,740
1965	9,818	10,037	21,871
1970	10,588	10,378	23,461

Source: EIA.

The Beginning of the Japanese Invasion

As the transistor began to supersede the electron tube for radio receivers in the 1950s, the Japanese began an all-out effort to master its technology and to use it as the basis for an assault on the world electronic markets. Akio Morita, the founder of Sony, reported:

> Miniaturization and compactness have always appealed to the Japanese. Our boxes have been made to nest; our fans fold; our art rolls into neat rolls; screens that can artistically depict an entire city can be folded and tucked neatly away, or set up to delight, entertain, and educate, or merely to divide a room. And we set as our goal a radio small enough to fit into a shirt pocket. Not just portable, I said, but "pocketable." Even before the war RCA made a medium-sized portable using tiny "peanut" vacuum tubes, but half the space was taken up by an expensive battery, which played for only about four hours. Transistors might be able to solve that power and size problem.
>
> We were all eager to get to work on the transistor, and when word came that it would be possible to license the technology, I went to New York to finalize the deal in 1953.[8]

---

[8] Akio Morita, *Made in Japan* (New York: E.P. Dutton, 1986).

Morita then described the difficulties he encountered with Japan's Ministry of International Trade and Industry (MITI) in obtaining permission to remit $25,000 to Western Electric for the initial license fee. But he prevailed, and Japan was set on a course that would eventually lead to its near domination of the world's commercial electronics industry. At first, its engineers and scientists copied the West, but dedication, determination, and a great deal of natural talent combined to create a highly educated cadre of Japanese engineers and scientists who soon became among the world's leaders in solid-state technology.

The translation of technology into salable products was not accomplished overnight, and it was not until the late 1950s that transistorized radios were first offered in the Japanese market. Significant quantities of exports to the United States began in 1961 with 5.2 million units. U.S. imports grew rapidly thereafter to 17.6 million in 1965 and 34.7 million in 1969 (not all of these came from Japan). Since 1969, Japan and Third World countries have virtually monopolized the U.S. radio market.

American electronics manufacturers were not greatly alarmed at first by the Japanese penetration of the radio market. It was much smaller than the television market, and radio was regarded as a product in which the Japanese excelled because of their talent for miniaturization. History has shown that this was a vast underestimation, since it was only the opening wedge. From radio and audio equipment, the Japanese progressed to monochrome television, color television, video recorders for the home, and finally complex capital equipment for television broadcasting stations and production houses.

The Decline of AM and FM's Revenge

AM radio, which had succeeded so brilliantly in adapting itself to the competition from television, was not as successful against its sister medium, FM. The emergence of FM as the primary radio medium after decades in the economic wilderness is described in Chapter 3. At this point, it suffices to say that the main reasons were technical. FM's capability for high fidelity and stereo, its freedom from static, and the superiority of its nighttime coverage were the dominant factors. FM's gain was AM's loss, and in 1988 AM commanded only about 25 percent of the radio audience.

The role of high-fidelity stereo in FM's competitive position is underscored by demographic studies of the audience. One study reported that FM had three times as many teenage listeners as AM and that AM had 20 percent more listeners in the over-55 age-group. The presumption is that teenagers listen to rock music, which demands high fidelity and stereo, while older listeners prefer news and quieter music, which demands less fidelity.

The pioneers in FM broadcasting found sweet revenge in its triumph. For many years after World War II, it was regarded somewhat contemptuously by members of the AM broadcasting establishment. For at least two decades, their contempt seemed justified. The FM listening audience grew with painful slowness, and FM-only stations, considered as a group, operated at a loss until 1976.

In the meantime, the pioneers endured seemingly endless years of losses and the jibes of the AM broadcasting establishment. Their anguish was increased by the honest doubts that FM would ever succeed. It is unfortunate that many of these pioneers did not live to enjoy FM's final victory.

For those who persevered, the rewards have been great. FM profits are at an all-time high, and the market value of FM stations continues to rise. As a final irony, in 1986 the FCC repealed the rule that limited simulcasting, the simultaneous broadcasting of the same material on FM and AM. Simulcasting was permitted in the early days of FM to help *it* survive. As FM became more profitable, the FCC required increasing amounts of nonduplicated programming. In 1986, complete duplication was permitted—this time to help *AM* survive.

■ A LOOK AHEAD

It is dangerous to play prophet in the broadcasting industry. The history of broadcasting is replete with bad forecasts by the industry's foremost authorities. Some of the worst have been made by zealous advocates of new technologies who were overly anxious to predict that earlier technologies would become obsolete. Their favorite analogy is the buggy whip.

It was predicted that radio would eliminate phonographs; it did not. It was predicted that television would elimnate radio; it did not. It was predicted that satellites and cable systems would eliminate over-the-air broadcasting for the delivery of television programs; they have not. Finally, it was predicted that FM broadcasting would supersede AM. Since the future of AM will be shaped in large part by its competitive position in relation to FM, the accuracy of this prediction is the key issue in evaluating the future of AM.

FM has won its position by virtue of its technical superiority, and proponents of AM are trying to meet this challenge by improving AM technology. The problem of AM stereo standards is being solved, although the results will not be as good as those for FM stereo. The recommendations of the NRSC, when adopted by the FCC, will improve AM performance. A more questionable change, a reduction in channel width from 10 kHz to 9 kHz, would increase the number of channels but at the expense of increased adjacent channel interference.

All these developments will be helpful and are to be encouraged, but given the inherent propagation and bandwidth characteristics of the AM broadcast band and the vast investment of the public and broadcasters in the present technology, it appears that future technical improvements will come slowly and be limited. The quality of AM radio may improve, but it will not equal that of FM.

If one accepts this premise, it follows that the future, perhaps even the survival, of AM will depend on an overall shortage of radio channels, the clear channel stations' ability to serve rural areas at night, and nontechnical factors such as FCC regulation and programming.

Even though there are more than nine thousand commercial stations (about forty-nine hundred AM and forty-one hundred FM) and more than one thousand educational stations on the air, there is still a shortage of channels in some areas. The great diversity in public tastes and program materials, along with the low cost of operating a station, have created a demand for station licenses. This demand cannot possibly be met by FM stations alone under current assignment policies—even if they were modified in accordance with proposals in FCC Docket 80-90 that could add as many as seven hundred new FM stations. The channel shortage is so great that the FCC is proposing to extend the AM band upward an additional 100 kHz to 1,700 kHz. This expansion was authorized in principle by the 1979 World Administrative Radio Conference (WARC) for North and South America. The assignment rules have not yet been established, but one proposal is that all stations operate with 1 kilowatt of power, nondirectional, day and night. In any case, the survival of a few high-power clear channel stations with their unique ability to cover vast unserved rural areas seems assured under almost any set of assumptions.

The FCC is seriously considering a change in its ancient duopoly rule (common ownership of two stations in the same market) that would permit a single individual or corporation to own and operate more than one station in a market. This could lead to economies of scale and other efficiences that would improve AM's economic condition.

Perhaps the greatest opportunity for AM radio lies in using program formats that minimize the effect of its technical deficiencies. High-fidelity stereo is not needed for a news broadcast or for many types of music. Many AM stations are currently using such formats.

The overall outlook for AM radio, then, is one of cautious optimism. It has serious competitive problems, and its glory days are over. But it has survived major challenges in the past, and perhaps innovative management operating in a free enterprise system can continue to make AM radio profitable and provide an important additional service to the public.

3

■ ■ FM RADIO BROADCASTING

On January 31, 1954, Edwin Howard Amrstrong, a distinguished inventor and professor of electrical engineering at Columbia University, returned to his Manhattan apartment, dressed himself meticulously in formal clothes, then stepped out the window to his death thirteen floors below. This tragic act culminated the bitter conflicts that marked the introduction of FM broadcasting in the United States.

The controversial issues were

1. Who invented FM?
2. Should FM broadcasting be permitted?
3. What portion of the radio spectrum should be allocated to it?

Although these issues stemmed from the conflicting economic interests of the participants, they were greatly aggravated by pride, envy, and anger—three of the seven deadly sins. Armstrong's suicide resulted from his despondency over the interminable patent litigation he was carrying on against RCA. He wished to win this litigation not so much for the financial rewards (he was already a modestly wealthy man, and the legal costs of the litigation approached or exceeded the judgment he hoped to receive) but to satisfy his pride by obtaining formal recognition that he was the inventor of FM. His antagonist, David Sarnoff, was similarly motivated, and the result was an impasse that could not be resolved by money.

■ MAJOR EDWIN HOWARD ARMSTRONG

If one were to rank the engineers and scientists who contributed most to the development of broadcasting technology, Edwin H. Armstrong (known as Howard to his family and friends and shown in Figure 3–1) would be among those at the top of the list. His inventions displayed a high order of creativity, were commercially important, and had a lasting effect on the industry. Broadcasting owes much to his inventive genius.

From a technical standpoint, his invention of the superheterodyne circuit for radio receivers may have been the most significant (see Chapter 1), but he is best known for the development and advocacy of wideband frequency modulation for radio broadcasting. He was more responsible for the intro-

Courtesy Smithsonian Institution, with permission.

■ **Figure 3–1**   Major Edwin Howard Armstrong.

duction of FM broadcasting in the United States than any other individual. It is appropriate, therefore, to introduce this subject with a short biography of this distinguished engineer.

Armstrong was born in New York City in 1890 to cultured and modestly affluent parents. His father was manager of the U.S. branch of the Oxford University Press, and his mother was a schoolteacher. He enjoyed a happy childhood first in New York and then in suburban Yonkers where his parents moved in 1902.

His interest in becoming an inventor began in 1904 when his father, returning from England, brought him a book titled, *The Boys' Book of Inventions*. Soon after, he began experiments in wireless telegraphy, using an attic room in his Yonkers home. Still later, his high school physics teacher allowed him to construct a transmitting and receiving station on the roof of the school.

He entered Columbia University in 1909, majoring in electrical engineering. As in most electrical engineering departments of the time, the curriculum was heavily weighted toward power, a subject that did not interest him in the least. This lack of interest, together with his irreverent attitude toward the power engineering faculty, made him unpopular and from time to time jeopardized his academic standing.

The attitude of the electrical power faculty toward the communications curriculum, which they sometimes described as "the study of piddle currents," was condescending. Nevertheless, the communications faculty included a number of distinguished pioneers in communications, including John Morecroft and Michael Pupin. These men defended Armstrong against the attacks of the power engineering faculty and were an important influence in his development as an inventor.

His extraordinary talent for invention was displayed while he was an undergraduate, for it was during this period that he conceived the regenerative receiver circuit. His father, fearing that his preoccupation with invention was interfering with his studies, refused to loan him the $150 necessary for the patent filing fee until he graduated. Armstrong graduated in June 1913, his father made the loan, he filed the application, and the patent was granted in October 1914. In the meantime, his supporters on the faculty were successful in having him appointed as an instructor at a salary of $600 per year.

Unfortunately, this invention initiated the continuous dispute over patents and inventions that was to mar Armstrong's life and embitter his personality. Three other inventors, Lee De Forest (patent assigned to AT&T), Irving Langmuir (patent assigned to GE), and Meissner (patent assigned to Telefunken) claimed priority on the patent for the regenerative circuit. Litigation dragged on for two decades, and finally, in 1934, the U.S. Supreme Court decided in favor of De Forest.

It was a terrible blow to Armstrong, and shortly thereafter, in a highly emotional scene at an annual meeting of the Institute of Radio Engineers (IRE), he publicly returned the medal he had received from the Institute for this invention many years previously. Despite the Supreme Court ruling, however, many engineers believed the honor belonged to him, and he continued to have strong support from the broadcast and communications technical communities.

When the United States declared war on Germany in World War I, Armstrong entered the Army as a signal officer. He was sent almost immediately to the front lines in France to observe the communications systems being used by the French and U.S. armies and to update the wireless communications system for the U.S. Army. He received the French Chevalier de la Legion d'Honneur for his wartime service, but his most important accomplishment during the war years was the invention of the superheterodyne circuit, an activity only indirectly related to his official duties.

His application for the superheterodyne patent was challenged, but he successfully withstood this, and the patent was granted on June 8, 1920. Soon afterward, on October 5, he sold it to Westinghouse for $350,000 but retained the rights for amateur equipment. These rights were lucrative, and for a time his royalty income was more than $10,000 a month.

Shortly after the armistice, he was promoted to the permanent rank of major, and ever afterward he was known in the industry as Major Armstrong. His prolific efforts as an inventor continued after he returned to civilian life, and his next major invention was an improved version of the regenerative circuit, the superregenerative circuit. Although it had a number of important applications, it could not withstand competition from the superheterodyne. Sarnoff, however, was sufficiently impressed to convince RCA to purchase the patent for $200,000 plus sixty thousand shares of RCA stock (later increased to eighty thousand shares).

At this time, Armstrong and Sarnoff were good friends. In the course of his visits to Sarnoff's office, Armstrong fell in love with Sarnoff's secretary, Marian MacInnes, whom he married on December 1, 1923. Thus years later, when he was engaged in a venomous FM patent battle with RCA, he was not only one of RCA's largest stockholders but also was married to its chairman's former secretary.

The income from his patents allowed him to live in modest affluence and to establish his own research laboratory in Alpine, New Jersey. There, he and a small staff of loyal and enthusiastic employees carried out research in radio circuitry and systems for nearly three decades. In addition, he rejoined the electrical engineering faculty at his alma mater, Columbia University.

In the estimation of Armstrong and his body of supporters, the culmination of his career as an inventor came with his development of frequency modulation (FM) as an alternative to amplitude modulation (AM), the method then used universally in radio broadcasting systems. After filing his FM patent applications during the early 1930s, his enthusiasm for it soon grew far beyond the normal paternal instinct of an inventor. He became a zealot with a mission to bring the benefits of the new medium to the public. He was FM's evangelist, and for the remainder of his life, he devoted his considerable talents as an engineer and salesman to its promotion and the defense of his patent position. In the end, these activities led to his death.

Armstrong was so convinced of FM's superiority that he could not conceive that an objective engineer or broadcast executive could fail to share his enthusiasm for it. When some elements of the industry did not, he immediately assumed that it was the result of their greed and their desire to enhance their private economic interests by maintaining the status quo, even at the expense of the public interest. This was his appraisal of the attitude of RCA and his old friend David Sarnoff when the latter refused to support the development and promotion of FM after choosing to concentrate his company's resources on television.

Armstrong's final battle with RCA came in 1948, when he filed a lawsuit charging infringement of his FM patents. The resulting litigation, which ended with Armstrong's suicide, is described later in this chapter.

Armstrong's life was an archetype of the classical Greek tragedy in which the hero is destroyed by a single fatal flaw—the sin of excessive pride. But his achievements live on for the rest of the world to enjoy.

■ THE BASIC TECHNOLOGIES OF FM BROADCASTING

Invention and Development

The question "Who invented FM?" and the circumstances surrounding its invention were debated at length in the acrimonious patent litigation between Armstrong and RCA during the post–World War II years. This complex issue is discussed later in this chapter.

The next phase in the evolution of any new technology is its development, the transformation of an invention into a practical product or system. There can be no doubt that much of the credit for the early development of FM belongs to Armstrong.

The concept of FM is simple, perhaps even obvious. It began to appear in the technical literature in the early 1920s, and a number of technical papers on the subject were published during the next decade. But a recognition of its unique properties and the application of these properties to a practical broadcasting system required a high degree of creativity and years of patient study and experimentation. It was in performing this function that Armstrong made his greatest contribution to FM technology.

Frequency Modulation

As its name implies, FM, or frequency modulation, differs from AM, or amplitude modulation, in that the *frequency* rather than the *amplitude* of the radio frequency signal is varied or modulated in accordance with the audio signal being transmitted.

There are two distinct types of FM, narrow-band and wideband. Narrow-band systems, in which the frequency variations are small, are widely used for radiotelephone communications and other applications that do not require high-fidelity sound reproduction. Wideband systems, with large frequency swings, are used for broadcasting. (See illustration on next page.)

Armstrong's most important contribution to FM technology was the discovery of the dramatic improvement in performance that resulted from the use of wideband FM. This also was the basis for his patent position.

FM WAVEFORM AND SPECTRUM

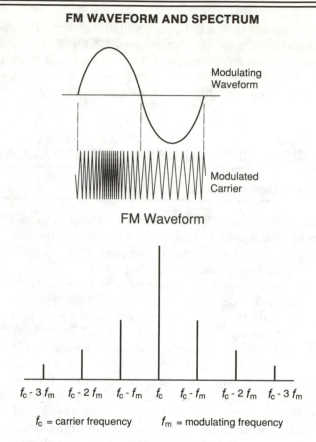

FM Waveform

f_c = carrier frequency f_m = modulating frequency

FM Spectrum

This drawing shows the waveform and frequency spectrum that result from the frequency modulation of a radio frequency carrier with an audio signal. The distribution of energy in the radio frequency spectrum is a representation that is more useful to the engineer than the waveform.

To conserve spectrum space, radio communication systems typically use small deviations and narrow channels. This is known as narrow-band FM. For example, land mobile systems are limited to a deviation of ±5 kHz.

To achieve optimum performance, broadcast FM stations use wideband FM with large deviations and wide channel bandwidths. In the United States, the deviation is ±75 kHz, and the bandwidth is 200 kHz. Wideband FM is also used for the television sound channel, although the deviation of ±25 kHz is not as great as that used for FM broadcasting.

The spectrum of a wideband FM waveform is complex. In theory, the sidebands extend indefinitely above and below the carrier. In practice, nearly all the sideband energy is contained within a well-defined frequency band that extends above and below the carrier by an amount equal to the frequency deviation plus the frequency of the modulating signal.

Specifically, he claimed the following advantages of wideband FM over AM and narrow-band FM:

- Virtual freedom from atmospheric interference or static from electrical storms. Wideband FM is inherently more immune to this type of interference.
- Greater freedom from man-made interference, such as that produced by arcing power lines and electrical appliances
- Greater immunity to interference from stations operating on the same frequency. With AM, a weak interfering signal from a distant station having a strength of as little as one-twentieth that from the desired station can still be heard in the background. With FM, an interfering signal must have a strength of more than half the desired signal to be heard at all.
- The capability for higher fidelity reproduction, particularly with respect to dynamic (volume) range and frequency response.

Having discovered these unique characteristics of wideband FM, Armstrong became its missionary. His goal was the creation of a broadcasting system that was technically as perfect as possible. To achieve this, he was unwilling to make any compromises. His most important recommendation, which the FCC ultimately adopted, was 200-kHz spacing between channels—twenty times the 10-kHz spacing in AM broadcasting. Only the major improvement in performance resulting from wideband channels could justify such prodigal use of a limited resource.

With the benefit of hindsight, one might conclude that the choice of a 200-kHz channel width was overkill for monaural broadcasting. On the one hand, excellent performance, even for the most demanding program material, can be obtained with narrower channels. On the other hand, 200-kHz channels provide the capacity for high-quality stereo broadcasting, now in common use but not contemplated during the prewar years when the 200-kHz standard was established. Armstrong and the FCC probably chose the optimum channel bandwidth, although possibly for the wrong reason.

Very High Frequencies

The requirement for spectrum space resulting from the choice of the 200-kHz channel width dictated the location of FM in the VHF region of the radio frequency spectrum—30 to 300 MHz. There simply was not enough space at lower frequencies. Even at VHF, FM had to compete for spectrum space with television, and this became another point of contention during the birth and growth of FM.

The necessity of locating FM in the VHF portion of the spectrum was fortuitous because the advantages of the broadcast system Armstrong had proposed did not result solely from its use of FM. In part, they were the consequence of the bandwidth of the FM channels and their location in the spectrum. If AM had wider channels and a similar spectrum location, it would have enjoyed many of the advantages of Armstrong's FM system, and it would have lacked some of its disadvantages. In their zeal to sell FM, Armstrong and his supporters were not always careful to explain this.

While the necessity for locating FM in the VHF region was obvious and uncontested, its optimum position within this band was not. The establishment of this position led to one of FM's most acrimonious disputes.

VHF Propagation   The propagation mechanisms of the VHF radio waves used for FM broadcasting, 88 to 108 MHz, are quite different from those of the medium-wave frequencies, 550 to 1,600 kHz, used for AM broadcasting. The relative advantages of AM and FM broadcasting are determined as much by these differences as by the inherent differences between AM and FM modulation. VHF transmission is more nearly like that of light rays, and it essentially follows a line-of-sight path with a range of thirty to sixty miles depending on the terrain and the antenna height. Like light, it is subject to the phenomena of *reflection*, *refraction*, and *diffraction*.

VHF waves are reflected from both the earth and terrestrial objects, and the received signals are a complex mixture of direct and reflected waves. Under some conditions, reception of this mixture results in *multipath distortion* and a serious distortion of the audio signal.

Refraction bends the path of VHF waves around the earth's curvature as they pass into the less dense layers of the upper atmosphere. This increases the distance to the radio horizon to about one and one-third times the distance to the geometric horizon.

Diffraction bends the paths of waves as they pass over sharp obstacles. It often makes reception possible behind buildings and other obstacles in the direct path of the wave.

A third characteristic of VHF is that long-distance nighttime ionospheric, or sky-wave, transmission is eliminated or greatly reduced. This has both advantages and disadvantages. The advantages are that stations are not subject to nighttime interference from distant stations operating on the same frequency and their day and night coverage areas are essentially the same. The disadvantage is that the useful range of FM stations is more limited. They cannot duplicate the nighttime coverage of remote rural areas that clear channels provide in the AM broadcast band.

Fourth, the severity of atmospheric static is less at VHF than in the AM broadcast band. Similarly, the severity of man-made interference is less at VHF frequencies.

The Effect of Antenna Height A typical FM antenna is a structure twenty-five feet high mounted on a supporting tower. Unlike an AM antenna, the tower is not a radiating element but merely serves as a support.

The height of the antenna has a major effect on the station's coverage for two reasons. First, increasing the height of the antenna increases the distance to the horizon and hence the range of the station. Second, increasing the antenna height, even at locations between the station and the horizon, results in a more favorable phase relationship between the direct and reflected waves and thus increases the strength of the received signal. The effect is quite large. For example, increasing the antenna height from five hundred to one thousand feet produces the same effect as increasing the power by a factor of 5.

■ THE EARLY DEVELOPMENT OF FM

Armstrong's Initial Research

From 1928 to 1933, Armstrong worked diligently at his Columbia University lab (Figure 3–2) to construct a working model of an FM system. With the limitations imposed by the primitive state of vacuum tube technology at the time, his experimental circuits required as many as one hundred tubes. He was a thorough investigator, and he made more than a hundred thousand measurements of the performance of FM modulated signals. It was during this period that he discovered the unique properties of wideband FM. From July 1930 to January 1933, he filed four FM patent applications; all were granted simultaneously on December 26, 1933.

Two years later, on November 5, 1935, he made the first public demonstration of FM to the New York chapter of the IRE together with a paper titled "A Method of Reducing Distrubances in Radio Signaling by a System of Frequency Modulation." An amateur station in Yonkers owned by his friend Randy Runyon transmitted the signal for the demonstration. Curiously, it made little initial impact on the assembled engineers.

In the years prior to 1933, RCA and AT&T engineers also experimented with FM. They were generally unimpressed and were unable to note any advantages over AM. This may have resulted from their failure to recognize the great difference in the characteristics of wideband and narrow-band FM.

RCA Turns Its Back

From the outset, Armstrong recognized that the support of a substantial number of radio manufacturers and broadcasters was required if FM were to

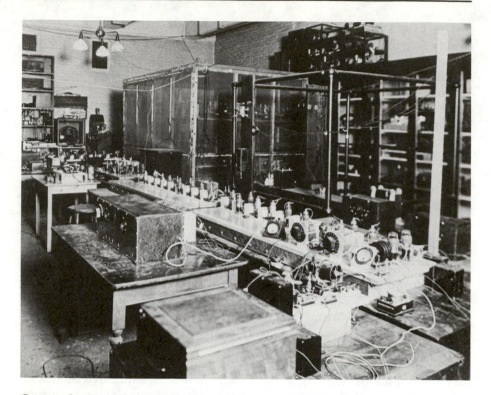

Courtesy Smithsonian Institution/Columbia University, with permission.

■ **Figure 3–2** Armstrong's Columbia University Laboratory.

be successful. He had no desire to be a manufacturer or broadcaster. His rewards were the patent royalties and the honor and prestige of being recognized as the inventor of FM. But someone had to build and market the receivers and transmitting apparatus, someone had to construct and operate the broadcasting stations, and he needed as much support as he could muster to persuade the FCC to authorize FM broadcasting.

Armstrong was so convinced of the superiority of FM that he anticipated no difficulty in obtaining the industry's support. It was with this attitude that he approached David Sarnoff to enlist the aid of RCA and NBC. It was a logical step. Sarnoff and Armstrong were friends, RCA was the leading manufacturer of radio transmitters and receivers, and NBC was the leading radio network. With their support, the success of FM was almost guaranteed.

For a time, RCA cooperated. Armstrong first demonstrated FM to Sarnoff in December 1933. This was followed by visits from RCA engineers, who were sufficiently impressed to recommend field tests. In response to this recommendation, Armstrong, with RCA's agreement, installed a test transmitter on the Empire State Building in 1934. It shared facilities with RCA's television field-test transmitter, and for two years it carried out tests of FM broadcasting.

The reactions of RCA's engineers to these tests were mixed. Some were favorable, but other were cautious or clearly unfavorable. One of the more optimistic reports was issued by Harold Beverage, a highly regarded engineer, who stated that frequency modulation with a deviation of 100 kilocycles would increase the service radius by 3 to 5 times as compared with AM. RCA's sales, patent, and legal departments, however, were concerned about FM's threat to the corporate position and opposed RCA's support.

Faced with conflicting advice and wishing to save RCA's resources for television development, Sarnoff made his decision. In 1936, he advised Armstrong that RCA would cease any further activity on behalf of FM. (RCA's engineers, however, quietly continued a modest program of experimental work to protect the company's patent position.) He stated that, whatever advantages it might have over AM, it was only another form of radio broadcasting. In contrast, television was a completely new and potentially much more powerful medium, and Sarnoff had decided that RCA should devote all its research efforts to the development of television. As a tangible sign of this decision, he asked that Armstrong's experimental FM antenna be removed from the Empire State Building and replaced with a television antenna.

Armstrong was amazed and outraged. He could not believe that Sarnoff's desire to give television a higher priority was his real reason for turning his back on FM. Sarnoff's real motive, he decided, was to protect NBC and other AM broadcasters from FM competition and RCA from a threat to its patent position. AM broadcasters were prospering greatly in spite of the depression, and much of their success resulted from their licenses for the limited number of high-power AM stations. FM would increase the number of available channels, thus increasing competition, and it would provide approximately equal coverage for all stations.

Sarnoff's stated position seems reasonable enough today, but in trying to understand Armstrong's reaction, it is important to remember that television's ultimate success was by no means assured. It had major unsolved technical and economic problems, and the latter appeared almost insurmountable in light of the depression. As Armstrong saw it, television was a costly and highly speculative venture with a long-term payout at best, while radio was a proven medium and FM seemed to be a sure thing. He concluded, therefore, that the decision to abandon FM in favor of television must have resulted from ulterior motives. The long-standing friendship between Armstrong and Sarnoff came to an end.

Like every coin, this one had two sides. If Armstrong was correct in his belief that Sarnoff's motive was to protect AM broadcasters, how can

Sarnoff's costly and single-minded development of television, a medium that was a far greater threat to AM radio than was FM, be explained? The broadcasting establishment feared television more than FM, and Sarnoff was the object of vicious attacks by many radio broadcasters.

Finally, if fear of competition was an element in RCA's decision to withdraw its support of FM, the motives of Armstrong and other FM enthusiasts were not entirely pure either. The success of the AM broadcasters who were licensed to use the scarce AM channels aroused envy in FM's proponents. Throughout the literature of FM's early years, evidence of this envy is never far beneath the surface and is sometimes quite obvious. There was an enormous desire to "get the fat cats," and this was surely one of the reasons for the messianic zeal that developed in the FM establishment.

The FM Establishment Takes Form

Having failed to obtain RCA's support, Armstrong pressed ahead, aided by some of RCA's principal competitors, notably Zenith and GE. RCA was not the only game in town, and soon a strong and vocal FM establishment formed. At times its members seemed to live to oppose RCA as much as support FM. Whatever their motives, Armstrong had no dearth of supporters in the manufacturing industry.

The Development of FM Technology

A number of major technical tasks had to be performed before a proposal could be made to the FCC to allocate a block of frequencies for FM broadcasting. It was necessary to persuade a skeptical engineering community that wideband FM would do all that Armstrong claimed. Extensive field tests had to be carried out to establish performance under real-life conditions. Transmission standards had to be developed. A number of critical electric circuits had to be developed as a basis for receiver and transmitter designs. In short, the process of bringing FM out of the laboratory and making it a practical operating system was a major engineering task.

To carry out this task, Armstrong enlarged his laboratory at Columbia to engage in an expanded program for the development of FM. In 1936, following RCA's decision to remove his FM field-test antenna from the Empire State Building, he began construction of a test transmitter in Alpine, New Jersey, that included a four-hundred-foot tower to support the antenna. (Ironically, the transmitter was manufactured by RCA.) He was able to test the range and performance of FM signals over a wide variety of terrains with this installation. Armstrong also developed limiter, discriminator, and preemphasis/deemphasis circuits, three of the basic circuit types required in an FM system. And he established the 200-kHz channel width for FM broadcasting.

Unfortunately, Armstrong' popularity and influence were not enhanced by his self-righteous pronouncements and combative disposition. An example of Armstrong's attitude was his response to a mathematical paper that, having drawn the conclusion that narrow-band FM would not be practical for broadcasting, failed to make a distinction between narrow-band and wideband. Angered, Armstrong wrote a scathing paper titled "Mathematical Theory vs Physical Concept" that attacked mathematics as a tool of physical research.

Nevertheless, the technical development of FM proceeded rapidly. In 1939, the Yankee Network erected an experimental FM transmitter on Mount Asnebumskit in New England. Two weeks later, RCA quietly applied for an experimental license. In all, some twenty experimental stations were constructed during this period, and by the middle of 1939, sufficient engineering information had been collected to provide the technical basis for a petition to the FCC for an allocation of a substantial portion of the spectrum for FM broadcasting.

■ THE FCC ACTS

The FM Rule-Making Hearing

Late in 1939, the FCC convened a Rule Making hearing to determine whether to allocate VHF spectrum space for the purpose of commercial FM broadcasting. Frequencies for experimental broadcasting had been assigned since 1935, but licensees of experimental stations were not permitted to accept advertising or to derive any income from them. FM's advocates believed the time had come to permit regular commercial broadcasting and were lobbying vigorously for this authorization.

The FCC addressed three questions in this hearing:

1. Should VHF frequency bands be allocated for radio broadcasting?
2. If so, should it be AM or FM?
3. Where and how much of the spectrum should be allocated for broadcasting? This quesion was directly related to the second, since the amount of spectrum needed would depend on the choice of FM or AM.

The first question was almost no contest. There was a legitimate argument that an allocation of an additional frequency band for radio broadcasting would make a greater variety of programming available to the public. In addition, the regulators believed that opening up a new band would increase the competition to AM broadcasters who were waxing fat, protected from free competition by the scarcity of channels. These broadcasters, while they did not welcome additional competition, were in no position to oppose it. To have done so would have been seen as opposing the public interest. Thus, there was no question that the Commission would establish a VHF broadcasting service.

The answer to the second question, AM or FM, was not as simple. Some of the advantages claimed for FM were the result of its position in the VHF spectrum rather than of the type of modulation. Furthermore, it was prodigal in its use of spectrum space, and FM receivers were expected to be more expensive.

These arguments could not withstand the onslaught of Armstrong and his supporters. For them, the promotion of FM was a crusade. They played skillfully on the dissatisfaction of the academic, literary, and artistic communities with the programming available on standard AM radio. In 1938 and 1939, Armstrong addressed innumerable groups, clubs, and technical societies touting the advantages of FM. This public relations effort was highly effective, as he promised a continuous flow of high-quality programs transmitted with superb technical quality. After a time, FM broadcasting acquired something of a mystique in the public's mind.

The Position of RCA and CBS

The official position of RCA and CBS was that FM was an interesting technical development that did not have much commercial significance because the public was not interested in high fidelity. This position particularly angered Armstrong. It was their opinion, he said sarcastically, that the public wanted to hear conductor Leopold Stokowski in "low fidelity" (under his premise that vast numbers of the public wanted to hear Stokowski).

RCA's ability to oppose FM at the FCC, assuming it might have wished to do so, was limited. Given the temper of the times and RCA's dominant position in the industry, aggressive opposition to FM would have been counterproductive. RCA and its subsidiary, NBC, chose, therefore, to assert their request strongly for VHF spectrum space for television but to maintain a low profile with respect to FM. Late in 1939, RCA petitioned the commission to allocate VHF spectrum space for thirteen 6-MHz television channels. This would have reduced the space available for radio broadcasting, but RCA suggested that this problem could be solved, even with FM, by reducing the frequency deviation and the channel width.

Outside of the regulatory arena, RCA opted to encourage FM as little as possible, to maintain a modest level of FM engineering development to protect its patent position, and to challenge Armstrong's patents, if necessary, in court.

The FCC Chairman

The cause of FM's proponents was aided by the predilections of the FCC's strong-willed chairman, James Lawrence (Larry) Fly. He honestly believed that FM was a superior medium for aural broadcasting, he was irritated by RCA's aggressive demand for spectrum space for commercial television, and he had a populist resentment of the huge profits being made by AM broadcasters. Accordingly, his response to RCA's request for thirteen television channels was that there could be no allocation of television channels until the needs of FM had been considered and met.

Toward the end of the hearing, RCA, sensing the inevitable, softened its position toward FM. It stated publicly that FM was ready for commercial development. In addition, Sarnoff offered to buy Armstrong's patent rights for $1 million, an offer Armstrong rejected.

The FCC's Report and Order

With a pro-FM chairman, a strong and effective FM lobby, and FM's opposition muted, it is not surprising that the FCC acted with unprecedented speed. On May 20, 1940, it issued a Report and Order stating in summary that "Frequency modulation is highly developed. It is ready to move forward on a broad scale and on a full commercial basis."

The Report further stated that FM was technically superior to AM for broadcasting at frequencies above 25 MHz, citing "unanimous" industry opinion as evidence. It accepted Armstrong's recommendation for 200-kHz channels.

The Report noted that the opening of a new high-frequency band would correct the inequalities of the standard broadcast band; it stated, "These inequalities result from the scarcity of frequencies, their technical characteristics and the early growth of broadcasting without technical regulation." The new band would allow some communities to have their first station and to increase the number of services available in others. The disparity in service areas between stations would be eliminated: "Competitive broadcast stations in the same center of population will insofar as possible be licensed to serve the same area."

The decision was to allocate forty 200-kHz channels in 8 MHz of spectrum space—from 42 to 50 MHz—for commercial and educational FM broadcasting. Commercial broadcasting was authorized on thirty-five channels and educational broadcasting on five. It was a complete victory for Armstrong.

The choice of FM over AM for VHF broadcasting was the right one, but it did not result from an impartial hearing. Industry opinion favoring FM was not as unanimous as the FCC indicated in its Report. But, with the exception of RCA, there were no vested interests that had reason to support AM. The AM broadcasting establishment did not welcome any VHF competition, but given its inevitability, AM broadcasters perceived (correctly for the next two decades) that FM would be difficult to introduce. Accordingly, they gave it lukewarm support.

For RCA, the decision was a defeat. For the first time since the earliest days of radio, its technical leadership in broadcasting was challenged. This affected its potential royalty income, and it piqued Sarnoff's pride. Also, the frequency band allocated to FM overlapped a portion of the channel that had been tentatively planned for television channel 1. It was one of the factors that ultimately led to the elimination of this channel and a reduction in the number of VHF channels from 13 to 12.

The report included a number of caveats. First, it noted that the service area of FM stations would be limited though greater than the primary service area of many standard broadcast stations. Second, it pointed out that FM would not have long-distance coverage in the VHF portion of the spectrum. Thus there would be a continuing need for AM in the standard broadcast band to provide rural service. The report also refused to forecast that FM would supplant AM: "The extent to which in future years listeners will be attracted away from the standard band cannot be predicted."

Finally, the report raised a serious question as to whether the 42- to 50-MHz band was optimum for FM broadcasting because of potential sky-wave interference: "The effect of sky-wave interference will not be known until additional stations are placed in operation in various parts of the country. If later developments should favor the use of higher frequencies, the Commission will consider the facts at that time." This simple statement, which seems to have been ignored by most of the industry, was a premonition of a battle that would tear the industry apart a few years later.

Commercial FM Broadcasting Begins

The FCC decision was received enthusiastically by the FM establishment. There was a rush for licenses by both newcomers and established AM broadcasters. The intensity of competition for licenses was enhanced by the terms of the FCC Order, which established three classes of stations by power level based on the clear, regional, and local station hierarchy of AM broadcasting. Licenses for the higher power stations, particularly in major markets, were especially desired. FM broadcasting soon had its own would-be fat cats.

By the time station construction was halted by the United States' entry into World War II, Construction Permits (the preliminary to licenses) had been granted for 58 stations—seven to educational broadcasters, twelve to new broadcasters, and thirty-nine to the licensees of AM stations. The AM licensees, generally not eager to promote FM but not wanting to be left behind, constructed stations as a defensive measure and for the most part duplicated the programming of their AM stations.

RCA's lukewarm support of FM was evidenced by the fact that NBC constructed only one station, in New York. CBS was a little more optimistic and built stations in New York, Chicago, and Philadelphia.

A number of manufacturers, most aggressively Zenith, began the promotion and sale of FM receivers. The initial response of the public was good, and when receiver production was halted by the war at the end of 1941, nearly 400,000 sets had been produced and sold. FM appeared to be off to a good start.

■ THE WORLD WAR II HIATUS

Soon after the United States entered World War II, the FCC ceased authorizing new FM stations. The construction of stations authorized but not yet completed became difficult or impossible because of government restrictions on the use of critical materials. The manufacture of FM receivers also was halted. Thus FM broadcasting was essentially frozen at its December 7, 1941, level.

The hiatus in the growth of FM broadcasting did not extend to its technology. On the contrary, just as World War I had provided a strong stimulus for the development of AM broadcasting technology, World War II provided the stimulus and funds for an enormous expansion in radar and radio communications research, much of which was directly applicable to FM broadcasting.

At the beginning of the war, VHF technology was in its infancy. Vacuum tubes and circuit components that could operate in this region of the spectrum were just being developed. Knowledge of the properties of VHF radio waves, particularly their propagation characteristics, was limited. Few engineers or technicians had formal training or experience in the design and maintenance of VHF equipment.

Wartime research changed this. Like FM, long-range search radar operated in the VHF region of the spectrum. Short-range tactical radio communications moved upward in the spectrum to VHF and shifted from AM to FM. Much of the technology developed for these critical military products could be transferred directly to FM broadcasting equipment and systems. An enormous array of vacuum tubes, circuit components, and system technologies developed under forced draft during the war for radar and communications equipment became available to the designers of FM transmitters and receivers.

Equally importantly, thousands of young men came out of the armed services trained in the maintenance of VHF equipment. FM broadcasting no longer had to develop its own technicians.

As a consequence, while the short-term effect of the war was to halt the growth of FM broadcasting, its long-term effect was to accelerate technical development.

The war greatly enhanced the size and stature of the electronics engineering profession that evolved from radio engineering. The IRE later merged with the electrical engineers' professional society, the American Institute of Electrical Engineers (AIEE), which had once treated it with condescension, to form the Institute of Electrical and Electronics Engineers (IEEE).

World War II also provided a field test of public acceptance of FM broadcasting. The results were inconclusive. No overwhelming public preference for FM could be detected. FM stations were not profitable, in vivid contrast to AM stations that were making more money than ever. FM enthusiasts pointed out, correctly, that it was not a valid test. The limited number and type of receivers in the field did not encourage programming that would demonstrate the high-fidelity capabilities of FM. For the most part, programming was similar to or a duplication of that available on AM. In addition, the owners of FM-only receivers had the choice of only a limited number of stations. The common chicken-and-egg problem of a new broadcasting service was not yet solved.

Because of these ambiguous results, industry opinion was divided as to the future of FM. For the most part, AM broadcasters, possibly engaging in wishful thinking, were not optimistic. But the prewar FM enthusiasts did not lose their zeal, and they eagerly resumed the task of building a profitable FM industry.

■ THE GREAT SPECTRUM BATTLE

But for a problem that resulted indirectly from wartime research, the wartime hiatus might have ended shortly after May 7, 1945, when Germany capitulated. With Germany defeated and Japan's position becoming increasingly desperate, the War Production Board announced that it would soon lift its restriction on the manufacture of radio receivers. But the manufacture of FM receivers could not be resumed until the uncertainty concerning the location of FM broadcasting in the spectrum was resolved. Would it remain in the 42- to 50-MHz band, or would it be moved to some higher frequency? This issue was raised as the result of War Department measurements of VHF sky-wave propagation, and it was the basis of the great spectrum battle. This conflict began in 1944 and was not settled until March 1946. In the meantime, manufacturers could not begin the sale of receivers, and no new stations could be authorized or built.

Norton's Research

The catalyst for the spectrum battle was Kenneth Norton, a quiet and highly competent engineer on the FCC staff during the prewar years. In an unspectacular way, he had made important contributions to the broadcasting industry. His equations for calculating the ground-wave coverage of radio stations in the AM broadcast band were described in Chapter 2.

At the outbreak of the war, Norton was transferred temporarily to the War Department. There, he became privy to the results of classified experimental programs in the sky-wave propagation of VHF signals.

The lower limit of the VHF range, 30 MHz, is also the upper limit of frequencies at which long-distance sky-wave transmission normally occurs (see Chapter 1). But nature abhors sharp discontinuities, and under abnormal ionospheric conditions, sky-wave transmission sometimes occurs in the lower portions of the VHF spectrum. There was sufficient prewar evidence that this might be a problem for FM broadcasting to cause the FCC to include a warning in its authorization of the 42- to 50-MHz band. Wartime field measurements added greatly to the bank of available data, and Norton reexamined the question.

He combined the War Department results with data obtained from the Bureau of Standards and from measurements on a commercial FM station, WGTR, in Paxton, Massachusetts, to develop new predictions of the extent of VHF sky-wave transmission. He then calculated the amount of cochannel interference in the 42- to 50-MHz FM band that would result. These calculations indicated that objectionable interference would occur much more frequently than had previously been estimated. They also showed that this interference would occur much less frequently at higher VHF frequencies.

Norton reported his calculations confidentially to the FCC, and he strongly urged that the FM broadcast band be moved to a higher frequency. His forecast of the number of hours per year that listeners at the edge of the service area would receive E-layer sky-wave interference after all the channels were occupied was indeed sobering:

Frequency (MHz)	Hours per Year
43	830–2,410
48	475–1,400
66	55–190
104	0.65–2.6

The FCC took Norton's report seriously, partly because of its respect for his competence and partly because he confirmed the FCC's earlier suspicions voiced in its original allocation of the 42- to 50-MHz band. It responded by convening a hearing late in 1944 to consider this new evidence. Since classified information was disclosed, the hearing was not open to the public, but industry representatives with proper clearance were admitted. Norton was the FCC's principal witness. He presented his data and repeated his recommendation that the FM band be moved.

The FCC's Dilemma

The commission had developed its rules for the spacing between cochannel FM stations on the basis of the signals propagated through the troposphere, the lower level of the atmosphere. Their range was limited, and spacing of less than two hundred miles was permitted. This spacing made it possible to assign a large number of stations on the same channel. But if sky-wave propagation had to be considered, it would be necessary to increase the spacing and reduce the number of cochannel stations drastically.

In addressing Norton's report at the hearing, the FCC had to answer two questions. First, how often would objectionable interference occur in the 42- to 50-HMz band? This was a technical question. Second, how much interference could be tolerated before it became objectionable? This was a judgment call. Norton's recommendation presented the FCC with a serious dilemma. On the one hand, the FCC did not wish to saddle the new service permanently with an inferior spectrum allocation. On the other hand, it had to recognzie the serious impact of adopting Norton's recommendations on FM broadcasters and equipment manufacturers and on the development of the industry. Among the effects were the following:

- Some 400,000 receivers in the hands of the public would become obsolete.
- Transmitter and receiver manufacturers would be faced with costly redesign and retooling programs.
- The time required to carry out these programs would delay the postwar resumption of FM's growth.
- Broadcasters would be faced with the expense of replacing their transmitting equipment and would lose their existing audience base—the owners of the 400,000 receivers.
- The service areas of stations operating at higher frequencies might not be as great, particularly in hilly or mountainous terrain.

The Industry's Position

Given these adverse effects, it is not surprising that the manufacturers and broadcasters who had pioneered FM reacted violently to the Norton recommendation. Their concern for the problems resulting from moving the band was aggravated by a terrible sense of injustice. As they saw it, they had risked considerable resources to initiate this new service. The war had interrupted its growth before they could recoup their investment. And now companies such as RCA, which had invested little or nothing in the industry, could compete with them on an equal basis.

The consequence was that Norton, one of the least combative of individuals, found himself embroiled in an acrimonious dispute with most of the FM broadcasting industry, both manufacturers and broadcasters. As time passed and it appeared that his views carried a lot of weight with the Commission, the debate became venomous. Norton stuck to his position, and in the end it prevailed.

In a broad sense, the dispute was not over technical issues but over policy. The FCC wanted to provide FM service to all parts of the country, including the most sparsely populated. It was in these areas on the fringes of the stations' coverage that sky-wave interference would be the most serious. Hence the commission wished to minimize it.

The industry's priorities were different. Providing totally interference-free service to a few listeners on the remote edges of their service areas was of minor commercial importance to broadcasters. Their primary interest was service to the population centers in the metropolitan areas they served. They had located their transmitters to provide strong signals in these areas, where even Norton did not claim that interference would be a problem.

Politically, however, the FM industry could not explicitly oppose the Commission's desire to provide interference-free service to rural areas. The industry's political problems were aggravated by a curious reversal of roles. A few years before, it had portrayed itself as a white knight attacking the monopoly of the greedy AM establishment. Now it was a new establishment defending the status quo and its own monopoly with equal vigor.

Since its political position was weak, the industry had to base its opposition on technical grounds. Norton's calculations, which admittedly were based on a number of assumptions and extrapolations, were forcefully attacked. Determining the extent of sky-wave interference was a statistical problem requiring measurements over a long period of time—ideally over an eleven-year sunspot cycle—and at a number of distances and latitudes. It was impractical to take sufficient measurements to provide a definitive answer, however, and any forecasts were subject to some degree of uncertainty.

Norton's assumption that the desired signal must be at least ten times as strong as the interfering signal was a clear weakness in his calculations. Armstrong pointed out that this ignored the unique ability of wideband FM signals to override interference and that a $2:1$ ratio was more appropriate. This was the weakest link in Norton's calculations, and industry engineers probably should have challenged it more strongly.

In addition, the FM industry alleged that the performance of higher frequencies would be inferior in a number of important respects. Lower frequencies would have greater coverage, particularly in hilly terrain. The higher frequencies were said to be more vulnerable to interference resulting from long-distance atmospheric transmission under certain weather conditions. Severe reservations were expressed concerning the difficulties in designing and manufacturing transmitting and receiving equipment to operate at higher frequencies. Finally, great emphasis was placed on the delay in bringing this new service to the public should the spectrum shift occur.

The FCC Decides

In the end, these arguments were not persuasive with the FCC and particularly its engineering staff, which had great confidence in Norton's knowledge and judgment. By contrast, the industry's position was believed to be not only technically weak but also self-serving and therefore suspect.

The industry's technical evidence concerning propagation was even sparser and less conclusive than Norton's. The concerns it expressed as to the manufacturability of higher frequency equipment were not credible in view of the rapid advances in technology resulting from wartime research. As for the delay, the Commission's engineering staff viewed this as a temporary problem that did not outweigh the problem of permanently allocating an inferior portion of the spectrum.

The recommendations of the FCC's engineering staff were crucial in establishing the Commission's position. A change in its chairmanship in late 1944 from Fly to Paul Porter had little effect on the momentum that had built up among the staff favoring a change. And so, on January 5, 1945, the FCC issued a report that recommended moving the FM band to 84 to 102 MHz. This subjected existing FM broadcasters not only to the expense and temporary loss of audience resulting from the change in frequency but also to the potential of twice as much competition because of the number of available channels.

In desperation, the industry proposed compromises. The Radio Technical Planning Board, an industry advisory group, recommended extending the upper limit of the band from 50 to 58 MHz. None of the stations then on the air would have to move, and they could be heard by existing receivers. Armstrong recommended moving the band to 48 to 66 MHz, thus preserving at least a portion of the original allocation. Neither of these proposals was persuasive, and on May 9, 1945, the FCC issued a tentative and conditional order in which it proposed to move the FM band to 84 to 102 MHz, the frequency range recommended in its January Report.

The Commission's order recognized that the propagation data upon which it had based its decision were not completely adequate. Partly to strengthen its position but mostly to appease the industry, it ordered one more measurement program. The decision to move the band was made contingent on the completion of another set of measurements on sky-wave transmission that were to be made during the summer of 1945.

This contingency was soon overtaken by events. On May 7, 1945, just two days before the FCC issued its conditional decision, the War Production Board announced its plans to permit the manufacture of radio receivers.

This action put tremendous pressure on the FCC for an expeditious final decision. Manufacturers could not begin the production or even the engineering design of FM receivers as long as there was any uncertainty as to FM's location in the spectrum.

The Commission did not wish to hold up receiver production, so on June 5, without waiting for the results of the additional measurements, it ordered another oral argument on the location of the FM broadcasting band, a necessary procedural step before making a final decision. The oral argument was held on June 22 and 23. The result was predestined, and less than a week later, on June 27, the Commission issued a final decision to move the FM band. This was surely something of a speed record for bureaucratic action, and it suggested that the decision had already been made. The oral argument was only a formality.

In its written decision, the Commission weighed the evidence meticulously. On the basis of the available technical data, it stated, stations at distances of one thousand to two thousand miles from other cochannel stations would suffer substantial sky-wave interference at the outer edges of their service areas in the 42- to 50-MHz band. In the 84- to 102-MHz band, the interference would be negligible.

Because of the incompleteness of the data, the precise amount of interference was uncertain. In addition, the amount of interference that could be tolerated was strictly a judgment call. But the Commission was determined that every possible step should be taken to ensure interference-free service to rural areas. In its view, this required that FM be moved in spite of the many problems it would cause. Norton's recommendations had prevailed over those of the entire industry.

The Industry's Reaction

Very few FCC decisions have been so unpopular. The pioneer broadcasters, who had established FM stations before the war and had no profitable AM stations to support a costly program schedule, had to begin from scratch. They felt betrayed by the same Commission that had so enthusiastically supported and encouraged FM in 1939 and 1940.

The pioneering manufacturers were in the same position. They had made major investments in the design of 42- to 50-MHz eqipment and receivers. Now they had to start over on an equal footing with manufacturers that had spent and risked nothing on lower frequency designs.

Zenith was the most bitter opponent of the FCC decision. As both an FM broadcaster and manufacturer, it had regarded FM not only as a profitable business but also as a means of breaking the dominance of its hated competitor, RCA. With the movement of the FM band upward and with the increase in the number of channels, Zenith lost most of the advantages it had hoped to enjoy as a result of its costly pioneering effort.

On January 2, 1946, Zenith filed a petition with the FCC based on additional measurements that purported to discredit Norton's studies. The petition requested that dual-band operation be permitted. The measurements were analyzed by an FCC–industry committee in a series of informal meetings. They were marked by a heated attack on Norton by Zenith engineers. Norton withstood the abuse quietly.

Two months later, on March 5, the FCC rejected Zenith's petition on the grounds that its measurements were incomplete and inconclusive. Additionally, it pointed out the disruptive effect of dual-band operation. And, in a statement that must have been particularly galling to Zenith, it pointed out that dual-band operation would give an unfair advantage to manufacturers, such as, Zenith, that already had low-band product designs.

Zenith responded angrily, but it was useless. The location of the FM band was set at 88 to 108 MHz.

Throughout the battle, RCA had maintained a low profile, but Sarnoff could not have been unhappy with the decision.

What If?

With the perspective of time, it is interesting to speculate about the wisdom of the FCC's decision. The best guess is that changing frequencies made very little difference in the growth of the industry but that increasing the number of channels was highly desirable.

Undoubtedly, the FCC overestimated the seriousness of sky-wave interference in the lower band. Still, although moving the band may have caused some delay in FM growth, there were far more fundamental factors, to be described subsequently, that inhibited the initial growth of FM. In the long run, the effect of the spectrum change was minor.

As for the number of channels, they are in short supply today in many areas, even with the availability of two and a half times as many channels as in the original allocation. This is evidence enough of the wisdom of increasing their number. Of course, this may not have been the view of the FM pioneers who would have preferred less competition!

■ TECHNICAL REGULATIONS

The main purpose of the FCC's Report and Order was to move the FM band upward in the spectrum, but it also included specifications and rules for channel widths, power, antenna height, and frequency assignments. On the whole, these were less controversial.

Channel Widths

The Order preserved the 200-kHz bandwidth of each channel in spite of some testimony that questioned it as being overly prodigal of spectrum space. Spectrum space was originally allocated for ninety channels, twenty educational and seventy commercial. A few months later, on August 24, the commission added ten more channels, and the band was extended to its present limits, 88 to 108 MHz. This gave FM one hundred channels, two and a half times as many as in the pre-war allocation.

Power and Antenna Height

Three classes of stations were established: Class A local stations with a maximum radiated power of 3 kilowatts, Class B regional stations with a maximum radiated power of 50 kilowatts in the crowded northeast section of the country, and Class C regional stations with a maximum radiated power of 100 kilowatts for the remainder of the country. (Radiated power is the product of the transmitter power, minus transmission line losses, and the antenna gain.)

The limits on power were significantly lower than had been previously established in the 42- to 50-MHz band. There would be no superpower fat cat stations in FM broadcasting. Some of the pioneers who had hoped to be fat cats grumbled that they would be at a disadvantage in relation to AM clear channel stations, but their complaints fell on deaf ears.

The power and antenna height limits have undergone a number of revisions since the Commission's 1946 order. In part, the changes were an attempt to compensate for the rigidities in the frequency assignment principles that had been adopted. Three subclasses of stations were created to provide greater flexibility in station assignments, and southern California was added to the Northeast as a congested area requiring lower power. Current station classes with their permitted power levels and antenna heights are as follows:

Class	Min. Radiated Power[a]	Max. Radiated Power	Max. Height
A	100 w	3 kw	328 ft.
B1	3 kw	25 kw	328 ft.
B	25 kw	50 kw	492 ft.
C2	3 kw	50 kw	492 ft.
C1	50 kw	100 kw	961 ft.
C	100 kw	100 kw	1,968 ft.

[a] The power is the product of the transmitter power (less transmission line losses) and the antenna gain.

Class A, B, and B1 stations may be authorized in the Northeast and southern California. Class A, C, C1, and C2 stations may be authorized in the remainder of the country. The maximum antenna height may be exceeded if the transmitter power is reduced accordingly.

Circular Polarization

The original FCC rules for FM broadcasting specified horizontal polarization—that is, the electric field in the wave was to be horizontal. The original *FM Standards of Good Engineering Practice,* issued in September 1945 immediately after the end of the war, permitted circular polarization, but this was not incorporated in the rules until January 1956. Horizontal polarization generally requires a horizontal receiving antenna for optimum reception (although in some cases the apparent plane of polarization may be rotated by reflections). A circularly polarized wave has both vertical and horizontal components, and it frequently produces better results with indoor rabbit ears and other nonhorizontal antennas.

Another important advantage of circular polarization is that the FCC permits the radiated power of both the vertically and horizontally polarized components to equal the maximum values specified in the table. This permits stations to double their total radiated power.

In spite of the advantages of circular polarization, it was seldom used in the early years of FM broadcasting because of its cost. In the early 1960s, a few FM stations began to feel sufficiently affluent to afford it, and in more recent years its use has become common.

Preemphasis/Deemphasis

The Rules contained a requirement for preemphasizing the high-frequency components of the audio signal—that is, amplifying them more than the low frequencies before transmission. The amount of preemphasis was specified precisely so that complementary deemphasis circuits with reduced gain at higher frequencies could be incorporated in FM receivers. Since the background noise in FM systems is greatest at the higher frequencies, reducing the receiver gain reduced the noise in the output.

Frequency Assignment Principles

As with AM, FM frequency assignments were originally made on a demand basis. An applicant for a new FM station would engage an engineering consultant to make a "frequency search" to identify a channel that met the FCC's minimum separation criteria for cochannel and adjacent channel stations. He or she would then file an application for that channel, and, assuming there were no competing applications and that the applicant was otherwise qualified, the FCC would grant the application.

This policy was changed in November 1964. The rule changes were motivated by the FCC's experience with its AM assignment policies, which it felt were too lax and led to excessive levels of interference. The Commission was determined not to repeat the same mistake with FM.

Following the practice it had established for the assignment of television channels at the conclusion of the 1949 freeze (see Chapter 4), the Commission established a Table of Allotments that assigned specific channels to each city. The basis of this table was a set of rigid mileage separation criteria for cochannel and adjacent channel stations. In the interest of administrative simplicity, the commission ignored the laws of nature and defined the distance between cities for radio interference purposes as the distances between their post offices.

Other provisions included the following:

- The Table of Allotments is a part of the FCC's Rules and can be changed only as the result of a formal Rule Making proceeding. It cannot be changed by a simple administrative order.
- The required mileage separations cannot be violated under any circumstances, not even for extraordinary terrain features such as a mountain range between the cities. If the post offices in two cities fail to meet the minimum mileage separation criteria by even a few feet, an assignment of cochannel or adjacent channel stations is forbidden.
- Transmitter sites also must meet the minimum separation criteria.
- Directional antennas are permitted for the purpose of improving coverage but not to reduce spacing between stations. The effective radiated power in any direction cannot exceed the maximum specified for the station class.

These Rules were so rigid that the FCC subsequently introduced some degree of flexibility by defining the station subclasses B1, C1, and C2 with a selection of minimum mileage criteria. Further flexibility has been added by allowing low-power drop-ins if they meet the FCC's interference criteria. Some members of the industry are concerned that yielding to the pressure for more channels in the FM band will eventually lead to the same problems of excessive mutual interference as AM has experienced.

■ THE POSTWAR FM BROADCASTING INDUSTRY

Initial Postwar Enthusiasm

FM started smartly enough soon after the FCC issued its Order moving the band to higher frequencies. As with its prewar introduction, there was a rash of applications for licenses, some from newcomers seeking to make a fortune and some from AM broadcasters seeking to protect their positions. During the second half of 1945, 456 stations were authorized. Three years later, by January 1, 1949, 976 stations had been authorized and 687 were on the air.

The rush for licenses was matched by the rush of equipment manufacturers to design and produce transmitters and antennas. Contrary to the dire predictions of those who had opposed moving FM upward in the spectrum, the difficulties in producing higher band equipment did not materialize, and FM transmitters and antennas were soon available.

The design of FM receivers that were as simple, foolproof, and inexpensive as AM receivers was more of a problem. Early FM receivers were hard to tune and had a tendency to drift. Unlike AM receivers, they could not be accurately tuned by adjusting for maximum volume. While these defects were seemingly trivial to engineers, they were serious barriers to the public's acceptance of FM. The engineers solved these problems later by the addition of automatic frequency control (AFC), but this was not available in FM's early years.

FM Stagnates

Although the initial enthusiasm of would-be FM broadcasters was great, something very disturbing happened as stations went on the air. There were very few listeners. Few receivers were being sold, and most FM stations were operating in the red. The forecasts of the skeptics were confirmed. FM acquired an aura of failure in the broadcasting establishment, and its enthusiasts were regarded as quixotic idealists. Worse yet, it became the subject of ridicule.

The failure of FM to fulfill the expectations of its enthusiasts was reflected in a reduction in the number of new stations. At the beginning of 1948, the number of stations authorized (granted construction permits) reached a peak of more than one thousand, but many of these were never built. The number of on-air stations reached a peak of slightly more than seven hundred in 1950 and then began to decline. Many authorizations were allowed to expire, and a number of stations went dark as the result of intolerable operating losses.

Armstrong and the FM establishment naturally blamed FM's woes on the

FCC decision to move the frequency band. But with the passage of time, it appeared that there were more fundamental reasons.

Neither the technical deficiencies of AM nor the virtues of FM were particularly important to most of the public at that time. The majority of the population lived in urban areas where strong local signals overcame static and nighttime sky-wave interference. Most rural listeners lived beyond the range of VHF stations, and they continued to rely on AM for radio service. As for high fidelity, the broadcasting station was only one link in the system. All the links—studios, recordings, intercity lines, and receivers— had to be upgraded to take advantage of FM's potential. (Only since about 1980 have satellites provided the potential for nationwide distribution of high-fidelity network programs). In addition, high-fidelity receivers were expensive, and few members of the public were willing to invest in them.

Furthermore, the only type of programming on the air at that time that could benefit from high-fidelity reproduction was classical music. High fidelity was of little value to news and sports programs, situation comedies, and soap operas. Most of the popular music of that era lacked the dynamic range and the emphasis on percussive effects that require high-fidelity reproduction and characterize the music of the 1970s and 1980s.

Classical music, then as today, appeals to only a small minority of the listening public. Armstrong and his supporters, most of them well educated and culturally sophisticated, seemed unwilling or unable to grasp this fact. It was a bitter pill for them to swallow, and they tried to place the blame on everything but this fact. As evidence, only forty-four of the thirty-eight hundred commercial FM stations on the air today describe their program format as Classical.

FM had another serious problem: a lack of programs with wide popular appeal. Radio was still in the network era, and the major networks with their extensive affiliate groups had all the highest rated shows. They were so costly that FM stations could not compete for them. The most popular programs on FM were those of the network affiliates, which were already available on AM.

FM also had some technical problems. The problems of receiver tuning and drift have already been described. In addition, multipath propagation in hilly and congested terrain, often leads to signal distortion, a problem that continues today.

Finally, as television grew in popularity, the automobile audience became increasingly important to radio broadcasting. Few automobiles had FM receivers.

The result was a dark decade for FM from 1948 to 1957. After the number of on-air stations reached a peak of 733 in 1951, it declined steadily to a low of 530 in 1957. Few FM stations could generate sufficient revenues to cover the costs of even modest program schedules, and many simply gave up. Others obtained a small incremental revenue by adding a wireless "wired

music" service on a subcarrier, thus taking advantage of some of the generous capacity of the 200-kHz bandwidth. Most AM broadcasters with FM
stations reduced costs as much as possible by duplicating programs. With
the perspective of time, this ten-year period does not seem long, but for the
FM broadcasters who lived through it, it seemed like an eternity.

FM Blossoms

But "don't ever say never!" Beginning in 1957, something happened.
The decline in the number of FM stations was halted, and it began to climb,
very slowly at first and then more quickly. What caused this turnaround?

First, there were steady technical improvements, particularly in receivers, which became more sensitive. The tuning problem was solved with
AFC, and the development of the transistor permitted a major reduction in
the size, weight, and portability of receivers. AM/FM radios became standard equipment in most new cars.

In addition, the public demand for stereo broadcasting grew. The 200-
kHz FM channels provided ample capacity for high-quality stereo transmission, and the FCC authorized stereo broadcasting in 1961.

Finally, high fidelity came into its own as a commercially important
service. This came about not because of an order-of-magnitude change in
the number of classical music lovers, but because of a new form of popular
music generically described as rock. Rock music lovers like it loud and
with plenty of percussive effects, and both of these qualities require high-
fidelity reproduction.

An equally important factor in the growth of FM was the basic change in
the nature of radio broadcasting after television became popular. This
change, described in Chapter 2, made it possible for radio broadcasters to
operate their stations more economically and to make a profit with a frag-
mented audience. This led to a demand for more channels that could not be
satisfied with AM. Beginning in the mid-1950s, it became virtually impossible to find an available full-time AM channel, even with elaborate directional atnennas, except in the most sparsely populated parts of the country.
By 1965, it became difficult to find a channel even for daytime-only operation. Thus FM was the only way new stations could go on the air on a
full-time basis. It is fortunate, perhaps, that Armstrong did not live to see
that the initial success of FM was due to a scarcity of AM channels, not to its
superiority as a form of broadcasting.

With an increasing number of FM stations coming on the air, the public
had more incentive to buy FM receivers, and this in turn made FM broadcasting more profitable. The chicken-and-egg cycle was restarted, and it led
to a growth in the numbers of both.

The Renaissance of the Networks

One development that was helpful to both AM and FM broadcasting, but particularly to FM, was the renaissance of the radio networks. Their economic influence is far less than it was during the network era, but networks provide an attractive additional source of programming for their affiliates.

The renaissance, which began in the 1970s, resulted from the confluence of a number of favorable developments. First, the FCC loosened its regulation of radio networks, which were permitted to offer multiple program services to more than one affiliate in a single city. In addition, the technical developments that helped radio stations adapt to the postnetwork era (see Chapter 2), such as magnetic tape recording, made it possible for networks to offer program services at much lower costs. Network managements also learned to adapt their programming to appeal to radio audiences in the television era. Finally, the advent of satellite distribution of radio programs provided a flexible and cost-effective method of distributing high-fidelity programs to affiliates throughout the country.

The result of these favorable developments has been a proliferation in the number of networks. ABC has been the most aggressive with seven program services—Contemporary, Rock, Information, Entertainment, FM, Talkradio, and Direction. CBS offers two program services, CBS Radio–Radio. Westwood One purchased the Mutual Broadcasting System and, in 1988, the three services of the NBC network. A newer network, the United States Radio Network, offers three program services. Other networks serving the general public include Fox Broadcasting Company and the Sheridan Broadcasting Network.

In addition, a number of national networks serve special audiences. Among these are the National Black Network and National Public Radio. The 1988 *Broadcasting Yearbook* also lists ninety-five regional networks. Clearly, networks are again an important part of radio broadcasting, although they do not enjoy the dominance of its earlier years.

FM's Growth Record

Share of Radio Listening Audience   The most dramatic indicator of FM's growth is its ever-increasing share of the radio audience (Figure 3–3). It reached parity with AM in 1979, and by 1988 its share had risen to 75 percent.

On-Air FM Stations   As the FM listening audience increased, the number of on-air FM stations grew correspondingly (Figure 3–4). As of mid-

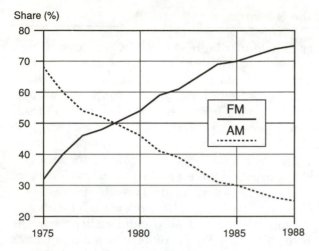

■ **Figure 3–3** AM–FM share of the radio listening audience. Note the dramatic growth of FM radio and the corresponding decline of AM. (Data from NBC radio research.)

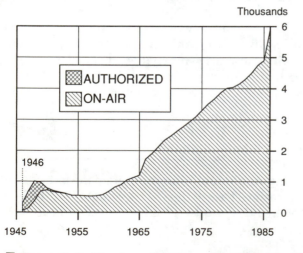

■ **Figure 3–4** Number of commercial and public FM Stations on the air, 1946–1986. This illustrates the rise, fall, and rise of FM broadcasting in the post–World War II era.

1988, there were 4,085 commercial stations and 1,339 public stations on the air. In total, this exceeded the number of AM stations.

Revenues   The ultimate measure of the growth of FM broadcasting has been the increase in its revenues (Figure 3–5). FM broadcasting has become a healthy and prosperous industry. FM-only stations as a group became profitable in 1976, and their profitability continues to grow.

FM offers the public a wide choice of program material, including high-fidelity stereo service to lovers of both rock and classical music. The number of classical stations is small, but at least one service is available in most large cities and in many college and university towns. After a rocky start, FM is a financial success, although not exactly what Armstrong had expected. He did the right thing, albeit for different reasons, and broadcasting and the listening public owe him much.

■ POSTWAR TECHNICAL DEVELOPMENTS

Transmitters and Antennas

With the rapid progress in the development of vacuum tube technology during and after World War II, the design of FM transmitters did not pose any extraordinarily difficult problems. FM transmitters (Figure 3–6), with constant output power, had none of the problems of AM transmitters resulting from the variation in power output throughout the modulation cycle.

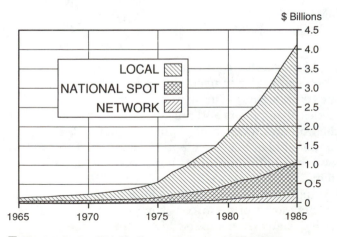

■ **Figure 3–5**   FM Station revenues, 1965–1985.

Courtesy RCA Corporation.

■ **Figure 3–6**   An FM transmitter.

The most complex FM transmitter component is the *exciter,* which generates the FM signal. Exciters originally used electron tubes, but they were replaced by solid-state components as suitable transistors became available. The exciter drives a series of amplifiers and finally a power amplifier, which delivers power to the antenna. With typical antenna gains of 6, 20- and 40-kilowatt power amplifiers became the standard sizes used to produce effective radiated powers of 100 kilowatts for horizontal and 200 kilowatts for circular polarization.

Several antenna configurations were offered by manufacturers in the immediate postwar years, some of them consisting of complex arrays of dipole antennas. These designs were superseded by the simple pylon antenna, which consists of a slotted vertical steel cylinder acting as both radiator and mechanical support (Figure 3–7, left). More complex structures

Courtesy RCA Corporation, with permission.

Courtesy RCA Corporation.

■ **Figure 3–7** FM antennas. (*Left*) A pylon antenna. (*Right*) A circularly polarized antenna.

(Figure 3–7, right) were required to radiate circularly polarized waves after this type of transmission was permitted by the FCC.

Receivers

As with transmitters, the escalation of electron tube development in World War II helped greatly in the design of postwar receivers. Their biggest problems were difficulty in tuning and a tendency to drift. Both these problems were solved by the addition of AFC, which locks the receiver to the station once it is manually tuned to approximately the correct frequency.

As with AM receivers, transistors revolutionized FM receiver design, and they are now universally used. They were particularly useful in the design of combination AM/FM receivers.

Stereo

The growth in the number of rock music listeners, supplemented by the small but influential classical music audience, led to the increasing importance of stereophonic transmission. Fortunately, the 200-kHz FM bandwidth, one of Armstrong's imperatives, accommodates high-fidelity stereo transmission, and the need for stereo was seen long before the demand became great.

STEREO TRANSMISSION SPECTRUM

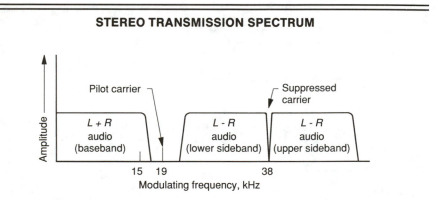

This diagram illustrates the spectrum of the signal transmitted for FM stereo broadcasting. The left and right channels, L and R, are added and transmitted as the base band. This signal can be received on a monaural receiver, thus providing compatibility. The L − R signal, usually of low amplitude, amplitude modulates a 38-kHz subcarrier, which is then suppressed. A pilot carrier operating at one-half this frequency is transmitted to reconstitute the subcarrier at the receiver. The L and R signals are derived from the L + R and L − R signals. Stereo transmission degrades the signal-to-noise performance of the system somewhat.

FCC Rules permit the authorization of a second subcarrier, a Subsidiary Communications Authorization (SCA), at a frequency above the stereo subcarrier for the transmission of auxiliary services, such as background music for restaurants and offices. This service was originally allowed as a financial aid to FM stations when most of them were unprofitable.

The National Stereophonic Radio Committee was formed in 1959 to examine various methods proposed for transmitting stereo sound. After extensive studies and field tests, the system proposed by GE was recommended and subsequently approved by the FCC in 1961. The approval of a single system ensured compatibility between transmitters and receivers, and a high percentage of FM broadcasts are now in stereo. Unfortunately, as noted in Chapter 2, the FCC has refused to approve a single standard for AM stereo.

Laser Recording

The laser, an acronym for light amplification by stimulated emission of radiation, is an exciting new technology for generating infrared and visible radiation having unique properties (see Chapter 10). Laser beams are used in radio and television for audio and video recording (see Chapter 9) and fiber-optic cable transmission (see Chapters 7 and 10). For audio and video recording, their most important property is their ability to be focused on extremely small spots.

The technical quality of audiodisk recordings has been improved enormously by the invention of laser recording. Because of their small size, the records are usually called *compact disks*, or simply CDs. The frequency and dynamic range of the recorded sound, together with the total absence of record noise, result in extremely high quality sound.

For laser recording, the sound signal must first be digitized (see Chapters 1 and 10). The digital signal then modulates a very sharp laser beam, focused on a photosensitive master, and records a pattern of dots. The playback disk has a shiny reflecting surface, and the dot pattern on the master is transferred to it as a series of tiny nonreflecting pits.

On playback the process is reversed (see Figure 3–8). A laser beam is focused on the disk and is made to track the spiral path of the recording. The nonreflecting pits on the disk modulate the reflected laser beam with the recorded digital information, which is then reconverted to analog form. The size of the spiral track on a CD is only about 1.5 millionths of a meter wide, comparable to the wavelength of infrared light and 3 thousandths of the diameter of a human hair.

Satellite Program Distribution

Communications satellites (see Chapter 8) were available for radio program distribution by the networks in the early 1970s, but they did not begin to be used widely for this purpose until 1980. At first, little need for satellites was perceived by radio broadcasters. The networks were in a decline, and

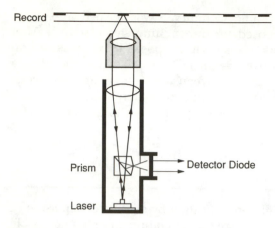

Record

Prism

Detector Diode

Laser

■ **Figure 3–8**   CD playback system.

medium-fidelity distribution facilities were readily available through the telephone system.

But innovative radio networks, particularly ABC, recognized that satellites could mean their revival. Networks could not meet the growing demand for high-fidelity stereophonic programs with the technical limitations of the telephone system's intercity network, and satellites provided an economical solution. The cost of receive-only earth stations steadily declined as a result of the relaxation of FCC requirements, competition, technical advances, and economies of scale until it reached an affordable figure for most stations (less than $5,000 in 1988).

ABC was the first major radio network to use satellites. It took advantage of the bandwidth available on a satellite channel by transmitting the signal in digital form, thus improving the quality further. ABC was followed by other networks, and satellite distribution is now widely used.

■ THE RCA–ARMSTRONG PATENT BATTLE

No history of FM broadcasting would be complete without a description of the epic RCA–Armstrong patent battle. While FM was struggling to achieve commercial acceptance in the decade following the end of World War II, an equally intense struggle was in progress in the nation's courts over the question "Who invented FM?"

The Contestants

The issue was as much pride as money, and the contestants, David Sarnoff and Edwin Armstrong, were two very proud men. Sarnoff had an enormous pride in RCA, which to a large extent was his company. A small indication was that he never referred to it as "RCA"; it was always "The RCA." RCA had been the leader in radio technology since its early days, and Sarnoff was upset by FM's challenge to its leadership. In addition, as described in Chapter 2, the RCA patent pool was a major source of its profits, and much of RCA's future depended on obtaining and maintaining most or all of the basic patent for radio systems and apparatus. The pool was under severe attack from the Justice Department and RCA's competitors on antitrust grounds, and Armstrong's FM patents, if determined to be valid, would be another crack in the structure.

The patent issue was joined near the end of World War II. RCA, although notably unenthusiastic about FM, felt it necessary to market FM transmitters and receivers in order to maintain its position as the leading radio manufacturer. Sarnoff refused to pay patent royalties to Armstrong on the ground that his FM patents were invalid.

Much of the rest of the industry, accustomed to following RCA's lead in patent matters, followed suit. At the same time, however, another contingent of manufacturers, led by Zenith, agreed to pay. Among other effects, this put them at a cost disadvantage as compared with RCA and its followers.

Armstrong continued to be infuriated by RCA's unwillingness to give FM its total support, and RCA's refusal to pay him patent royalties made him even more furious. In this state of mind, he was unwilling to seek a compromise and instead insisted on unconditional surrender, which meant RCA's unqualified recognition that he was the inventor of FM.

Armstrong could ill afford to take this position. His income from his earlier patents was dwindling, and his FM royalty income had been greatly reduced by the refusal of RCA and others to pay. His legal expenses were enormous, and he had the continuing costs of his research programs. His lawyers repeatedly urged him to negotiate a settlement, but he refused.

Michael Pupin, a distinguished engineer and one of Armstrong's mentors at Columbia, remarked that when one's claim to an invention is disputed one will fight for it just as a tigress would fight for her cub. In a similar vein, Carl Dreher stated, "Beyond a certain point, devotion to a cause, however admirable, enters the realm of pathology."[1] Both statements accurately reflected Armstrong's state of mind.

With Armstrong's refusal to be satisfied with nothing less than RCA's complete capitulation and Sarnoff equally unwilling to yield, a court battle was inevitable. On July 22, 1948, after months of preparation with his lawyers, Armstrong filed suit against RCA and NBC, charging them with willfully infringing and inducing others to infringe on five of his basic FM patents.

---

[1] Carl Dreher, "E. H. Armstrong: The Hero as Inventor," *Harper's*, April 1958.

Patent law is an arcane subject, full of subtleties and fine distinctions that are almost incomprehensible to the layperson. The arguments in this case were extremely complex, and rival those of medieval theologians discussing the number of angels who can dance on the head of a pin. This history will not attempt to describe them in depth but rather to give a simplified summary of the case.

The Issue

The fundamental issue was whether wideband FM, the essence of Armstrong's patents, was sufficiently different from narrow-band to be separately patentable. Was there a quantum difference between the characteristics of wideband and narrow-band systems, or did they differ merely in degree? If the former was held to be the case, Armstrong could obtain a basic patent, and his position would be extremely strong. If the latter was found to be true, only specific circuits or concepts for the generation and reception of FM signals could be patented, and these could probably be circumvented with new and different designs.

RCA naturally took the latter position and based much of its defense on the development of new FM apparatus—the Beers receiver by George L. Beers of RCA, a new transmitter by Murray G. Crosby of Bell Laboratories, and the ratio detector by Stuart L. Seely, also of Bell Laboratories. RCA had patent rights to all these circuits and devices through its cross-licensing arrangements. They were alleged to be entirely new developments, dissociated from Armstrong's system.

This allegation enraged Armstrong. He was equally enraged by the quality of the FM receivers being manufactured. Philco, for example, offered a receiver without a limiter, a basic element in the FM system as perceived by Armstrong.

The Trial

The trial began on February 14, 1949, with depositions in the offices of Armstrong's legal counsel, the prestigious (and expensive) Cravath, Swain & Moore. Armstrong was the first witness, and RCA's strategy of attrition was soon disclosed. He was questioned about every aspect of his career, his accounting practices, his relationship with Columbia University, his use of the university's letterhead, and on and on, almost ad infinitum. After several months, his attorneys appealed to the court to end his examination. The court responded by ordering questioning of Armstrong to cease by the end of the year—more than ten months after he had taken the stand.

Armstrong's attorneys retaliated by obtaining a ruling requiring RCA to produce from its files all documents having any reference to FM. This huge mass of data was collected and analyzed under a special master, Judge P. J. McCook, a process that took two more years. The trial had become an endurance contest.

In February 1953, four years after the trial had begun, Sarnoff took the stand. He was admittedly a nontechnical witness, and his testimony did little to shed light on the technical issues of the case. He stated that his decisions had been made on the basis of advice from his technical experts, of whom he said he "had more than a dog has fleas."[2] He made one statement that infuriated Armstrong anew: "I will go further, and I will say that the RCA and the NBC have done more to develop FM than anybody in this country, including Armstrong."[3] This hardened Armstrong's resolve to refuse a negotiated statement, even in the face of his attorneys' advice and his wife's urgent pleas.

And so the trial dragged on. The basic issues of the case had long since been buried under mountains of irrelevancies and minutiae. A new generation of RCA lawyers employed at the end of World War II had spent their entire careers on the Armstrong case. Armstrong himself devoted virtually every waking hour to it. His wife became increasingly concerned, and one story has it that she swallowed her pride and took the desperate step of calling on Sarnoff, her former boss, to settle the case, pleading that it was ruining her husband's health. She urged, so the story goes, that this could be done at very little cost to RCA. All her husband wanted, she said, was RCA's public recognition that he had invented FM. Sarnoff refused. She also repeatedly begged her husband to settle the lawsuit, but he also refused.

Armstrong's Suicide

Armstrong was depressed not only by the burden of the trial but by other unfavorable events. In 1953, Zenith, his strongest supporter in the manufacturing community, notified him that it was ceasing FM royalty payments. FM broadcasting, contrary to his hopes and expectations, was still a commercial failure. Matters finally came to a head on Thanksgiving night in 1953. Faced with her husband's intransigence and ill herself, Marian Armstrong left her husband and went to Connecticut to live with her recently widowed sister. Two months later, on January 31, 1954, Edwin Armstrong committed suicide.

---

[2] Kenneth Bilby, *The General* (New York: Harper & Row, 1986).
[3] Lawrence Gessing, *Man of High Fidelity: Edwin Howard Armstrong* (New York: J. P. Lippincott, 1956).

The Settlement

The adversaries, sobered by the tragedy and drained by the years of litigation, were soon able to negotiate a settlement, and RCA and Armstrong's estate signed an agreement. The basic question of who invented FM went unresolved, but RCA paid the estate $1 million for infringement of Armstrong's FM patents, all of which by then had expired. This was an insignificant amount compared with the direct and indirect legal costs to the contestants, and but for the excessive pride of the two principals, the same result could have been achieved years earlier.

The hostility between the RCA and Armstrong camps continued for many years. There was no doubt in the minds of Armstrong's supporters as to who invented FM, regardless of the outcome of the case.

■ A LOOK AHEAD

The future of FM broadcasting looks bright indeed. The major technical problems that plagued its early years have been solved. Receivers are stable, reliable, and easily tuned. AM/FM automobile radios are readily available and commonly sold or supplied in new cars. The FM broadcasting system is well adapted to the transmission of stereo programs, and a high percentage of FM broadcasts are in stereo. Program production has benefited from the same advances in studio equipment that have been so important in AM radio and the sound channel for television. Laser disks have brought audio recording to a high degree of quality. Transmitters are stable and efficient. While advances in audio technology are by no means at an end, there are no major technical weaknesses in today's FM broadcasting system.

As a result of these technical advances, FM now has complete public acceptance. The chicken-and-egg cycle of receivers and programming has been broken. As an industry, it is profitable. It probably will not supplant AM, but FM will continue to be the primary radio broadcast medium.

4

■ ■ MONOCHROME TELEVISION

It was September 1947, and NBC's affiliates had convened for their annual meeting with the top network executives during the National Association of Broadcasters (NAB) convention in Atlantic City. It was an especially important meeting because the NBC contingent was headed by David Sarnoff, who was scheduled to make a major address.

The affiliates were a happy lot. They were basking in the aftermath of ten years of unprecedented prosperity in radio broadcasting. NBC was the clear leader among the networks, and the affiliates were sharing in its prosperity. For the most part, the prospects for the future looked equally bright.

There were a few clouds on the horizon, but most of them were no bigger than a man's hand. CBS, under the able leadership of William Paley, was becoming a competitive threat, but it would be more than a year before Paley would score the greatest coup of his career by luring NBC's top entertainers such as Jack Benny and Amos 'n' Andy to CBS. FM radio was receiving much publicity, but it was not considered to be serious competition. The affiliates' only real concern was television.

Sarnoff had been ardently espousing the cause of television since his interest was aroused by a historic meeting with a young Westinghouse engineer, Vladimir Zworykin, in 1927. He had committed a substantial portion of RCA's research and engineering budget to its development. He had lobbied vigorously with the FCC for the approval of transmission standards and the allocation of a major segment of the radio frequency spectrum for television broadcasting. This effort had come to fruition in two FCC actions: (1) an Order in late 1945 reinstating the standards and channel allocations approved just before the outbreak of World War II, and (2) a denial of the CBS petition to permit color broadcasting with the incompatible field sequential system. Now Sarnoff's task was to get the broadcasters' support—the encourage them to build stations for the NBC network and to provide programs that would be an inducement for the public to buy television sets.

Television broadcasting is now so profitable that it is difficult to imagine that there was once considerable doubt about its future. Some radio broadcasters were blatantly hostile toward it. They had a good thing going in radio; why risk spoiling it with this very costly new medium? Sarnoff's promotion of television had not made him any friends among this group. CBS's actions confirmed his opponents' opinions. It had decided not to apply for any stations other than its New York flagship, and it had publicly advised its affiliates to wait for color. This was the situation facing Sarnoff as he addressed the NBC affiliates.

He began by citing industry statistics. There were 13 television stations on the air and 150,000 television sets in use. He forecast that there would be 50 stations and 750,000 sets by the end of 1948. (There were actually to be 51 stations and more than 800,000 sets.) He also forecast a television audience of 5 million viewers at that time.

Sarnoff then reviewed the status of the industry and outlined his fore-casts for its future—the prospects for local and network television, its programming, its future as an advertising medium, its manufacturing prob-lems, and its potential for profits. He ended with a remarkably prescient peroration, which was one of his finest hours. He did not talk down to his audience, did not threaten or bully them, and did not give them a slick sales speech. Instead he spoke to them as one businessman to another, urging them to consider the potential of television in their future:

> Affiliates of the NBC: This is the message I would like to bring to you. I do not want to ask you to buy television stations, or to erect them, or to urge you to enter television beyond your own convictions, or to promise you immediate profits. But I feel that I should be less than frank if I did not on this occasion, particularly when you are all assembled, share with you the thoughts I hold, not only about the future possibilities of television—and my enthusiasm is unlimited as to them—but also about the possible effects that television may have upon the present broadcasting business.
>
> I have lived through several periods of development in the fields of com-munication and entertainment. I remember the day when wireless as a service of transoceanic communication was regarded by some as a joke. In the days when I worked as a wireless operator, a cable company could have acquired the Marconi Wireless Telegraph Company of America for a few million dol-lars. Those who owned cables could not see wireless as a competitor of cables. Who, they asked, would send message that were secret through the air? Who would entrust important messages to a medium that was filled with static?
>
> Today, the law says to Western Union [International]: "You must divest yourself of the cables." But now it is difficult to find a buyer for cables. Today, radio is the medium of international communications and can reach every country directly.
>
> I lived through the day when the Victor Talking Machine Company—and they did a great job in their day—could not understand how people would sit at home and listen to music that someone else decided they should hear. And so they felt that the "radio music box" and radio broadcasting were a toy and would be a passing fancy. What was the result? Not many years after their fatal dream, RCA acquired the Victor Talking Machine Company, and the little dog changed its master.
>
> I saw the same thing happen in the field of talking motion pictures. It was urged by many that people would not go to a movie that made a lot of noise

and bellowed through an amplifier and disturbed the slumber of those who enjoyed the silent movie. That, they said, was a preposterous idea! The very virtue of a silent movie, they contended, was its silence. And then—in 1927— came Warner Brothers with "The Jazz Singer" and Al Jolson. Almost overnight a new industry was born. The silent actor became vocal, and the silent picture was given an electronic tongue. Today, who goes to a silent movie?

Now, I should like to impress upon those of you engaged in radio, that for the first time in its history, radio has a stake in the present. It must be careful not to act like the cable companies, which looked upon the new children of science as ghosts of obsolescence that might affect their established businesses. In their desire to perpetuate and to protect their existing businesses, some of them stubbornly resisted change and progress. Finally, they suffered the penalty of extinction, or were acquired by more progressive newcomers.

Let me assure you, my friends, after more than forty years of experience in this field of communications and entertainment, I have never seen any protection in merely standing still. There is no protection except through progress. Nor have I ever seen these new scientific developments affect older businesses, except favorably, where those who were progressive gave careful thought and study to the possibilities of new inventions and developments for use in their own businesses.

Despite the fact that the Victor Talking Machine Company passed into radio hands, more phonograph records are made and sold today than ever before. And so it is with the entertainment industry. Talking pictures saved that industry at a time when it needed saving and has kept it prosperous ever since. Television in the theater may be as much of a stimulant to an industry which at the moment, at least, needs a new stimulant as sound was to the silent movie. [Forty years later, theater television has failed to have a significant impact on the entertainment industry. One reason is the lack of a satisfactory large-screen television projector.]

Therefore, may I leave you with this final thought: I am not here to urge you to enter the field of television beyond the point where you yourself think it is good business for you to do so; nor to urge that you plunge in all at one time. Rather I would suggest that you reflect carefully and thoughtfully upon the possible ultimate effects of television upon your established business if you do nothing, and of the great opportunities for your present and future businesses if you do the right thing![1]

Not everyone in the audience liked Sarnoff, but they all respected him. Most of them followed his advice and applied for television stations in their cities, and those that did became wealthy. His speech was the catalyst the industry needed to start it on its way to becoming one of the free world's major industries after several decades of technical progress and the resolution of contentious regulatory issues.

---

[1] David Sarnoff, "Television Progress," *Broadcast News*, December 1947, 26–27.

■ THE BASIC MONOCHROME TELEVISION TECHNOLOGIES

Early History

Like most major technologies, television was developed rather than invented. Many scientists and engineers contributed to its development, and it was a multinational effort.

In 1873, Joseph May, a British engineer engaged in the operation of the transatlantic cable, noted that selenium produces an electric current when exposed to light. He published this information in a letter to the Society of Telegraph Engineers that was disseminated widely in the technical community.

In 1875, an American engineer, G.B. Carey, proposed a method of transmitting pictures that would use a camera consisting of a rectangular array of selenium cells, each connected to a light bulb in a corresponding array in the receiver. The large number of circuits required to connect the cells with the light bulbs made such systems impractical. To overcome this difficulty, the *scanning principle* was invented. This made it possible to transmit a television signal over a single circuit.

The Scanning Principle

The scanning technique (Figure 4–1) is fundamental to all television systems. It was an indispensable initial step in the development of television technology, and it was a major breakthrough. It did not have a single inventor but appears to have occurred to a number of engineers simultaneously.

The picture is viewed as an array of small picture elements (pixels), and brightness information from these elements is transmitted sequentially rather than simultaneously. This method takes advantage of the retentivity of human vision, which perceives the rapidly repeated sequential display of picture elements on the receiver tube as a single continuous image.

Television Cameras

The technology of cameras and other devices for generating television signals has progressed through three phases: *scanners, storage tubes,* and most recently *charge-coupled devices* (CCDs), which use solid-state components.

The earliest scanners, such as the *Nipkow disk,* were mechanical. They were followed by the *flying spot scanner* and the *image dissector,* which used electronic scanning.

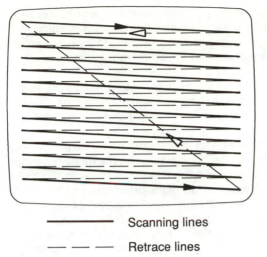

——————— Scanning lines

— — — Retrace lines

■ **Figure 4–1** The scanning principle. In the television camera, an optical image of the scene is scanned vertically and horizontally to form a raster. A signal-generating device converts the variations in brightness along the scanning pattern into an electrical signal. The picture tube or other display device reverses the process. The variations in the electrical signal are reconverted to brightness variations, thus reconstituting the picture. Synchronizing the scanning patterns in the camera and picture tube is an essential requirement of this technique.

BANDWIDTH REQUIREMENTS AND INTERLACED SCANNING

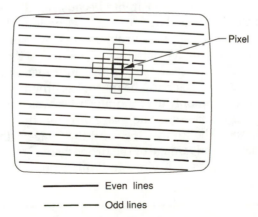

——————— Even lines

— — — Odd lines

The scanning technique in effect divides the picture into a grid of picture elements (pixels). The height of each element is equal to the spacing between scanning lines. Its width is determined by the bandwidth of the system and is equal to the amount by which the scanning beam moves during one-half cycle of the highest frequency of the band. In order to resolve fine detail in the picture, it is necessary that the picture elements be very small; this leads to the requirement for a wide transmission band and the voracious spectrum requirements of television systems.

The number of picture elements that can be transmitted is $2Cf_c/R_f$, where C = the fraction of the frame time devoted to transmission of the picture information; f_c = the bandwidth of the system; and R_f = the number of complete pictures, or frames, transmitted per second.

The number of picture elements and hence the sharpness of the picture can be increased without increasing the bandwidth by reducing the frame repetition rate, R_f. If R_f is reduced too far, however, annoying flicker results. Also, the frame rate should be an integral multiple or submultiple of the primary power frequency to minimize the visibility of hum bars in the picture. The effect of these constraints was minimized by the use of interlaced scanning, in which odd and even numbered lines are scanned in alternate fields. (This major breakthrough in television technology was invented by an RCA engineer, Raymond C. Ballard, in 1932.) This makes it possible to reduce the frame rate by a factor of 2 without increasing the flicker.

In the United States and other countries in which 60 Hz is the standard power system frequency, the field rate is 60 Hz and the frame rate is 30 Hz. In Europe and other areas that use 50-Hz power systems, the field and frame rates are 50 Hz and 25 Hz. The lower frame rate of European television causes some sensation of flicker, particularly with very bright pictures observed in the periphery of the viewer's vision.

For U.S. broadcasting systems, $C = 0.75$; $f_c = 4.25 \times 10^6$; $R_f = 30$; and the number of picture elements = 210,000.

Scanners suffered from low sensitivity because all the signal-generating information had to be obtained during the brief period when the scanning spot swept across the scene. This problem was overcome by the use of storage tubes in which an image of the scene was focused continuously on a photosensitive surface. The signal-generating information was stored there until it was removed by an electron beam that scanned the surface.

The earliest of the commonly used storage tubes was the *iconoscope*. It was followed by the *image iconoscope*, the *image orthicon*, the *vidicon*, the *Plumbicon*, and the *saticon*.[2] The iconoscopes and the image orthicon are *photoemissive* devices that take advantage of the property of some materials to emit electrons when illuminated. The vidicon, Plumbicon, and saticon are *photo-*

---

[2] Several other pickup tubes were developed, some of them intermediate steps in the development of more advanced tubes and others designed for specialized applications. They include the *orthicon*, the *CPS emitron*, the *multiplier orthicon*, the *image isocon*, the *silicon intensifier tube*, the *chalnicon*, and the *newvicon*.

conductive, which means they rely on the property of other materials to change their electrical resistance when illuminated.

The development of CCDs was made possible by rapid advances in solid-state technology along with innovative circuit designs. They are currently displacing storage tubes for some types of color television cameras and may ultimately replace them completely.

Scanners, iconoscopes, image orthicons, and vidicons were widely used in monochrome television systems, and the history of their development is described in this chapter. Plumbicons, saticons, and CCDs came later, after color television had replaced monochrome for broadcasting, and their history is related in Chapter 5.

Television Display Devices

Like television cameras, display devices progressed from mechanical to electronic. Mechanical displays typically consisted of a light source whose brightness could be varied by applying the signal voltage to it. The light from the source was then projected in a scanned raster onto a screen or other viewing area by one of a number of optomechanical configurations. Mechanical viewers had serious limitations of resolution and brightness, and during the 1930s they were superseded by the CRT, or kinescope (see Chapter 1). This device was nearly ideal for the purpose, and its use has been almost universal.

■ MECHANICAL TELEVISION SYSTEMS

Inventors anticipated the public demand for television even before the beginning of radio broadcasting. So many scientists, engineers, and experimenters devoted their talents to its development during the late nineteenth and early twentieth centuries that it is impossible to answer the question "Who invented television?" The efforts of a few, however, were so important that they deserve recognition as television's pioneers. Since the basic components required for an electronic television system had not yet been invented, all the early systems were based on mechanical components.

Paul Nipkow

Paul Nipkow proposed the first practical mechanical scanner (Figure 4–2) in Germany in 1884. The basic scanning device was a rotating disk with perforations arranged in a spiral around its periphery. Light passing through the perforations as the disk rotated produced a rectangular scan-

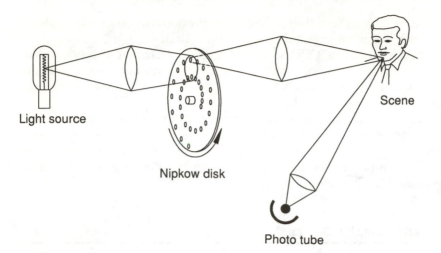

■ Figure 4–2 Nipkow disk television system. A disk with perforations arranged in a spiral around its periphery, later named the Nipkow disk, was the basic component of Nipkow's system. In Nipkow's conception, the scene to be televised was brightly illuminated and its image focused on the plane of the disk. As the disk rotated, the image was scanned by the perforations, and light from different portions of it passed through to the photocell. The number of scanned lines was equal to the number of perforations, and each rotation of the disk produced a television frame.

In a later version, the process was reversed. The photocell in the scanner was replaced by a light, and the disk perforations caused the scene to be scanned by a bright spot. Light reflected from the scene was then picked up by a photocell located outside the scanner.

In the receiver, the light source was a neon glow tube whose brightness was controlled by the signal voltage. The light from the tube passed through a synchronously rotating perforated disk and formed a raster in the field lens or on a projection screen.

ning pattern, or raster, which could be used either to generate the electrical signal from the scene or produce an image from the signal at the receiver.

Nipkow's concept existed only on paper, for he never built a prototype. Its basic principle—the use of a rotating mechanical device to produce the scanning effect—was, however, used in all the subsequent television systems until all-electronic systems became practical.

John Logie Baird and Charles Francis Jenkins

In the years following the publication of Nipkow's concept, a number of experimenters dabbled with it in an attempt to develop a practical working system. It was not until the 1920s, however, that two independent inventors, Baird in England and Jenkins in the United States, and two major research laboratories, GE and AT&T, constructed systems that performed well enough to have commercial promise.

John Logie Baird was a Scottish engineer-inventor who became interested in television in 1923 at the age of thirty-five. His first apparatus, based on the Nipkow principle, was extremely crude. It had only 8 scanning lines, each of which had 50 picture elements, resulting in a total of 400 picture elements per frame. (Compare this with the 210,000 picture elements provided by the current U.S. broadcasting standards.) Further, it was not capable of transmitting hafltones and was limited to silhouettes. With improvements, however, it was good enough to attract venture capital, and a new company, Television, Ltd., was formed in 1925 to exploit Baird's invention.

That year also marked a successful demonstration of the system to Gordon Selfridge, the American-born owner of Selfridge's department store in London. Selfridge was so intrigued by television's promotional possibilities that he engaged Baird for a sum of 20 pounds per week to demonstrate his apparatus three times a day in the store's front windows.

These demonstrations not only publicized Baird's system but also attracted more venture capital, and the Baird Television Company was formed to exploit it commercially. With additional funds available for research, Baird was able to improve the system further, and on June 26, 1926, he sought the imprimatur of the scientific community by demonstrating the system in his Soho Street laboratory to a select audience that included members of the prestigious Royal Institution. The demonstration was successful, and it encouraged him to continue to develop the system.

By 1929, it had been improved sufficiently to attract favorable attention from the British Broadcasting Corporation (BBC), and it began regular experimental broadcasts on September 30. Although the system had been greatly improved since the early 1923 experiments, it was extremely crude by today's standards. The number of scanning lines had been increased from 8 to 30, and the system was now capable of reproducing halftones. To keep the bandwidth within the 7,500-Hz audio standard, the frame rate was only 12 per second. This picture rate produced highly objectionable flicker with receivers of reasonable brightness, but it was tolerable with the dim pictures of Baird's receivers. In spite of these limitations, public interest was high, and Baird's copany sold a fair number of receivers.

In the meantime, laboratories in both the United States and Great Britain were working to develop all-electronic systems. In Britain, the effort was led by Electric & Musical Industries, Ltd. (EMI). EMI formed a joint venture with the Marconi Wireless Telegraph Company to demonstrate to the BBC a system using an electronic storage tube (the *emitron*, a version of the iconoscope) in the camera. The EMI system, with 405 interlaced scanning lines per picture and 25 frames per second, approached current scanning standards.

To meet this competition, Baird proposed three alternatives. One was a refinement of his original Nipkow disk system with 240 lines and 25 noninterlaced frames per second. A second proposal was designed to overcome the low sensitivity of the mechanical system when used for live pickups by photographing the scene on motion picture film, developing it rapidly (in about one minute), and scanning the film mechanically. A third alternative yielded on the issue of an all-mechanical system and proposed the use of the Farnsworth image dissector tube in the camera.

The BBC subjected the Baird and EMI systems to competitive on-air tests beginning in November 1936. The test period was originally planned to last two years, but the superiority of the EMI system was so overwhelming that it was chosen after only three months. The last BBC transmissions using the Baird system were on February 13, 1937.

Baird was devastated by the BBC decision, for it effectively repudiated his lifework. In the eyes of the British public, however, he was the inventor of television. His role and his reputation with the public were described much later by a Marconi historian:

> It is ironic that Baird, who contributed not a single invention to television as we know it today, should be regarded by the general public as its Father figure, while the names of those who were truly responsible should be known only in electronic circles. But let no one begrudge Baird his niche in the public mind; he was, after all, the first man to produce true television pictures and even though his success led many along the blind alley of mechanical scanning, he was, as P.P. Eckersley once remarked, the aphrodisiac which stimulated others to research, and ultimately to produce a more rewarding system.[3]

While Baird was inventing, developing, and promoting his mechanical system in England, an American inventor, Charles Francis Jenkins, was following a parallel course with another mechanical system in the United States. Its basic component was a flat rotating ring. The thickness of the ring increased gradually around its circumference, thus forming a rotating prism. As a light beam passed through the ring, its path was bent in proportion to the ring's thickness at the point of passage. Thus the beam could be made to scan a scene or a frame of film with each rotation of the ring. By using two rings overlapping at right angles, the beam could be made to scan both horizontally and vertically, thus forming a raster. Jenkins called these components *prismatic rings*.

Like Baird, Jenkins had a flair for publicity and an entrepreneurial nature. He began by using his prismatic-ring system for facsimile, the transmission of still images to a receiver where they are recorded photographically. In October 1922, he demonstrated the transmission of photographs and maps

---

[3] W.J. Baker, *A History of the Marconi Company* (London: Methuen & Company, 1970).

between a number of locations in Washington, D.C., to representatives of the Navy and Post Office. The quality of the transmission was good enough to be used by the Navy for sending weather maps to ships at sea.

In 1923, he was able to demonstrate the transmission of moving pictures in a laboratory environment, and by 1925 he had perfected it sufficiently for a public demonstration. The source of the pictures was a short motion picture film of a Dutch windmill. The film was scanned by a light beam from a neon tube that passed through the rotating prismatic rings to a photocell. The receiver also used a beam from a neon light source that varied in intensity with the signal and was made to form a raster on a small screen, six by eight inches, by another set of prismatic rings rotating synchronously with those in the transmitter.

The picture quality was marginal to say the least—the images were small, dim, and fuzzy, and they had no halftones—but the novelty of television was so great and speculative fever so intense that Jenkins had no difficulty in raising a very substantial sum to develop and promote his system. In 1927, he obtained the first license for an experimental shortwave television station, W3XK, from the FRC and began broadcasting a series of film shorts that could be received at distances of several hundred miles. Investment bankers were impressed, and they were willing to underwrite a stock offering worth $10 million for the Jenkins Television Corporation. He established a receiver manufacturing plant in Jersey City and a studio for producing programs in New York. Since his license was experimental,[4] he could not receive advertising income, and his profits had to come from the sale of receivers. He sold several thousand units at prices ranging from $85 to $135, clearly luxury items at the time. This demonstrated the novelty value of television, since buyers received only small, blurred pictures for their money.

Jenkins's system and his company ultimately failed for the same reason as Baird's: The picture quality was not good enough to compete with the electronic systems that were becoming available. Jenkins died heartbroken in 1934 at the age of sixty-seven after a long illness.

AT&T, GE, and RCA

The 1920s were the era of the radio trust (see Chapter 2), and three of its members—AT&T, GE, and RCA—took an intense interest in the development of television. Although they provided potentially strong competition

---

[4] The distinction between "experimental" and "commercial" stations was fundamental. Experimental stations were permitted to engage in broadcasting for the sole purpose of field tests. They were not allowed to sell advertising or otherwise charge for the use of their facilities.

to Jenkins, their activities actually helped him because they added an aura of respectability to the medium.

AT&T first demonstrated a television system in 1927 under the direction of one of Bell Laboratories' scientists, Herbert E. Ives. It used Nipkow disks at the transmitter and receiver and sent eighteen 50-line pictures per second. Two receiving devices were shown, one a projection system with a 2- by 2½-foot screen and the other a 2- by 2½-inch personal viewer intended as an adjunct to the telephone. The system was capable of handling rudimentary halftones. Video transmission was demonstrated both by wire and with a 1,575-kHz radio carrier.

About a year later, GE demonstrated a new television system developed by Ernst Alexanderson of Alexanderson alternator fame. Its receiver used a Nipkow disk that was two feet in diameter, and it produced a three-inch-square picture that had to be observed through a viewing aperture. The video signal was transmitted from the GE laboratory in Schenectady, New York, by a 7.9-MHz shortwave transmitter, and the sound was broadcast over GE's AM broadcasting station, WGY.

During this period, RCA had no laboratory or manufacturing facilities, and its role was to act as a sales agent for GE and Westinghouse radios (see Chapter 2). Sarnoff, however, followed television's technical developments with deep interest, correctly forecasting that it would one day become a major communications medium. The timing of his forecast was in error, however, as he predicted that television would be well established by the late 1930s. In fact, it took ten years longer.

Sarnoff was outwardly enthusiastic about the AT&T and GE demonstrations, and RCA obtained licenses for three experimental stations in the New York–northern New Jersey area. Privately, he had growing doubts about the ultimate success of mechanical systems. This was consistent with his faith in electronics, which remained with him throughout his life. Accordingly, when the GE–Westinghouse–RCA trust was dissolved in 1929 and RCA obtained its own research facilities, one of his first acts was to initiate a long and costly research program in electronic television systems.

FRC Regulation

In view of the hopelessly inferior quality of the television pictures produced by mechanical systems, the number of entrepreneurs and major corporations that wished to become broadcasters was extraordinary. In 1927, this forced the FRC, which had just been formed and had its hands full with AM radio problems, to take action.

The FRC recognized the inadequacies of the available mechanical systems, and it wisely refused to license stations for commercial operation until much-improved picture quality could be achieved. It also came to grips with television's bandwith problem. Its 1928 annual report quoted a memorandum from RCA's chief engineer, Alfred N. Goldsmith, as follows:

> A 5-kilocycle [kHz] band will permit the television broadcast of a crude image of a head with comparatively little detail. A 20-kilocycle band will permit the broadcasting of head and shoulders. . . . An 80-kilocycle band will permit transmission of the picture of two or three actors in fairly acceptable detail.
>
> The allocation of bands 100 kilocycles wide for television is strongly advocated, since this is clearly the minimum basis for a true television service of permanent interest to the public.

Television channels now occupy 6 MHz, sixty times the 100 kHz Goldsmith advocated as a minimum. It is doubtful that television service would be "of permanent interest to the public" today if it were constrained to 100-kHz channels, and even by 1928 standards it was marginal.

Goldsmith's analysis pointed out the dilemma that faced the FRC. There was not enough spectrum space in the medium-wave band occupied by AM radio, even for 100-kHz channels. The shortwave spectrum was wider, but stations operating there were subject to severe interference from distant stations. Finally, satisfactory transmitting and receiving equipment for the VHF portion of the spectrum was not yet available. It was clear that the limitations of mechanical systems were not the only problems inhibiting the progress of television.

Under the circumstances, the FRC did the best it could. It authorized limited service in the AM broadcast band with 10-kHz bandwidths. It allocated four 100-kHz channels in the 2,000- to 3,000-kHz range, and it allocated three much wider bandwidth channels in the 43- to 80-MHz range. Unfortunately, the three wider bandwidth channels were barely usable because of the primitive state of the VHF equipment.

The establishment of scanning standards was also an FRC responsibility, and the Radio Manufacturers Association (RMA) established a committee to make recommendations. Motivated by its commercial desire for immediate approval of some system, no matter how inferior its performance, the committee unanimously recommended fifteen frames, or pictures, per second (noninterlaced) and forty-eight lines per frame. Fortunately, the FRC refused to approve this standard, which would have resulted in pictures of unacceptable quality.

Broadcasting with Mechanical Systems

In spite of television's many technical difficulties and the prohibition against the sale of advertising by experimental stations, the number of stations continued to grow. Licensees had a variety of motives—to create a market for receivers, to gain technical experience, to obtain publicity, and to achieve grandfather status for the assignment of commercial licenses when they became available.

In 1929, fifteen experimental stations were licensed to operate. In addition to RCA's three stations in the New York City area, GE had three in Schenectady, Jenkins had one in Washington, Westinghouse had one in Pittsburgh, the trade publication *Radio News* had two in New York, the Chicago Federation of Labor had one in Chicago, and entrepreneurs in New York and Chicago had four others.

During the next three years, more than fifteen additional stations were licensed in all areas of the country, but by 1932 the growth in the number of stations had ended, and many of them had gone off the air. None of them was obtaining any immediate financial benefits, either directly or indirectly, and as time progressed, it became obvious that the inherent limitations of mechanical systems were so serious that they had no long-term future. Although the situation was exacerbated by the depression, it was the poor performance of these systems that caused them to fail.

In the meantime, great strides were being made in the development of all-electronic systems, both in England and the United States, and it was here that the future of television lay.

■ EARLY ALL-ELECTRONIC SYSTEM DEVELOPMENT

Alan A. Campbell-Swinton's Proposal

All-electronic television was not a new idea in the 1920s. As early as 1908, Alan A. Campbell-Swinton, a prominent and respected consulting electrical engineer in London, proposed a system that used CRTs for generating the electrical signal at the transmitter and displaying the picture at the receiver.

In his concept, the receiver CRT would operate as a conventional kinescope, or picture tube, with electromagnetic scanning to form a raster. CRT technology had progressed far enough to make this an achievable proposal.

The transmitter tube was another matter. Campbell-Swinton postulated a tube in which an electron beam would electromagnetically scan a special screen mounted behind the tube's faceplate. The screen consisted of an array of small cubes of a metallic photosensitive material. He described the screen's operation when an image of the scene was focused on it as follows: "On one side of this screen the cathode-ray beam impinges and on its other side is a chamber filled with sodium vapor or any other gas which conducts electrons more readily under the influence of light than in the dark. The metallic cubes readily discharge electrons when the scanning beam strikes, the discharge being proportional to the amount of illumination on the particular cube."[5] The discharged electrons were then conducted through the gas to be amplified and generate the video signal.

Although Campbell-Swinton continued to describe his proposal both in lectures and in the literature, he never carried it to the point of a working system. One reason was that he questioned its commercial value. In 1924, he addressed the Radio Society of Great Britain on the possibilities of television with wire and wireless. In the discussion period that followed, a Mr. L.B. Atkinson expressed the view that his proposal could be developed into a working system with "perhaps one or two years research." In response, Campbell-Swinton stated, "I wish to say that I agree entirely with Mr. Atkinson that the real difficulty in regard to this subject is that it is probably scarcely worth anybody's while to pursue it. That is what I have felt all along myself. I think you would have to spend some years of hard work, and then would the result be worth anything financially?"[6]

In spite of Swinton's lack of enthusiasm for the commercial possibilities of television and his failure to pursue his proposal to the point of a working system, it included the essence of modern all-electronic systems. Some of the concepts he proposed for the signal-generating tube, while never used in precisely the form he described, were incorporated later in practical tubes. Synchronized electromagnetic scanning of the signal-generating and display tubes is an integral part of modern television systems. And CRTs are used for picture display exactly as he proposed.

Swinton's proposal also highlighted the major missing link in all-electronic systems—the lack of a suitable electronic signal-generating device that could convert optical images into electrical signals. Solving this problem was the key to the successful development of all-electronic television.

Television Signal-Generating Tubes

Credit for inventing the first practical television signal-generating, or pickup, tubes must be shared by two engineers, Vladimir K. Zworykin and Philo T. Farnsworth. Zworykin's invention was the iconoscope, which evolved into the image iconoscope, and Farnsworth's was the image dissector.

---

[5] Alan A. Campbell-Swinton, "Presidential Address" (Speech to the Roentgen Society, 7 November 1911).

[6] George H. Brown, *and part of which I was* (Princeton, N.J.: Angus Cupar, 1982).

Zworykin and the Iconoscope   Vladimir K. Zworykin (1891–1982) was born in czarist Russia, the son of an affluent ship-owning family. He studied electrical engineering at the University of St. Petersburg, where he attracted the attention of Professor Boris Rosing, one of the pioneers in television picture tube technology. Rosing disappeared during the Bolshevik Revolution of 1917–1919, but Zworykin was able to escape and emigrated to the United States in 1919.

Zworykin was employed by Westinghouse as a research engineer, and for a time he was able to continue his work in all-electronc television. The result was the invention of the iconoscope (Figure 4–3), which he was first able to demonstrate to Westinghouse executives in 1923.

Courtesy RCA Corporation.

■ **Figure 4–3**   Vladimir Zworykin and the iconoscope.

The quality of the pictures produced by the first iconoscopes was quite poor, although it was a quantum leap ahead of pictures from mechanical scanners. It was not good enough, however, to persuade Westinghouse's executives to fund further research. Discouraged, Zworykin left Westinghouse for a time but returned after he was unsuccessful in finding a corporate sponsor for his work elsewhere. It was at this time, while RCA was still partly owned by Westinghouse, the he received an audience with Sarnoff.

Sarnoff listened enthusiastically to Zworykin's heavily accented but impassioned description of the commercial potential of the iconoscope for television broadcasting, and he immediately decided to find a way to finance its development. He asked Zworykin how much it would cost to develop a television system based on the iconoscope. Zworykin estimated that it would require about $100,000. After RCA had spent more than $50 million, Sarnoff repeatedly told this story in mock humility as an example of Zworykin's salesmanship and his own gullibility.

In 1929, about a year after this meeting, the radio trust was dissolved,

THE ICONOSCOPE TUBE

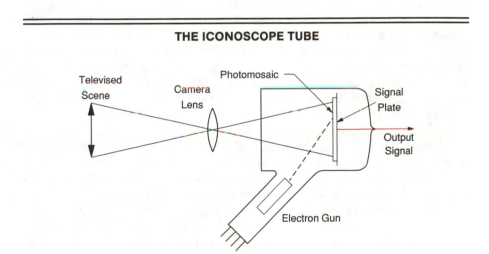

The photomosaic was the critical element in the iconoscope tube. It consisted of tiny particles of a photoemissive material mounted on a mica insulator, which in turn was mounted on a metal signal plate. When light from the scene was focused on the mosaic, electrons were emitted from the particles in proportion to the intensity of the light, leaving them positively charged. When the mosaic was scanned by the electron beam, the number of electrons absorbed by each particle was determined by the amount of its positive charge. The resulting current was transferred to the signal plate by the capacitance between the particle and the plate.

and RCA's relationship with Westinghouse was ended. Many Westinghouse engineers, including Zworykin, were transferred to the Camden, New Jersey, plant, which RCA had just purchased from the Victor Talking Machine Company, and Sarnoff was able to keep his promise by giving Zworykin the financial support he needed.

Zworykin proceeded enthusiastically, his progress closely followed by Sarnoff. By 1931, Sarnoff was convinced. He made a corporate decision to cease all further work on rotating disks and to commit major resources to the development of electronic systems. He hastened the demise of mechanical systems by buying Jenkins's troubled company for $500,000, much to the relief of Jenkins's bankers. In return, he received the rights to certain patents and the services of some of Jenkins's talented engineers. It was a courageous decision to embark on this program in the depths of the depression, but Sarnoff never lacked the courage of his convictions.

By 1932, Zworykin had perfected the iconoscope sufficiently to begin field tests of an all-electronic system at RCA's Camden plant. Improvements continued, and the iconscope was clearly the winner of its intense prewar contest with Farnsworth's image dissector.

Its major advantage over the image dissector was its substantially greater sensitivity, although it was still inadequate for many broadcasting applications. In 1934, Zworykin and RCA's German licensee, Telefunken, developed a more sensitive model, which Zworykin called an image iconoscope. This gave the iconoscope a further competitive advantage over the dissector, but it confused the patent situation because the modification was the addition of an imaging section that closely resembled the dissector.

Compared with modern tubes, the sensitivity of the image iconoscope was low, it was difficult to operate, and its performance was only fair. Still, its performance was good enough to be the basis for an all-electronic television system that received tentative FCC approval for commercial broadcasting just before the United States entered World War II.

After the war, the image iconoscope was replaced by the image orthicon, a product of wartime research. The original iconscope design continued to be used for film pickup, where plenty of light was available, until it was replaced by the vidicon in 1954.

Zworykin spent the remainder of his career basking in the glow of his success with the iconscope and Sarnoff's fulsome praise. He was furnished a private laboratory in RCA's research center in Princeton, New Jersey, and he served as a gadfly—not always welcomed but always heeded because of his relationship with Sarnoff—to the company's technical community. He continued to work there until his death in 1982.

Farnsworth and the Image Dissector   Philo T. Farnsworth was a Mormon, born on a farm in 1906 and educated at Brigham Young University, where he majored in engineering. He was forced to leave college before graduating because of the strain on his family's finances resulting from his father's death. In 1926, he opened a radio store in Salt Lake City, and he began work on television pickup tubes in its shop. The store soon failed, but he was able to persuade two groups of financial backers, first in Los Angeles and then in San Francisco, to support his research. He displayed a creativity that bordered on genius while working with incredible speed, and in 1927 he applied for a patent on his invention, which he called an image dissector.

The image dissector was a nonstorage scanner, and it suffered from the inherent problem of all devices of this class—extremely low sensitivity. Farnsworth was losing the competitive battle with the iconoscope because of this problem, and he developed a number of modifications and improvements in an effort to improve its sensitivity. Some of them were moderately successful, but none of them could bring the sensitivity up to the level of the image iconoscope or even the iconoscope. To obtain more capital for this endeavor, he entered into an alliance with Philco, one of RCA's radio competitors, in 1929, but this partnership was dissolved after two years as the result of Philco's dissatisfaction with the results.

Although the image dissector was never a commercial success, Farnsworth's patents caused big trouble for RCA. In 1932, he filed a patent interference suit alleging that the iconoscope infringed on his dissector patents. His case was strengthened when Zworykin added an electron imaging section in the image iconoscope that was very similar in concept to this component of the dissector. In 1935, both the patent examiner and the appeals court ruled in favor of Farnsworth, and Sarnoff's entire program was at risk, particularly the use of the image iconoscope, whose sensitivity was almost essential for live pickup.

Rather than appealing the case further, Sarnoff decided to make a deal. He first tried to buy Farnsworth's patents, but Farnsworth refused. He was finally forced to enter into a cross-licensing agreement that involved the payment of significant license fees. Agreeing to pay others for the use of their patents was totally foreign to the RCA patent department, and legend has it that one of RCA's lawyers had tears in his eyes as the agreement was signed.[7]

Farnsworth went on to found a successful electronics company, the Farnsworth Television and Radio Corporation in Fort Wayne, Indiana (later merged with Federal Radio), which he served for a number of years as vice president for research. He was well liked, even by Sarnoff, who was not known for charitable feelings toward his competitors. Years later, Sarnoff paid him a gracious tribute in testimony before a Senate hearing: "It is only fair that I should mention this—an American inventor who I think has contributed, outside of RCA itself, more to television than anybody else in the United States, and that is Mr. Farnsworth of the Farnsworth television system."[8]

---

[7] Kenneth Bilby, *The General* (New York: Harper & Row, 1986), 128.
[8] Ibid., 128.

Television Display Tubes

The technology of television display tubes is much simpler than that of pickup tubes, and its history can be told quickly.

It was recognized at the outset that the CRT (originally called the Crookes tube) provided an almost perfect solution to the problem of an electronic display device, and it never had serious competition for home receivers. The technology of producing, focusing, and deflecting electron beams had been developed by Sir J.J. Thomson and others in connection with his measurement of the properties of the electron. The use of these techniques to form a scanning raster was suggested by a number of engineers, notably Alan Campbell-Swinton. There remained the problem of causing the electron beam to produce a visible image on the faceplate of the tube (which was called a *kinescope* by Zworykin, from the Greek words *kinein*, "to move," and *skopein*, "to watch").

The first scientist to use a phosphor coating on the inside of the CRT bulb, which would glow when struck by electrons, may have been Karl F. Braun in Germany in 1897. The technology was developed further by Max Dieckman in Germany (1906) and Boris Rosing, Zworykin's mentor at the University of St. Petersburg (1907).

With the basic technologies of kinescopes established, further progress required thousands of man-years of intense and meticulous engineering effort. Phosphors that would glow with white rather than green light had to be developed. Many incremental improvements were made in the electron gun, which generates the electron beam, the focusing and deflection components and circuitry, and the construction of tubes that were sturdy enough to withstand accidental implosion in a home environment. Above all, it was necessary to develop techniques and machinery for producing tubes in mass quantities at low cost.

The electronics industry has succeeded admirably in these efforts as the result of the dedicated work of a host of engineers who have worked in anonymity. The performance of tubes was brought to a high state of perfection, accidents involving the implosion of tubes have been extremely rare, and the cost of tubes has been steadily reduced, even in the face of an inflationary economy.

■ ALL-ELECTRONIC TELEVISION BROADCASTING BEGINS

By 1935, the key components of all-electronic television systems were sufficiently perfected to produce acceptable picture quality, although it was far below present-day standards. But there was a long road ahead. Complete systems of studio and transmitting equipment had to be designed and built.

Receivers had to be designed, manufactured, and sold. Programming had to be made available. In the United States, where broadcasting was handled by private industry, there was the ever-present chicken-and-egg problem for new broadcast services—no incentive to broadcast programs without receivers and no incentive to buy receivers without programs. Transmission standards had to be established and huge segments of radio spectrum space allocated to meet the enormous demands of television.

The three leading countries in the development of television were England, Germany, and the United States. Each country approached these problems differently in accordance with its business and governmental policies.

England and Germany

England and Germany, which were about to go to war, shared one common characteristic in their broadcasting systems: They were both run by quasi-governmental companies. This solved the chicken-and-egg problem because governmental organizations, undeterred by the need to show a profit, could begin broadcasting before any receivers were available.

The early broadcasts of the BBC, beginning in 1929 with Baird's mechanical system and progressing in 1936 to the EMI–Marconi electronic system, were described earlier in this chapter. The leading figure in the EMI program was Isaac Shoenberg, its chief engineer and like Zworykin a Russian émigré. (Shoenberg is still recognized today in England as the father of electronic television.)

In Germany, the two competing companies were Telefunken and Fernseh. Telefunken used a system similar to the BBC's with "ikonoscope" pickup tubes and 441 scanning lines (later superseded by the "super-ikonoscope," or image iconoscope). Fernseh offered a variant of one of the systems proposed by Baird in England. It used an image dissector, and to solve its sensitivity problem, the scene was first photographed on film and then converted to an electrical signal by the dissector. The battle between Telefunken and Fernseh was really no contest, and Telefunken won it hands down. (After the war, however, Fernseh emerged as Germany's leading supplier of television studio equipment. It was eventually acquired by Robert Bosch, a manufacturer of automotive components.)

After the war began in 1939, progress in television technology in both Germany and England nearly came to a halt, as their electronics industries were converted to the war effort. The BBC stopped television broadcasting entirely.

The United States did not enter the war until two years later, and this permitted it to continue developing its systems. The U.S. television industry received an additional boost from government-sponsored television research even during the war, which gave it a clear lead over Europe in television technology.

The United States

To a large extent, the prewar history of television in the United States is the history of RCA. Mechanical systems failed because of inadequate performance. Farnsworth never made it with his low-sensitivity image dissector. Other companies, especially Philco and Dumont, made important contributions, and CBS caused great confusion with its ill-fated and premature attempts to establish color television. But the driving force behind the development of television was RCA.

In May 1935, Sarnoff announced that RCA was prepared to spend $1 million for the development of a complete television system. That was an enormous sum in the middle of the depression and would include not only the cost of research but also an experimental transmitting station on the Empire State Building in New York, a studio in the RCA building, a program service, and test receivers located in the homes of VIPs and RCA and NBC employees. The press greeted the announcement with enthusiasm, although it correctly prophesied that Sarnoff's plan would cost far more than $1 million. One by-product of the plan was RCA's insistence that Edwin Armstrong remove his FM antenna from the Empire State Building to make room for the television transmitting station.

RCA's plan proceeded on schedule. The first test transmissions from the Empire State Building occurred in 1935 with a 4-MHz VHF channel and a carrier frequency of 49.75 MHz (just below the present channel 2). Initial tests were at 343 interlaced lines and 30 frames per second. Iconoscope pickup tubes were used for studio cameras. By 1937, the number of lines was increased to 441, and in 1939 the image iconoscope, with its much greater sensitivity, was introduced. Receiver kinescopes were small, typically five to ten inches, but this had the advantage of obscuring the system's relatively low definition.

The FCC cooperated by allocating blocks of the VHF spectrum for experimental television broadcasting. As had happened earlier with mechanical television systems, a number of broadcasters and manufacturers opted to operate experimental stations, and by the end of 1937, seventeen of them were broadcasting signals generated by electronic systems.

RCA's competitors were not idle. Although Philco had long since broken off its relationship with Farnsworth, it began demonstrations of its system in 1936 using an image dissector camera, and it announced its intention to manufacture television receivers. Farnsworth continued to promote the image dissector, and he increased its sensitivity by adding a secondary emission multiplier. Unfortunately, this caused new problems, and it was not successful. CBS built an experimental VHF station on the Chrysler Building in New York for experimental broadcasts with both iconoscopes and image dissectors. Its efforts were sidetracked, however, by the unsuccessful attempt by its principal scientist, Peter Goldmark, to develop a color system.

RCA also encountered a major problem in initiating a commercial television system. This problem, the regulatory delay that resulted from the violent controversy concerning transmission standards and channel allocations, arose as RCA sought FCC approval for these key parameters of the system. The great standards battle was about to begin.

■ THE GREAT STANDARDS BATTLE

The Issue and the Contestants

Television has been described as a lock-and-key system in which the scanning of the receiver kinescope must be synchronized precisely with the scanning of the optical image in the camera. The required precision of the timing is extreme—a fraction of one-millionth of a second. (The precision required for color television is even greater.) This characteristic makes it necessary to establish transmission standards so that the scanning of the receiver kinescope will follow that of the camera. The specification of these standards for television broadcasting in the United States, together with the related problem of frequency allocation, is the FCC's responsibility.

The major issue in the standards battle was not the content of the standards, although this was the subject of some dispute, but rather the timing of their approval. For television broadcasting to begin, it was necessary that the FCC establish these standards, allocate spectrum space, and approve the use of broadcasting stations for commercial purposes.

On one side of the battle were RCA, Farnsworth, and a few progressive broadcasters such as the *Milwaukee Journal*. Sarnoff, who in 1935 had committed $1 million to develop electronic systems, was understandably anxious to start reaping the financial rewards of this expenditure by the sale of receivers and broadcasting by NBC. By 1938, he believed that the technology of television had progressed far enough to enable the FCC to establish standards for commercial broadcasting, and he pressed for expeditious action. Although Farnsworth was competing with RCA on a technical level, he shared Sarnoff's desire for an early decision.

On the other side were most other manufacturers, especially Philco, Dumont, Zenith, CBS, and many broadcasters. They wanted a delay or an indefinite postponement of the standards for a variety of reasons. Some had honest doubts as to whether technology had progressed far enough to establish standards that could withstand the test of time. A stronger motive, however, was their commercial interests. RCA was ahead of them, and they needed time to catch up. They also were concerned that standards established with a strong RCA influence would put them in a poor patent position. Zenith was doing well in radios, and it would have been happy to have television postponed indefinitely.

The opposition by Philco and Zenith also was influenced by personalities. Their chief executives, Lawrence Gubb and Eugene McDonald, were bitter enemies of Sarnoff, and they opposed him as a matter of principle. CBS, acting on the advice of Peter Goldmark, advocated that monochrome television be skipped and that the country proceed immediately to color.

Radio broadcasters were prospering in spite of the depression, and they were reluctant to jeopardize this with a risky and costly new medium. Many of them strongly disliked Sarnoff for threatening their businesses with a competing service.

The FCC was ambivalent. On the one hand, it had a statutory obligation to encourage and foster new technologies. This should have put it on the side of RCA. On the other hand, it feared RCA's alleged monopoly, and it was reluctant to take any action that would strengthen RCA.

The FCC's concern over RCA's monopoly sharpened when the able and strong-minded James Lawrence Fly was appointed chairman by President Roosevelt in 1939 (see Chapter 3). Fly was an ardent New Dealer, and he fully subscribed to Roosevelt's description of big businesses as malefactors of great wealth. He was particularly critical of RCA, which he regarded as one of the worst of the monopolists. He had a populist's desire to act as a handicapper to give the "little fellow" a chance. His visceral reaction was to side with RCA's opponents and do nothing for a time. Only irresistible political pressure could persuade him to change his mind.

The Content of the Standards

The parameters of a television signal that require standardization include the picture repetition rate, the number of scanning lines, the bandwidth of the transmission channel, the locations of the sound and picture carriers in the channel, the type of modulation, and the form of the electrical signal.

Frame Rate, Number of Scanning Lines, and Bandwidth By the late 1930s, when the question of establishing official standards was first seriously addressed, most of the industry (with the important exceptions of Philco and Dumont) was in agreement that the frame rate should be thirty per second, half the standard power source frequency. The value of interlaced scanning for reducing the bandwidth without sacrificing picture sharpness also was recognized.

The major problem was to specify the number of scanning lines—a trade-off between picture sharpness and bandwidth. Increasing the number of lines improved the picture's sharpness, but the bandwith increased with the square of the number of lines (assuming horizontal and vertical sharpness were improved proportionally). The problem was to find the right compromise between picture quality and bandwidth.

The Video Waveform The video waveform (see below) must carry three types of information:

1. The *synchronizing signals,* which synchronize the scanning of the receiver and camera
2. *Blanking pulses,* which black out the picture at the end of each line and at the bottom of the picture so that the scanning beam is not seen as it returns to the left or top of the picture
3. The video information itself

Considerable ingenuity was shown in developing the video waveform, and it was not particularly controversial.

Method of Modulation This standard caused very little controversy. FM was chosen for the sound channel but with only one-third the frequency

---

THE VIDEO WAVEFORM

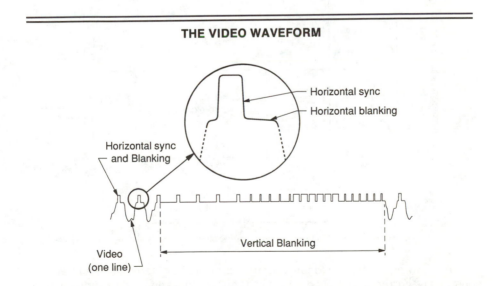

This is the video waveform specified in the current FCC standards. It shows the vertical and horizontal blanking signals, the synchronizing signals, and the video signals.

deviation permitted for FM broadcasting. FM was not even considered for the picture channel (although subsequent experience with FM in video recording showed that it might have been a better choice), but a form of AM known as *vestigial sideband* (see below) was chosen. Developed by Waldemar Poch and David Epstein of RCA, vestigial sideband modulation nearly cut in half the amount of spectrum required for the transmission of a given picture bandwidth.

The RMA Acts

In 1936, encouraged by Sarnoff and Farnsworth and with the FCC's tacit approval, the RMA established two committees to make recommendations to the FCC concerning television. One of them was to consider transmission standards, and the other was to consider the question of frequency allocations.

TELEVISION BROADCASTING CHANNEL

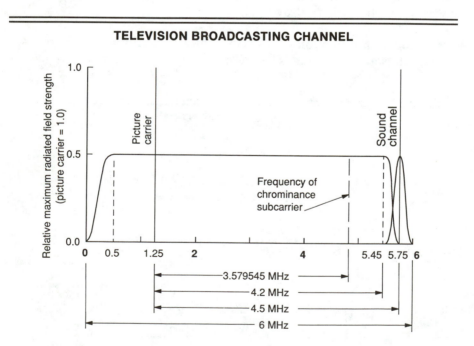

The drawing shows the location of the carriers and sidebands in a television channel as currently specified by the FCC for broadcasting. The sound channel uses FM with a maximum deviation of ±25 kHz (compared with ±75 kHz for FM broadcasting). The picture channel uses vestigial sideband AM. The dashed curve shows the spectrum space required if both sidebands were transmitted.

The allocations committee acted expeditiously and recommended that seven 6-MHz channels be allocated for television broadcasting. Late in the year, the standards committee recommended a preliminary standard of 441 lines and 30 frames per second with interlaced scanning. Video modulation was double sideband, which limited the video bandwidth to 2.5 MHz in the 6-MHz radio frequency channels. In 1938, the recommendation was modified to specify vestigial sideband modulation for the video signal, thereby increasing its bandwith to 4 MHz in the 6-MHz video frequency channel.

The FCC Responds

The FCC held a series of hearings to consider these recommendations, and the deep divisions in the industry soon surfaced, even though the RMA presumably spoke for all. To confuse the issue further, Dumont and Philco proposed two different standards. Dumont's was 625 lines and 15 frames, and Philco's was 605 lines and 24 frames.

The FCC's first actions related to frequency allocations, which were less controversial than the transmission standards. In 1936, it allocated 42 to 56 MHz, 60 to 86 MHz, and a number of channels above 110 MHz for experimental television broadcasting. These allocations were expanded on October 13, 1937, when the FCC issued its report and order concerning Docket 3929, a proceeding that considered the allocation of the spectrum from 10 kHz to 300 MHz very broadly. This order allocated nineteen 6-MHz channels for television broadcasting (but still limited to experimental broadcasting pending an agreement on transmission standards), the lowest being 44 to 50 MHz and the highst 288 to 284 MHz.

After a lengthy series of hearings, the FCC affirmed this action in an Order issued on March 19, 1939, stating, "The Commission believes that in order to permit television to be inaugurated on a nationwide basis a minimum of 19 channels should be reserved below 300 MHz." The Commission added that more channels would undoubtedly be required above 300 MHz.

These statements by the FCC were remarkably prescient, particularly since the technologies of both television and VHF were in their infancy and UHF (above 300 MHz) had scarcely been born. The 6-MHz channels are still the standard today, and nineteen VHF channels would not be sufficient to provide a nationwide competitive television service.

The Commission's orders did not solve the industry's allocation and regulatory problems, however. One of the nineteen channels was reallocated to FM in 1940 (the missing channel 1 in today's system) leaving television with eighteen. Moreover, the Commission steadfastly refused to establish transmission standards or authorize commercial broadcasting, and in 1938 it tightened the eligibility requirements for experimental licenses. In the future, they were to be granted only to organizations engaged in technical research. This put a number of stations that had been engaging in programming development off the air.

RCA Makes Its Move

Sarnoff, impatient with the repeated delays and concerned that RCA's competitors were catching up, decided he could wait no longer. In October 1938, he stated, "Television in the home is now technically feasible. The difficulties confronting this difficult and complicated art can only be solved from operating experience, actually serving the public in their homes."[9] He announced that RCA would begin manufacturing receivers according to RMA recommendations and that NBC would begin regular broadcasts, albeit experimental, in 1939. The inaugural program would be a broadcast of the opening of the New York World's Fair on April 20, 1939 (Figure 4–4). Sarnoff, Mayor La Guardia, and President Roosevelt would share the podium.

The plan was a calculated risk. Commercial broadcasting, including the sale of advertising, was not permitted, and all the costs of the program would have to be borne by the corporate treasury. It was certain to increase the animosity of RCA's competitors, and it was likely to antagonize the FCC. But Sarnoff was gambling that it would stimulate the public's desire for television and cause irresistible political pressure.

Sarnoff's gamble paid off. RCA's competitors were furious, and they protested its actions to the FCC, but public pressure did increase. An example was the action of the the *Milwaukee Journal*, a pioneer radio broadcaster. It filed for a commercial television station even though doing so was not allowed by the FCC's rules. In its November 6, 1938, issue, just two weeks after Sarnoff's announcement, the *Journal* stated that "experiments and investigation have shown that television has developed beyond the laboratory stage and is now ready for a service to the public." It also said that "adoption by radio manufacturers of RCA specifications for television equipment as trade standards is further justification for *The Journal's* new undertaking."

Sarnoff increased the pressure by keeping his promises. NBC increased its programming to twelve hours per week, and RCA receivers priced from $395 to $675 were offered on the market. More stations, all limited to experimental broadcasting, came on the air in New York, Philadelphia,[10] Schenectady, and Los Angeles.

---

[9] Bilby, *The General*, 132.

[10] The programming for the Philadelphia stations was informal and unprofessional and was frequently handled by engineers rather than skilled program personnel. There is a legend that the Philco station in Philadelphia installed a camera, often left unattended, at the University of Pennsylvania swimming pool. The engineers forgot to note that at certain hours bathing was limited to nude members of a single sex. The response from the viewers gave a good indication of the size of the audience.

Courtesy RCA Corporation.

■ **Figure 4–4** David Sarnoff demonstrating television at the 1939 World's Fair. This demonstration on April 20, 1939, which opened the 1939 New York World's Fair, marked the start of a regular schedule of television programming by NBC. It was a huge success. Only a few thousand were able to see it off the air or on monitors at the fair, but Sarnoff and television upstaged President Roosevelt and Mayor La Guardia in the press—no small accomplishment. The equipment remained at the fair, and hundreds of thousands of people were able to see television for the first time.

Not surprisingly, in view of the limited programming and the possibility of technical obsolescence, there was no rush to buy the high-priced receivers. Sarnoff had forecast that twenty thousand to forty thousand receivers would be sold in New York the first year. In fact, only eight hundred were sold. Rather than reducing the political pressure, however, the lack of receiver sales was taken as further evidence of the need for FCC action as a means of reassuring the public.

The FCC Compromises

The pressure for FCC action eventually became unbearable, and Fly reluctantly agreed to reconsider the question in a formal hearing in January

1940. The outcome was a compromise that satisfied no one and left the situation more confused than before. On February 29, the FCC issued an order that did not establish transmission standards but did authorize *limited* commercial broadcasting. *Limited* was defined as meaning that advertising sponsors could cover the cost of programming but that the stations would have to pay for operating costs. The new rules were to become effective in September. Like many other compromises, it turned out to be ineffective.

This order was only a small opening, but it was all that Sarnoff needed. With full-page ads in major New York newspapers, he announced that RCA would begin regular *commercial* broadcasts in September and that twenty-five thousand sets would be offered for sale at reduced prices. It was implied that this was the first step in a program that would result in the RMA recommendations becoming the de facto standards.

Again the industry was cast into turmoil. Philco engaged in a counter-publicity campaign advising the public *not* to buy RCA receivers because higher performance sets with more scanning lines would soon be available. It withdrew from the RMA, claiming that the organization was merely a puppet of RCA.

Fly was outraged. He took the unusual step of going on the air himself on the Mutual radio network. For an entire hour, he portrayed RCA as a bully and a monopolist and himself as the champion of the "little fellow."

In May 1940, the FCC rescinded its earlier order allowing limited commercial broadcasting and stated that the FCC would not approve full commercial broadcasting until the engineering opinion of the industry was agreed on standards.

Now it was Sarnoff's turn to be outraged. He fought back, using the potent political support he was able to elicit. He obtained favorable editorial comment from influential newspapers such as the *Philadelphia Inquirer* and the *New York Times*. The latter described the FCC decision as "absurd" and "unsound." The issue became so heated that it reached the desk of President Roosevelt.

It also reached the halls of Congress. Senator Burton Wheeler (D) of Montana chaired a hearing to determine whether the FCC had exceeded its authority. Fly accused RCA of "blitzkrieg" tactics and of openly defying the commission's intent. Wheeler's committee did not take a definite position, but it strongly suggested that the Commission settle the issue promptly and unambiguously.

The First NTSC and the Final Standards Decision

Faced with this criticism, Fly resorted to a technique that would be repeated ten years later in the color television battle. He appointed an ad hoc committee, the National Television System Committee (NTSC), to study the issue of television transmission standards and to make recommendations to the FCC. The chairman of the NTSC was W.R.G. Baker, a distinguished GE executive, who was able to add an aura of objectivity to the committee. One of his most important qualifications was that he had no connection with RCA.

The committee was formed in July 1940, and its final meeting was held on March 8, 1941. It submitted a set of recommended standards to the FCC specifying 525 scanning lines, a frame rate of 30 per second with interlaced scanning, and FM sound. These are essentially the same standards that are used for monochrome television today.

Fly and the remainder of the FCC, faced with growing political and public impatience and presented with the recommendations of a committee that included manufacturers, broadcasters, and other interested parties, acted with unprecedented speed. In April 1941, it adopted the NTSC recommendations as the standards of Good Engineering Practice Concerning Television Broadcasting Stations.

Thus ended the television standards controversy. Commercial broadcast television was free to proceed and would have done so had World War II not intervened. At the end of 1941, when the United States entered the war, thirty-two experimental stations had been authorized to begin commercial broadcasting. With receiver manufacturing and station construction forbidden by wartime regulations, only six of these were operating as commercial stations at the war's end.

■ THE WARTIME HIATUS

The following six stations continued commercial operation during the war:

1989 Call Letters	Location
WNBC-TV	New York
WCBS-TV	New York
WNEW-TV	New York
WRGB-TV	Albany/Schenectady
KYW-TV	Philadelphia
WBBM-TV	Chicago

With a shortage of parts and most of their skilled technical personnel transferred to defense work, staying on the air was a touch-and-go proposition. There were very few receivers in the hands of the public, and there was little incentive to offer a full-time program schedule. Most of these stations operated only a few hours a day and at a considerable expense to their owners. The war truly caused a hiatus in the growth of television broadcasting.

But the war did not create a hiatus in the development of television technology. Even more than for FM, it was a powerful stimulus. Television technology and the war effort had a synergistic relationship that benefited both.

One of the decisive technical developments of the war was radar, a British invention but one that owed much to American engineering. Radar and television use several of the same technologies—VHF and UHF transmission and reception, as well as nonsinusoidal, or pulse, signals—and engineers experienced in television were able to transfer their skills and knowledge directly to radar.

Similarly, the enormous financial support given by the Allied governments to the development of radar was beneficial to television broadcasting when it was resumed at the end of the war. And, as with FM, the engineers and technicians trained in radar proved immensely useful in the postwar development of television.

The greatest contribution of World War II to television technology was not related to radar, however. It was the development of the image orthicon pickup tube by Albert Rose of RCA.

The Image Orthicon

The image orthicon pickup tube was developed as the image transducer of the Block equipment, the code name for a family of unmanned radio-controlled flying bombs that were guided to their targets with the aid of on-board television cameras. The need for these bombs had arisen from the desire to destroy targets such as the German submarine pens along the French coast. Heavily protected by thick concrete walls and roofs, they were virtually impervious to bombing unless the bomber approached at water level and entered the pen at its opening. The Japanese could have solved this problem without advanced technology by ordering kamikaze pilots into the pens on suicide missions. But this solution was not acceptable to Western culture, and the Block equipment was developed instead.

As a weapon system, the Block equipment was only modestly successful, but the image orthicon camera was an enormous step forward in television technology. At the end of the war, it replaced the iconoscope as the standard pickup tube for studio and field cameras, and its position of dominance continued for twenty years until 1965, when it began to be replaced by the *Plumbicon*.

The image orthicon was a marvel of ingenuity. Its development required the services of skilled professionals in chemistry, electron optics, vacuum technology, and other sciences. Its manufacture was an intricate process involving a host of specialized skills and techniques. Its operation is com-

THE IMAGE ORTHICON

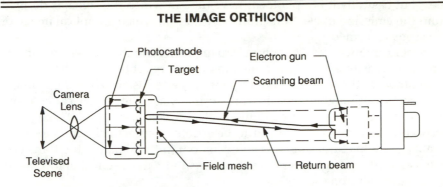

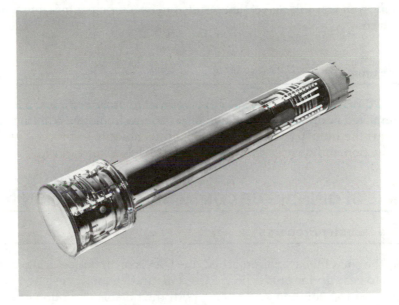

An optical image of the scene is formed on the front of the photocathode, and the electrons emitted from its rear form an electron image on the target mesh assembly, an extremely fine mesh spaced close to a thin glass membrane. Most of the electrons from the photocathode pass through the mesh and strike the glass membrane, causing secondary electrons to be emitted. These are collected by the mesh. This leaves the glass membrane positively charged, and electrons are absorbed by the scanning beam. The return beam provides a negative image—that is, its current is highest in dark areas of the picture where very few electrons are absorbed from the scanning beam.

plex, but basically it is a photoemissive tube in which the signal is derived from the emission of electrons from a surface on which an optical image of the scene is formed.

Its performance was a quantum leap forward from the iconoscope. It was far more sensitive, and good pictures could be obtained at light levels comparable to those used in photography. It was easier to operate than the iconoscope, although operators were normally assigned to the camera controls to make more or less continuous electrical adjustments. It also had a second-order effect that seemed to be a defect but actually enhanced the quality of the picture. This was an overshoot in the signal at the transition between dark and light areas in the picture and had the effect of increasing its apparent sharpness.

Its major defect was a high level of electrical "noise," or picture graininess. It was this defect that resulted in its ultimate replacement by the Plumbicon.

In Summary

World War II brought the growth of television broadcasting to a halt, but it stimulated enormous advances in television technology. With the end of the war, the industry was well positioned technically to proceed with the introduction and development of commercial television.

■ THE BEGINNING OF COMMERCIAL TELEVISION

The Industry Prepares

As the end of the war approached, the television manufacturing industry began to gear up for production. RCA, GE, and Dumont designed lines of studio and transmitting equipment. With the notable exception of Zenith, most major radio manufacturers joined the protelevision camp and began converting their plants for receiver production. NBC made plans to increase the number of hours of programming.

AT&T was busy constructing its L-1 coaxial cable network, primarily to handle an expected increase in telephone traffic, but, unlike the older open-wire lines and twisted pair cables, it would also be capable of distributing television network programming throughout the country. By the end of 1945, fifteen hundred miles of cable had been laid.

On the regulatory front, late in 1945 the FCC reinstated the prewar NTSC transmission standards and allocated a total of thirteen VHF channels for television broadcasting. It also announced that it would begin accepting applications for commercial television stations.

All the pieces seemed to be falling in place for the resumption of the growth of commercial television broadcasting after the wartime hiatus and for the beginning of a major new industry.

A Slow Start and Broadcasters' Concerns

At first, there was no rash of applications for station construction permits. Only six new stations went on the air in 1946 and only four more in 1947. Had there been a greater demand for stations, there would have been a hardware problem. The manufacturers of station equipment did not have the capacity to satisfy a huge demand in 1946 and 1947. But the reluctance of potential television broadcasters to apply for stations had a different basis— their fear that operating a television station might not be good business.

The construction cost of a television station was many times more that of a radio station. For example, the cost of a single image orthicon camera system, or "chain," was about $15,000. This sum was sufficient to buy the equipment complement for a small radio station. Construction costs for television stations, including studio equipment, transmitter, antenna and tower, buildings, and construction, were measured in the hundreds of thousands of dollars and could easily exceed one million. Radio broadcasters were not used to these sums.

Operating costs were equally high. Television stations required much larger staffs than radio, and other expenses were larger in proportion. According to FCC reports, in 1949 ninety-eight television stations reported annual operating expenses of $59.6 million, an average of $608,000 per station. In the same year, 1,824 radio stations reported expenses of $342.9 million, an average of $108,000 per station. The disparity in cost grew with the passage of time. In 1979, thirty years later, 723 television stations reported expenses of $6.2 billion, or $8.55 million per station, while 7,207 radio stations reported expenses of $2,624 billion, or $364,000 per station.

Programming was an equally serious problem. There were no facilities for the nationwide distribution of television programs. For a number of years, stations were forced to rely on locally produced programs or films. The production of live programs required techniques that were entirely different from those used in filmmaking. In addition, there was no large pool of talent, programming personnel, or technicians experienced in television production. New production techniques had to be developed, and everyone had to rely on on-the-job training. Perhaps the most serious problem in producing programs was the lack of a satisfactory method of making video recordings. Audio recording for radio stations was introduced shortly after the end of the war (see Chapter 2), but it would be ten years before a practical video recording system would become available.

Finally, there was the chicken-and-egg problem. According to the EIA, by the end of 1946, there were fewer than 100,000 television sets in use; 179,000 were added in 1947. This was far too small a base for a profitable television broadcasting industry. Clearly the broadcasters would have to break into the chicken-and-egg cycle and put programs on the air at a substantial loss until the number of receivers increased by several orders of magnitude.

These problems would have been overcome sooner, however, had there not been a doubt about the future of monochrome television. The radio broadcasting industry included many entrepreneurs who had become rich by taking risks, and they were not reluctant to take another one. But CBS, whose judgment in broadcasting matters was highly respected, took the position that the future of television was color and that monochrome was already an obsolete technology.

The Position of CBS

William Paley was a brilliant business leader, and his judgment in programming matters was unmatched in the industry. He had no technical background, however, and this made him vulnerable to misleading demonstrations and sales presentations by engineers who were also persuasive salesmen. In Peter Goldmark CBS had a senior engineering executive with an extraordinary talent for selling his ideas, even when they were bad ones.

Goldmark had started work on the *field sequential* color television system (see Chapter 5) at CBS in 1937, and he had managed to continue it during the war. Goldmark's conception was a mechanical system that used rotating color filter wheels at the camera and receiver. It was capable of producing attractive pictures with brilliant colors under controlled conditions, but it had inherent limitations that made it unattractive as an ultimate color system.

Goldmark was particularly skilled in staging demonstrations that emphasized the desirable features of his system and covered up its deficiencies. In 1940, in the middle of the bitter controversy over monochrome standards, he had demonstrated the system to the NTSC. Although it did not receive Committee acceptance because of its mechanical nature and other limitations, even CBS's competitors had to agree that the pictures were impressive. In 1945, he made formal demonstrations to two of CBS's top executives, Paul Kesten and Frank Stanton, who were running CBS while Paley was serving on General Eisenhower's staff. Had they been more experienced in technical matters, they might have recognized that the brilliance of Goldmark's pictures was superficial and it obscured the serious limitations of the system. But they were not, and Goldmark mesmerized them. Here is Paley's own description of subsequent events:

So there we stood at the end of 1945 poised for action with the RCA black-and-white system virtually assured of FCC approval and the CBS system not yet approved but, in our estimation, far superior to the black-and-white system.

Soon after my reutn to CBS [from service on Eisenhower's staff], Kesten, Stanton, and Dr. Goldmark made a major presentation to me of the CBS color system, including a demonstration of one of Dr. Goldmark's models. They all were absolutely enthusiastic about the system and our chances of FCC approval which would make CBS color television the standard for the entire industry. In color technology we were far ahead of RCA which had focused its research on black-and-white transmissions. . . . The dilemma, of course, was that the two systems were incompatible. But, argued Stanton and Kesten, everyone agreed that color television was vastly superior, technically and aesthetically, to black-and-white. It was just a matter of the public waiting perhaps one more year for the fully developed color system to be approved by the FCC rather than buying black-and-white sets immediately. In fact Kesten had indicated publicly eighteen months before, in April 1944, that CBS would soon be putting color television on the market and that the public should not be sold inferior black-and-white sets which would soon become obsolete. That had thrown the television industry, particularly the manufacturers of sets, into utter turmoil. How could they gear up for manufacturing television sets when RCA and CBS might be telecasting on two completely different systems?[11]

Receiver manufacturers made strenuous efforts to dissuade CBS from pursuing color at that time, but Paley refused to be persuaded. In the summer of 1946, as RCA was putting its black-and-white sets on the market, CBS petitioned the FCC to authorize commercial color broadcasting using the CBS system in the UHF band. It backed up its words with action by making a deliberate corporate decision *not* to apply for the four additional stations it was permitted under the FCC Rules. (CBS already had one station operating in New York.) CBS went a step further and advised its affiliates not to apply for black-and-white channels.

This decision turned out to be Paley's biggest mistake in all the years he ran CBS. CBS was ultimately forced to pay more than $30 million for VHF stations in Los Angeles, St. Louis, Philadelphia, and Chicago that could have been had for the asking in 1946. It was equally costly for the CBS affiliates who took his advice.

Although this CBS decision was risky and ultimately very costly, it had a certain rationale at the time. There was the prospect that millions of monochrome sets would be sold before the CBS system was approved. Because of the incompatibility of the color and monochrome standards, this would create an enormous and probably insuperable barrier to the introduction of

---

[11] William S. Paley, *As it Happened* (Garden City, N.Y.: Doubleday, 1979).

the CBS system. (The events of 1950, when an improved field sequential system was approved, showed that this concern was real.) The essence of the CBS strategy, then, was to receive speedy approval of its system, or, failing that, to block further growth of monochrome before large numbers of monochrome receivers were in the hands of the public. Its decision to discourage the growth of monochrome television by failing to establish its own network was part of that strategy.

The Second Standards Battle

The CBS petition set off another major industry controversy. All of the old shibboleths of monopoly and market dominance were dragged out. CBS cast itself in the role of a "little fellow" with a superior system being overwhelmed by the vast RCA, which was using its power to foist an obsolete system on the public.

For Sarnoff, who was eager to get television going, it began a period of intense frustration. To quote Yogi Berra, the New York Yankees' master of the malapropism, "It was déjà vu all over again." In fact, it was a twofold déjà vu. It reopened both the standards dispute, which had torn the industry apart before the war, and the mechanical versus electronic controversy, which Sarnoff thought had been resolved by the rejection of the Baird and Jenkins systems.

This time there were differences, however. RCA had the support of nearly all the television manufacturing industry. CBS found itself with no important allies among this group except Zenith, which claimed that it would never build a monochrome receiver. In addition, the FCC had a new chairman, Charles Denny, appointed in March 1945, who did not share Fly's antipathy toward RCA.

The FCC convened a hearing to consider the CBS petition. It was highlighted by RCA and CBS demonstrations of their color systems. Although the quality of the CBS color picture at that time was far superior, the FCC ruled in January 1947 that neither the CBS system nor UHF was sufficiently perfected to justify the authorization of their use for commercial broadcasting. It denied the CBS petition but permitted experimental color broadcasts to continue in the UHF band. CBS was stunned (see Chapter 5).

For Sarnoff, the FCC's decision was a total victory. One of the major uncertainties about the future of monochrome television had been removed. Now the ball was in his court, and he had to persuade broadcasters that television was a good investment.

■ TELEVISION BROADCASTING BURGEONS

The Rush for Licenses

In the months following the denial of the CBS color petition, the enthusiasm for monochrome broadcasting grew steadily. Sarnoff's September 1947 speech to the NBC affiliates, quoted earlier in this chapter, was the catalyst. By the middle of 1948, there were not enough channels to meet the demand in the larger markets, and a round of competitive hearings was begun to select the successful applicants. The growing number of construction permits granted by the FCC was reflected, after delays for construction, in the number of stations on the air:

Year's End	On-Air Stations
1945	6
1946	12
1947	16
1948	51
1949	98
1950	107

The Interference Problem and the "Freeze"

As the number of stations on the air increased, the FCC began to receive disquieting complaints of excessive cochannel interference. The Commission was very sensitive to this issue because there was a growing sentiment that it had been too lax in its assignment policies for AM stations, many of which were now receiving severe nighttime interference from cochannel stations. The FCC was determined not to repeat in television what was perceived as a mistake in AM radio.

An example of this interference problem was the situation in Princeton, New Jersey (the location of RCA Laboratories), which was about 45 miles from the NBC channel 4 transmitter in New York. Viewers in this area experienced considerable interference from NBC's channel 4 station in Washington, 255 miles from New York and 180 miles from Princeton. The effect was a venetian blind pattern of horizontal bars on the set resulting from interference between the two carriers.

With the problem affecting their home receivers, RCA engineers in Princeton had a personal motive in seeking a solution. In 1949, Alda V. Bedford and Gordon Fredendall found one in the technique of offset carriers. The carrier frequency of one of the cochannels was deliberately detuned from the other by one-half the scanning line rate, or by 7,875 Hz. The result was to make the "blinds" so narrow that they were barely visible.

The offset carrier technique was extremely effective and is still used today, but it came too late to dissuade the FCC from taking another step. On September 30, 1948, the FCC announced that it would take no action on pending applications until it had completed an exhaustive review of its policies for assigning new stations. This was the beginning of the famous (or infamous) "freeze."

When the freeze was imposed, 124 stations had been authorized and 45 were actually on the air. Sixty-three more of the authorized stations were built, and the number of on-air stations was capped at 108 until the end of the freeze, nearly four years later in April 1952.

Other Prefreeze Events

Receiver Manufacture   RCA was the pioneer in receiver manufacture, and it was quickly followed by a host of competitors. As the public responded enthusiastically to the new medium, manufacturers strained to meet the demand for receivers. The particulars of the progress in receiver technology and sales are described later in this chapter.

Broadcast Station Equipment   RCA dominated the broadcast equipment market in the prefreeze era. Its Broadcast Equipment Division had a mandate from Sarnoff to develop a complete line as a part of his plan to develop a television market for NBC and RCA's Receiver Division as quickly as possible. As noted earlier, RCA had benefited enormously from government-sponsored research during the war. It had an able and aggressive marketing organization headed by Walter W. Watts and Theodore A. (Ted) Smith, the former having learned the trade at Montgomery Ward under the irascible Sewell Avery.

There was also an intangible in RCA's success in the early days of television. Broadcasting was both a career and a way of life for its managers, salesmen, and professional personnel. For GE, RCA's major competitor at the time, broadcast equipment was just another business, and it did not benefit from the enthusiasm and singleness of purpose that marked RCA's operation. The results were reflected in market shares. RCA obtained 80 percent of the business during this period, while GE struggled to get most of the rest, about 15 percent.

Television Time Sales   Television time sales were negligible in 1946, grew slightly in 1947 and 1948, and reached a total of only $30 million in 1949, the first year of the freeze. Starting a television station at that time was an act of faith.

■ ALLOCATION AND ASSIGNMENT POLICIES

In FCC parlance, the term *allocation* describes the designation of blocks of spectrum space to specific services, such as television broadcasting. *Assignment* refers to the designation of channels within these blocks to specific localities or licensees.

Postwar Television Channel Allocations

Prefreeze Allocations The prewar allocations of television channels by the FCC were described earlier in this chapter. Television was originally given nineteen VHF channels, but one of these was lost to FM in 1940, and at the beginning of the war, it retained eighteen channels.

Television lost six more channels at the end of the war, five of them as the result of the relocation of the FM broadcast band (described in Chapter 3). On March 25, 1945, the FCC issued a proposed decision in which it reduced the number of television channels from eighteen to thirteen. Surprisingly, there was very little opposition from the television broadcasting industry.

On June 27, 1945, the FCC issued a decision allocating the VFH channels as they exist today except that it included channel 1. At the same time, it affirmed the NTSC transmission standards adopted in 1941. It also allocated a portion of the UHF band for experimental UHF broadcasting as a preliminary step for expansion of commercial television beyond the thirteen VHF channels.

On June 27, 1946, the Commission issued a Public Notice in which it reaffirmed its 1945 allocation but with a footnote that all channels except channel 6 must share with other services. The sharing concept turned out to be intolerable, and on May 5, 1948, after receiving much evidence, the Commission issued a notice of proposed rule making that eliminated sharing but reallocated channel 1 to other services. This proposal was adopted, and television broadcasting was allocated the exclusive use of the twelve VHF channels that it continues to occupy today.

Postfreeze Allocations The hearings that the FCC conducted during the freeze were lengthy and contentious, and it was not until April 11, 1952, nearly four years after the freeze was imposed, that it was lifted. Although the major issues addressed during the freeze were frequency assignment policies and color transmission standards, an indirect effect was that a major portion of the UHF spectrum was allocated to television broadcasting. This was because the new assignment policies reduced the number of assignments that could be made in the existing VHF channels—which was already inadequate before the freeze.

The number of VHF channels available in most major cities was so limited as the result of the post-freeze assignment policies that VHF channels alone could not satisfy the Commission's criterion of a nationwide competitive system. To provide a reasonable number of channels in these cities, the FCC decided to reallocate the television channels in the UHF spectrum from experimental to commercial. The April 11, 1952, Order that ended the freeze established seventy 6-MHz channels extending from 470 to 890 MHz and designated channels 14 to 83. Channels 66 to 83 were designated as "flexibility channels" and, unlike the lower channels, were not assigned to specific cities.

In allocating the UHF channels for commercial broadcasting, the commission was attempting to achieve two conflicting objectives. On the one hand, it was seeking to extend the interference-free coverage of the VHF channels and thereby improve service to rural areas. One of the FCC's strongly held views was that a resident of a remote rural village had as much right to television service as a resident of a large city. On the other hand, it wished to increase competition among broadcasters by authorizing more stations.

In attempting to achieve these conflicting objectives by allocating a portion of the UHF spectrum for commercial broadcasting, the FCC was obliged to indulge in wishful thinking and place hope before experience. Some of the provisions of the allocation and assignment notice appeared to be an attempt to overrule the laws of nature by executive fiat. The serious problems of radio wave propagation and receiver performance in this portion of the spectrum were well known, but the Commission, in its desperate attempt to resolve its dilemma, closed its eyes to them. The resulting trials and tribulations of UHF television are described in more detail later in this chapter.

One consequence of UHF's problems was that usage of these channels was considerably lower than the FCC had hoped, and the higher frequency flexibility channels were hardly used at all. In the meantime, the demand for channels in adjacent regions of the spectrum by the land mobile service grew astronomically, and the disparity between the crowded land mobile bands (used by public safety and commercial organizations for two-way vehicular communications) and the lightly used regions allocated to broadcasting became painfully apparent.

Faced with this disparity and extreme pressure from land mobile manufacturers and users, the FCC in 1971 agreed to permit the land mobile service to share channels 14 to 20 in eleven of the largest cities, where the crowding of the land mobile channels was the most severe. The areas were chosen because they were sufficiently removed from UHF television assignments to avoid harmful interference.

Pressure from the land mobile industry continued, however, and, noting that the higher UHF channels were hardly used at all, the FCC allocated channels 70 to 83 to the land mobile services. This allocation remains in effect today.

Assignment Policies for Television Channels

The allocation of spectrum space for television broadcasting, while complex and not without conflicts of interest, was relatively uncontroversial. No major and immediate vested interests were at stake, and allocations were made on a reasonably rational basis.

The assignment of television channels, however, involved major conflicts of interests, and it required significant public policy decisions. The result was that it commanded a large portion of the FCC's attention during the decade after World War II.

Prefreeze Assignments   At the outset, the FCC decided that channel assignments would be made on a demand basis but in accordance with a master plan that would be embodied in a table of city-by-city assignments. It issued the first table in 1945, which included assignments to 140 metropolitan districts. The assignments were based on preliminary engineering standards of radio wave propagation and tolerable interference levels derived from the best data available at the time.

As the demand for channels increased in late 1947 and early 1948, the number of assignments in the first table was inadequate, and the Commission issued a revised table in May 1948 that added assignments to some cities and provided assignments to others that were not on the original list. But adding assignments was contrary to the evidence that was appearing with respect to cochannel interference. To address this problem, the Commission scheduled a series of hearings during the summer of 1948. The evidence introduced at this hearing included new information on tropospheric propagation at distances from fifty to two hundred miles. It clearly showed that the existing engineering standards, which were based on average tropospheric conditions, did not accurately predict the interference levels that would exist for significant periods of time during abnormal tropospheric conditions.

This was a frightening discovery. The result could be chaos from cochannel interference if too many stations were licensed. It was in response to this concern that the Commission instituted the freeze in September 1948.

The Freeze and Postfreeze Assignments   During the nine months following the imposition of the freeze, the FCC considered the evidence available to it and on July 10, 1949, issued a notice of further rule making. This was a notice informing the public of proposed changes in the rules and invited comments.

The notice proposed a redesignation of UHF television channels from experimental to commercial and three classes of stations:

- Community stations (UHF only), limited in 20 kilowatts
- Metropolitan stations, limited to 100 kilowatts on VHF and 200 kilowatts on UHF with a five-hundred-foot limitation on antenna height
- Rural stations, with greater power and antenna height allowed under certain specified circumstances

In recognition of the more recent data on tropospheric propagation, the minimum cochannel spacing was increased to 220 miles for VHF and 200 miles for UHF. These spacings, however, were established as desirable objectives and not absolute requirements.

The FCC received many comments in response to this notice, and they were generally favorable. By a careful reshuffling of assignments, the number of VHF channels assigned to most of the major markets was reduced very little and, except for in Albany, New York, never by more than one. Furthermore, there was reasonable flexibility in the proposed rules, and it appeared that additional assignments could be made on a case-by-case basis with a proper engineering showing. It was anticipated that engineering consultants, who had shown great ingenuity in squeezing in more AM radio channels, would perform the same function in television.

Signs of the rigidity that was to characterize the final decision had begun to appear, however. The use of directional antennas to reduce the spacing between cochannel stations was forbidden, as was the assumption of any improvement from offset carriers. Most importantly, the FCC proposed a set of priorities that it would follow in making additional channel assignments. The first priority was to provide at least one television service to all parts of the United States. The fifth and last priority was to make additional assignments to cities that already had two or more. This sequence is precisely the opposite of the result of a free market. Major metropolitan areas would be denied additional television service so that channel reservations could be held for sparsely populated rural areas that would be unable to support stations economically.

The proposed set of priorities was a symptom of the dichotomy that has always characterized the FCC's attitude toward broadcasting. On the one hand, it is regarded as an essential public service, like the post office, to which every citizen is entitled. On the other hand, it has been regarded as a business that is regulated by the risks and rewards of the free market and is disciplined by the market to meet the needs and desires of the public. Broadcasting in the United States always has been suspended between these two worlds.

During the year following the issuance of the July 10, 1949, Notice, the

FCC's attention was almost totally taken up with the protracted color television hearings (see Chapter 5). Little attention was paid to the assignment problem except by a small group of engineers on the Commission's staff headed by its chief engineer, Curtis Plummer. These engineers were dissatisfied with the principles and policies proposed in the Notice, and they worked diligently on a series of revisions. The revisions were ultimately adopted, and they have had a profound effect on the development of television broadcasting.

The staff's dissatisfaction with the Notice stemmed from its experience with AM broadcasting. The policy with AM was to grant any new application provided it included engineering evidence that the station would not cause interference with existing stations above an amount specified as a maximum in the engineering standards.

This policy, the staff felt, led to undesirable excesses. As the demand for new licenses continued and the number of stations on each channel increased, the interference with existing stations became increasingly severe, even though it was held below the legal maximum for each new station. Another problem with this policy was that its administration was costly, contentious, and time-consuming.

The staff's solution was to develop a set of policies that were "simple and straightforward" and that would establish the Table of Assignments on the basis of rigid rules. To ensure the rules' rigidity, the staff went to extraordinary lengths to make sure there were no loopholes that an enterprising consultant might find to drop in additional assignments.

The mileage separation between cities to which cochannel stations were assigned was stated to be an absolute minimum with no exceptions. Mileage was to be measured between arbitrary points in the cities, usually the post offices. There was no flexibility; even a miss of a few feet would be disqualifying. Offsetting this rigidity somewhat was the minimum spacing for VHF stations, which was reduced from 220 miles to 180 miles on the basis that the offset carrier technique reduced the effect of interfering signals.

The three station classes were eliminated, and for assignment purposes all stations were assumed to be operating at the maximum operating power of 100 kilowatts for low-band (channels 2 to 6) and 200 kilowatts for high-band (channels 7 to 13) VHF. This policy led to some absurdities, which Commissioner Jones pointed out in an acerbic dissent. For example, Elko, Nevada—population 6,000—was presumed for assignment purposes to have stations as powerful as those in New York City. The staff recognized that this solution would greatly reduce the number of VHF channel assignments, but in its view this was not too high a price to pay for adminsitrative simplicity and the avoidance of creeping erosion of stations' service areas.

The staff's recommendations were adopted by the commission and issued on March 21, 1951, as the Third Notice of Proposed Rule Making. (In

1964, eight years later, similar policies were adopted for the assignment of FM stations; see Chapter 3.)

For the industry, the overall effect was a drastic reduction in the number of commercial VHF channels assigned to major metropolitan areas. This effect was exacerbated by the policy, assiduously lobbied by Frieda Hennock, the first female commissioner, of reserving one channel, VHF if possible, in every city for educational or public broadcasting. In thirty-seven of the fifty largest television markets, the number of commercial VHF channels was reduced. For example, in Chicago commercial channels were reduced from 7 to 4, in San Francisco from 7 to 4, in Boston from 5 to 3, in Cleveland from 5 to 3, in Dallas/Forth Worth from 6 to 4, in Houston from 4 to 2, in San Diego from 4 to 2, and in Scranton/Wilkes-Barre from 4 to 0.

The impact on the ABC network was particularly severe. It was then the weakest of the three major networks, and it was usually the third choice of local stations for affiliation. Thus, if there were only two VHF stations in a city, they would almost certainly choose NBC and CBS. In these markets, ABC was forced to beg for a few hours of released time from the NBC and CBS affiliates or to make do with a UHF affiliate. This further weakened its position vis-à-vis the other networks and delayed its becoming fully competitive.

Several hundred comments were filed in response to the Commission's third notice. They were directed both to the general assignment policies and to situations in individual cities. The thrust of most of them was a plea that the Commission assign more VHF channels to the major population centers.

The FCC's final response and the end of the freeze came on April 14, 1952, with the adoption of the Sixth Report and Order. In addition to the general statements of policy, each individual comment had to be addressed and a decision rendered. As a result, it was a massive document of more than five hundred pages.

The Sixth Report made a few revisions in the Table of Assignments, and it increased the maximum permissible power to 316 kilowatts for high-band VHF (316 is a multiple of the square root of 10) and 1,000 kilowatts for UHF. It amended the minimum spacings of cochannel stations to 170 miles for VHF and 155 miles for UHF in Zone 1 (the Northeast), 190 miles for VHF and 175 miles for UHF in Zone 2 (the South, Midwest, and West), and 220 miles for VHF and 205 miles for UHF in Zone 3 (Florida, Texas, and the Gulf Coast). Smaller spacing was allowed in Zone 1 in recognition of the closer spacing of its cities. Greater spacing was required along the Gulf Coast because of the greater range of tropospheric propagation in that area.

This Report was even more rigid with respect to policies that might allow deviations from the minimum mileage spacings. The prohibition of the use of directional antennas to reduce cochannel station spacing was reaffirmed. Although mountain ranges were recognized to be effective barriers against the propagation of VHF signals, terrain features were explicitly barred as a basis for reduced spacing. A related rule proscribed the use of field measurements to demonstrate the possibility of reduced spacing. The proposed rule forbidding a reduction in spacing by assuming less than the maximum power by one or both of the stations was affirmed. Finally, the Commission further tightened the Table of Assignments by increasing the number of assignments to small towns, thus making it even more difficult to locate stations in nearby large cities. In Nevada, for example, assignments were made to tiny hamlets such as Goldfield, population 336.

In short, the FCC made it clear that none of the techniques that might lead to a more flexible assignment policy would be acceptable, no matter how compelling the reasons. For practical purposes, the Table was cast in concrete, and since that time very few changes have been made.

Except for the fortunate holders of prefreeze licenses, who could scarcely contain their delight, the industry was almost universally critical of the Third Notice and its affirmation, the Sixth Report. With a little more flexibility, a number of additional assignments could have been made, and the Commission's action was widely regarded as a triumph of bureaucratic rigidity over common sense.

With a passage of time, it appears that this judgment might have been a bit severe. In support of the FCC's actions, one could state the following:

1. The assignment policies were clearly successful in achieving their major objective—the elimination of serious interference between cochannel and adjacent-channel stations.
2. The demand for channels in many metropolitan areas was far greater than could be satisfied by VHF stations, even with a more flexible set of assignment principles. It was imperative, therefore, that UHF be developed to supplement the VHF system. If more VHF stations had been assigned, there would have been fewer UHF stations, and its progress would have been even more difficult.
3. In spite of its early handicap resulting from the shortage of VHF channels, ABC was able to progress, and with good management and superior programs, it became an industry leader.

Perhaps the most questionable decision was the rejection of the principle of deintermixture—that is, the assignment of VHF-only or UHF-only channels to each city. This would have avoided competition between VHF and UHF in the same city, and the acceptance of UHF would have been greatly accelerated. It also would have made the use of VHF-only or UHF-only receivers possible in many areas, thus effecting savings in receiver costs. To have made such a decision, however, would have required considerable political courage, and the FCC lacked this. For its part, the industry had to face up to the fact that UHF was for real. The limitations on the number of VHF assignments were too severe to satisfy the needs of the marketplace.

■ THE GROWTH OF VHF TELEVISION

The Rush for Licenses

With the end of the freeze, the Commission was faced with the herculean task of selecting the successful applicants for the limited number of available VHF channels. Television had now proved itself as a profitable medium, and there was a rush of applications. Within six months after the end of the freeze, hundreds of applications had been filed and there were very few uncontested vacant VHF channels in the fifty largest cities. The FCC's engineering staff had been successful in easing its own burden by advocating a rigid assignment policy, but it had made the selection problem more dfficult for the Commission by limiting the number of available channels.

The successful applicants were selected as the result of administrative hearings at which their qualifications were examined. Given the rigid engineering rules, technical factors were rarely important in the decision. Instead, intangibles were emphasized, including program plans, local ownership, diversity of media control, and occasionally financial qualifications. The hearings were lengthy and costly, and success depended more on the skill of the applicant's attorney to present a case that fit the Commission's predilections than on his or her ability to manage a television station.

Denver had the honor of receiving the first postfreeze grants for VHF stations. It had no prefreeze stations, and it had retained its five prefreeze VHF channel assignments (although one of them was designated as an educational channel). Surprisingly, there were three uncontested applications, and these were granted on July 11, 1952, just three months after the end of the freeze. Grants of the contested applications trickled out more slowly, but by the end of 1954, nearly all of the most desirable channels had been taken, and the great rush was over.

VHF Station Growth

The objective of most of the successful applicants was to get on the air as quickly as possible. The growth in the number of VHF stations on the air, which had been halted by the freeze from 1949 to 1952, resumed with a spectacular increase from 1953 to 1955. In 1956, the growth rate of new stations began to decrease rapidly, as the most valuable channels were occupied. By 1960, all the commercial VHF channels in the larger markets were occupied, and most of the remaining growth came from educational stations and commercial stations in secondary markets (see Figure 4–5).

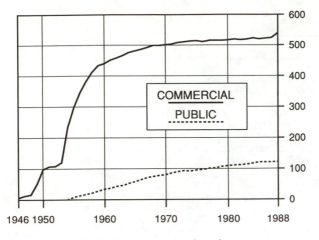

■ **Figure 4–5** VHF stations on the air.

The surge in new stations created an enormous market for studio and transmission equipment, and manufacturers, which had seen very lean years during the freeze, suddenly enjoyed great prosperity. Demand exceeded supply, everything sold at list price, and sales were limited only by the ability of the factories to produce. RCA was the dominant supplier with about 60 percent of the market. GE, then its major competitor, was a distant second with 20 percent, and the rest was divided among a number of smaller manufacturers. By the end of 1954, however, the rush to buy equipment was over, and the broadcasting equipment industry languished for two years until it was revived in 1956 by the introduction of the spectacular videotape recorder by Ampex.

Growth in the Number of Television Homes

The growth in the number of stations on the air was matched by an equally rapid growth in the number of television homes—homes having at least one television receiver (Figure 4–6).

Growth in VHF Stations' Revenues and Profits

The revenues and profits of the VHF broadcasting industry in the years preceding the freeze explain the reluctance of many broadcasters to apply for licenses that a few years later would be worth many millions of dollars. As the number of stations increased from 1946 to 1949, the total industry losses increased even more rapidly. A realistic analysis of television's cost and revenue prospects could easily have convinced one that it could never

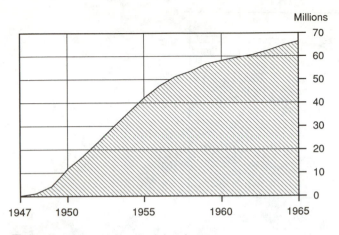

Millions

■ **Figure 4–6** U.S. television homes.

be profitable. It took courage and faith in the prophesies of respected industry leaders, such as Sarnoff's 1947 speech to the NBC affiliates, to persuade radio broadcasters and other entrepeneurs to proceed.

They did not have long to wait for their reward. Losses reached their peak in 1949, they were considerably less in 1950, and the industry became highly profitable in 1951. The 107 stations that had received their construction permits before the freeze were often described as lucky, but they were also courageous.

A longer range look at VHF stations' revenues and profits is even more impressive. Their total revenues (Figure 4–7) grew from $106 million in 1950 to $8.1 billion in 1980,[12] an average annual growth rate of 15 percent. Few industries have matched this rate for such an extended period.

The profit growth rate (Figure 4–8) was even greater, from a pretax loss of $19.2 million in 1950 to a profit of $1.6 billion in 1980, an enormous rate of return on sales. And since the book value of the tangible assets of most stations is comparatively low, the return on investment was even more impressive.

Industry profits continued to grow after 1980 until 1986 when they began to drop, partly as the result of competition from cable and VCRs and partly as the result of rapidly escalating program costs. The era of uninterrupted profit growth appears to have come to an end.

---

[12] The FCC stopped compiling revenue and profit totals for the industry after 1980. The most authoritative source of recent industry figures is the *Television Financial Report* published annually by the NAB. Since it is based on the voluntary disclosure of financial data by stations, it is not complete and the sample may not be representative. It showed that the average UHF station made a profit of $702,000 in 1984 but suffered a loss of $422,000 in 1986.

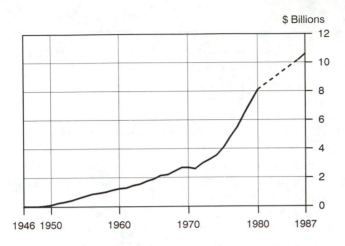

■ **Figure 4–7** VHF station revenues, 1946–1987.

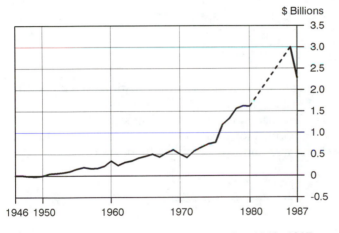

■ **Figure 4–8** VHF station pre-tax profits, 1946–1987.

■ THE TRIBULATIONS OF UHF

While VHF broadcasters were prospering beyond their wildest dreams, UHF broadcasters were beginning a long and costly struggle. At the end of the freeze, UHF left the world of wishful thinking and administrative fiats that had declared VHF and UHF practically equal and entered the real world. And a hard world it was! As UHF broadcasters went on the air, they found that everything was against them.

UHF's Problems

UHF had many problems, some inherent and others inherited. First, the propagation characteristics of UHF are not as favorable for broadcasting as are those of VHF. With its shorter wavelenegth, UHF is attenuated more rapidly by terrain, buildings, and vegetation. The small UHF receiving antennas also are not as effective in extracting energy from a passing radio wave as are the larger VHF units. In addition, the UHF sections of television receivers were not as sensitive as the VHF section. And UHF had the chicken-and-egg problem. All-channel receivers were more expensive than those for VHF only, and many viewers were not sufficiently interested in UHF programming to pay a premium to receive it.

UHF's most serious problem was programming. Networks would not affiliate with a "U" if a "V" were available, and most of the U's had to operate as independents. Today there are program sources for independent stations that approach or exceed the networks in popularity. But these sources did not exist in 1952 or for many years after, and most of the programs on independent UHF stations were second-rate or worse.

The Ebb and Flow of the UHF Station Population

The UHF station population went through four distinct phases (Figure 4–9). At the end of the freeze, many entrepreneurs, eager to participate in the profitable business of broadcasting and encouraged by the FCC's optimistic evaluation of UHF, applied for UHF construction permits. By the end of 1953, 112 UHF stations were on the air, more than the number of pre-freeze VHF stations.

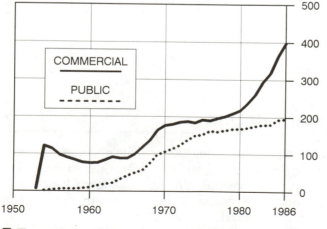

■ **Figure 4–9** UHF stations on the air.

This first phase ended very quickly. The new UHF broadcasters soon learned that VHF and UHF were not the same, and disillusionment set in. The number of UHF stations reached a peak of 125 in 1954 and then declined steadily as mounting losses caused many of them to go dark.

This was the beginning of the second phase, a period of losses and discouragement, which lasted from 1955 to 1965. During this period, fewer than one hundred UHF commercial stations continued to broadast.

The third phase began in 1965 with another triumph of hope over experience. Industry revenues began to increase, and, although expenses increased even faster, a number of favorable developments encouraged entrepreneurs to believe that the industry would ultimately become profitable. The number of commercial stations increased rather rapidly from 1965 to 1970, when it reached another plateau of about two hundred.

The industry finally became profitable in 1976, although the profitability

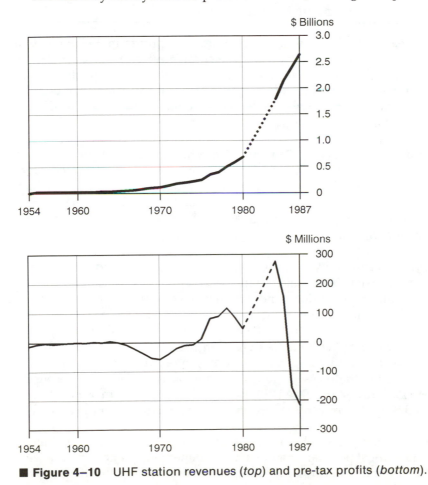

■ **Figure 4–10** UHF station revenues (*top*) and pre-tax profits (*bottom*).

of individual stations varied widely. This marked the beginning of the fourth phase, another period of rapid growth in the number of stations.

UHF Stations' Revenues and Profits

The revenues and profits of UHF stations through 1980 are shown in Figure 4–10. In comparison with VHF, their revenues were pitifully small. Total industry revenues did not exceed $100 million until 1969, sixteen years after the first stations went on the air. In that same year, VHF station revenues were $1.8 billion and radio station revenues were more than $1 billion.

Positive Developments for UHF

The breakthrough in industry revenues and profits in 1975 after two decades of stagnation resulted from the cumulative effect of a number of favorable developments. The most important of these were the overall growth of the broadcasting industry, the increasing availability of popular programs, the all-channel receiver law, FCC regulatory actions, the effect of cable television, and technical progress.

Broadcasting Industry Growth   The total revenues of the television broadcasting industry in 1952, the year in which the freeze was lifted, were $324 million. By 1986, industry revenues had grown to more than $21 billion, an annual growth rate in excess of 20 percent. The demand for television advertising became so great that VHF stations did not have the capacity to handle it.

Program Availability   As the result of the huge demand for programming by television broadcasters (and in recent years by cable television services), a multibillion-dollar industry has developed to produce and distribute television programs. The tabulation of program production and distribution companies in the 1988 edition of the Warren Publishing Company's *Television & Cable Factbook* gives an indication of its scope. It lists nearly eight hundred companies, including more than twenty news services. An active trade association, the National Association of Television Program Executives, Inc. (NATPE), has been formed to further the interests of the industry. Its conventions are an active marketplace for the industry's wares.

The major networks with their enormous buying power have first call on many of the most popular programs, but independent stations have a wide choice of program offerings, including reruns of programs that have proved their popularity by initial showings on the networks.

The All-Channel Receiver Law   The 1934 Communications Act was amended in 1963 to require all television receivers to be designed to receive both VHF and UHF channels. It gave the FCC the authority to reject sets for sale if the performance of the UHF channels did not meet satisfactory standards.

Its major proponent was Newton Minow, the FCC chairman who was best known for his description of television programming as a "vast wasteland." He was distressed by the competitive advantages enjoyed by VHF broadcasters, and he lobbied hard and successfully for the bill. It was opposed by most of the industry on the grounds that it interfered with the workings of a free market and forced the public to pay for a feature whether they wanted it or not. It was probably necessary, however, to solve UHF's chicken-and-egg problem.

The first all-channel sets did not completely solve the problem because the UHF band was difficult to tune, and they lacked sensitivity. The FCC, using its authority under the Act, tightened the specifications, and, manufacturers were required to improve the sets' performance and ease of tuning. A number of years elapsed before all-channel sets with satisfactory UHF performance were in general use, but the long-term trend was positive.

FCC Actions   The FCC was eager, in fact almost desperate, for UHF to succeed, and it looked with favor on any proposal that would be helpful in achieving that objective. One step was to increase the maximum permitted power and antenna height for UHF stations. The Commission's first proposal as the freeze was nearing its end was to limit the power to 200 kilowatts and the antenna height to 500 feet, the same as channels 7 to 13 on VHF. In the final Order ending the freeze, the limits were raised to 1,000 kilowatts and 1,000 feet. In 1957, they were raised to 5,000 kilowatts and 2,000 feet. The maximum powers and antenna heights now permitted are as follows (limits for VHF are lower in the Northeast):

	Max. Power	Max. Antenna Height
VHF channels 2–6	100 kw	2,000 ft.
VHF channels 7–13	316 kw	2,000 ft.
UHF	5,000 kw	2,000 ft.

The 5,000-kilowatt maximum for UHF was academic at first because transmitters that could achieve this power level did not become available until the early 1960s. The capital and operating costs could not be justified economically in many markets, and even this power level does not provide as complete coverage in most areas as lower powers at VHF. But raising the limits did provide UHF some degree of equality with VHF.

An equally important FCC action was to raise the limit of the number of

UHF stations that could be owned by one individual or corporation. Initially, this limit was seven stations, of which no more than five could be VHF. Currently, a single owner can be licensed to operate as many as fourteen stations, provided that the total population served by all the stations does not exceed 25 percent (30 percent for minority owners) of the nation's population. In making this calculation, only 50 percent of the households covered by UHF stations are counted.

Cable Television   Cable television was a mixed blessing for UHF. On the one hand, UHF signals could be received as well as VHF on cable systems. On the other hand, by importing the signals from distant broadcast stations and satellite-distributed programs (see Chapter 7), cable increased the amount of competition to UHF stations. On balance, cable has usually had a positive effect on local UHF stations in mixed UHF–VHF markets but a negative effect in all-UHF markets.

Technical Developments   The growth of UHF was encouraged and enhanced by major technical developments. Transmitter powers and antenna gains were steadily increased, thus making it possible for stations to approach or equal the maximum radiated power permitted by FCC Rules. Major improvements were made in the sensitivity and ease of tuning of UHF receivers. The invention of a method of recording video on magnetic tape (see Chapter 6) was a key factor in making the proliferation of program sources for independent stations possible. And the introduction of color (see Chapter 5) was one of the developments that led to the enormous increase in broadcasting revenues, which UHF shared.

Summary   In spite of the positive developments for UHF listed in the previous section, it still has an uphill battle in markets where it must compete with VHF. The 1978 reversal in profit trends shown in Figure 4–10 was followed by erratic results over the next decade. The drop in profits after 1984, which was experienced by VHF stations owing to competition from cable and increased programming costs (see above), was even more severe for UHF. As a result, the industry again fell into the red.

The Lost Opportunity of Deintermixture

Many of UHF's problems could have been avoided if the FCC had had the political courage to incorporate deintermixture in the order ending the freeze. The effectiveness of deintermixture is demonstrated by a comparison of the profitability of stations in Peoria, Illinois, and Fargo, North Dakota, two cities with an almost identical number of households in their Areas of Dominant Influence (ADIs). These are the figures for 1980:

	Peoria	Fargo
Number of stations	3 UHF	3 VHF
Revenues	$11,257,271	$8,791,942
Pretax profits	$1,615,531	$1,367,632

Subsequent to the freeze, the FCC took some small steps in the direction of deintermixture. For example, it deleted the single VHF channel assigned to Fresno, California, thus making it an all-UHF market. But by the time the effectiveness of deintermixture was recognized, it was too late to make major changes in the Table of Assignments.

UHF's Future

In spite of its recent progress, after more than 30 years, the cumulative sum of UHF stations' operating losses, the write-offs of facilities, and the losses incurred by equipment manufacturers that financed many of these stations have probably exceeded the profits. What then of the future?

In spite of all the technical advances that have aided UHF, it is inherently inferior to VHF as a broadcasting medium. Its short wavelength causes the signal strength to vary widely within very short distances, and the signal is so weak in the low points that no practical amount of transmitter power can bring it up to a satisafactory level. An ideal broadcast transmission medium would produce a uniform signal strength within each small region of its coverage area. UHF deviates the most widely from this ideal, and in competitive situations VHF stations will always have a significant advantage.

In addition, like all broadcast media, UHF will experience growing competition from cable-distributed programs. Being less affluent than VHF, UHF stations will not be able to afford the quality of programming to enable it to compete as effectively with cable.

Nevertheless, UHF has a promising future in markets where there is no VHF competition or where the number of VHF stations is insufficient to provide an adequate variety of programming services or outlets for advertisers. In these markets, UHF stations are prospering and will continue to do so.

■ INTERCITY VIDEO CIRCUITS

The lack of intercity transmission facilities posed a major barrier to the growth of television in its early years. It was obvious that networking would be as essential to television broadcasting as it had been to radio, and the need for a nationwide intercity video transmission system was urgent.

Intercity circuits for radio networks were essentially upgraded voice circuits, and they were a natural extension of AT&T's telephone network, both technically and economically. Intercity video circuits posed much more difficult problems. Their bandwidth, which was measured in megahertz rather than kilohertz, required the use of entirely different technologies.

AT&T readily assumed responsibility for providing these circuits. This was consistent with its perception that it had both the obligation and the right to be the sole supplier of intercity communication circuits of all kinds (except telegraphy). Since wideband circuits could also be used for the transmission of voice, it was also consistent with its obligation to carry out a major expansion of its intercity telephone capacity.

Coaxial Cable Versus Microwave

Two competing technologies, coaxial cable and microwave, had become available to AT&T for wideband intercity circuits. As its name implies, coaxial cable consists of an inner conductor—a solid or stranded wire—which is held in the center of a cylindrical outer conductor, typically 3/8 inch in diameter, by insulating spacers. The attenuation of the high-frequency components of a signal with coaxial cable is much less than with the twisted pairs used for voice circuits. This gives it the capability of transmitting broadband video signals over great distances. (Coaxial cable was first developed, however, for the transmission of a multiplicity of voice circuits and thus to replace twisted pairs for this purpose.) The attenuation is not zero, however, and it is necessary to insert amplifiers or repeaters at intervals to compensate for the cable losses. Increasing the bandwidth of the cable requires the installation of repeaters at more frequent intervals, and the bandwidth specification is a trade-off between cost and performance.

The technology of microwave transmission systems, typically operating at frequencies of about 4,000 MHz, had developed rapidly during World War II. These systems could handle the wideband requirements of television with ease.

The cable/microwave controversy in AT&T pitted its traditionalists against its Young Turks. The traditionalists had worked with cable all their lives, the nitty-gritty techniques for installing and maintaining it had been established through years of experience, and they felt more comfortable with a tangible connection between terminals than with an intangible radio link—"Nothing is as reliable as a wire!"

The traditionalists, quite rightly as it turned out, also were concerned that AT&T would lose its monopoly of intercity connections if the use of microwave became common. AT&T had an extensive existing network of rights-of-way for cable installations and, as a public utlity, could use the government's right of eminent domain to extend it. Private nonutility companies did not have this right, but it was unnecessary for microwave systems.

The Young Turks, schooled in the technologies spawned during the war, strongly urged the adoption of microwave, claiming that it was more economical than cable and was the wave of the future.

As befitted its conservative policies and deep pockets, AT&T settled this controversy by using both media. With the passage of time, a hybrid system evolved that used cable for short, high-density routes and microwave for longer hauls. Still later, satellites (see Chapter 8) became widely used for video transmissions, and in the future fiber optics will be a strong contender.

The L-1 Cable Network

Having made a decision to use both cable and microwave, AT&T proceeded to install a nationwide network (although not nearly fast enough to satisfy the impatient broadcasters). The first circuits, installed between New York and Washington, used the L-1 cable system. The cable was 3/8 inch in diameter, and repeaters were installed at intervals that resulted in a bandwidth of 2.7 MHz—adequate for monochrome but not wide enough for the 4.2-MHz bandwidth required for NTSC color transmission.

Once the massive AT&T organization began to move, it had the resources to act rapidly. As noted earlier, fifteen hundred miles of cable had been laid by the end of 1945, and cable circuits were soon extended to major cities throughout the South, Midwest, and East and finally to the West Coast.

The Microwave Network

While the L-1 cable network was being installed, AT&T and many electronics manufacturers were hard at work developing microwave equipment and systems for intercity transmission of voice, video, and data circuits. The fears of the AT&T traditionalists that microwave would result in competition were soon justified. Pipelines, electric utilities, and other bulk users of communication services began to install private microwave systems. This was the first crack in the AT&T monopoly and would eventually lead to divestiture and deregulation. Initially, however, the owners of private systems could not sell their services to others, and AT&T continued to be the major supplier of intercity service to the broadcast industry.

AT&T soon began the installation of its microwave system. It was capable of handling both voice and video, and the video bandwidth was adequate for NTSC color. Its routes often paralleled the L-1 system, but it provided the additional capacity required to handle the rapidly growing volume of communications traffic.

The AT&T system operated in the 3,700- to 4,200-MHz common carrier band. Its design was extremely conservative and costly. Repeater stations were installed at intervals of about twenty-five miles. They were massive structures, sometimes of masonry, and were built to withstand severe storms. Redundant (duplicate) facilities were provided so that no single failure would cause the station to cease operation. The cost of the system was considerably higher than that of private links, and this caused some grumbling by broadcasters, but they eventually felt that the added reliability was worth the cost. By 1960, the microwave network reached most of the major cities and had largely replaced cable for intercity video transmission.

■ MONOCHROME BROADCASTING SYSTEMS AND EQUIPMENT

The basic technologies and components of monochrome television—such as the interlaced scanning principle, vestigial sideband transmission, the iconoscope, the image orthicon, and the vidicon—were largely the work of single inventors or small teams of engineers and scientists. As the technology progressed from basics to products, the engineering effort shifted from invention to product design.

Product design is an exceedingly difficult engineering discipline. Its practitioners must have a thorough knowledge of the basic technologies, but they also must have an understanding of manufacturing processes and the demands of the marketplace. The design of a high-quality product that can be manufactured at a low cost, is reliable and easy to use, meets the needs of its application, and is appealing to customers in a highly competitive environment requires exceptional skill and experience. (As will be seen later, it is a skill that Japanese engineers possess in abundance.)

Product design is not as glamorous as invention, and it is unfortunate that its importance and difficulties are not recognized by the general public or often by the top managements of manufacturing companies. The design of the products described in the remainder of this chapter did not result from spectacular breakthroughs by individuals but from the dedicated work of hundreds of able and dedicated engineers working in the laboratories of manufacturing companies.

The Technical Components of a Broadcasting Station

The basic technical elements of a television broadcasting station are shown in Figure 4–11. The program sources are the network, one or more

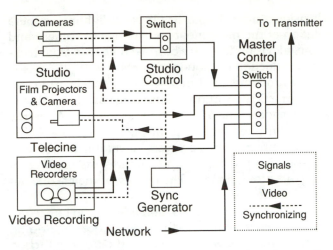

■ Figure 4–11 Television broadcast station system diagram. Note the synchronizing generator, which generates the timing pulses required to synchronize the scanning of the cameras and other program sources with the scanning of the receiver.

studios for live productions, film reproducing equipment for films and slides, and magnetic tape recorders that can be used for both recording and playback. The outputs of these sources can be combined for transmission.

The combination of program sources can be accomplished by a simple switch from one source to another or by a wide variety of *fades, lap dissolves,* or other *special effects*. The apparatus that produces these effects is often quite elaborate and is generally described by the generic term *terminal equipment*. The final program assembly is accomplished in master control, and its output is carried to the transmitter by cable or microwave.

Television Studios: Design and Production Practices

Designers of pretelevision radio studios paid great attention to their acoustic design. In contrast, television studio designers have given acoustics a rather low priority. There are several reasons for this. Radio studio designers were probably overly attentive to acoustics and had a tendency to surround their profession with a mystique that enhanced their prestige. There also had been a major improvement in microphones that had made them more tolerant of poor acoustics. But the most important reason was that television is primarily a visual medium, and viewers are not as conscious of the acoustical properties of the studio. As a result, the acoustical treatment of television studios is usually limited to lining the walls with sound-absorbent material and providing sound insulation from adjacent rooms and corridors.

The emphasis in television studio design has been in lighting, camera mobility, and the facilitation of the movement and storage of scenery and sets. In all these areas, the television industry borrowed heavily from techniques developed for the production of motion pictures.

There was, however, one fundamental difference between television and motion picture production practices. Motion pictures are usually produced with a single camera that shoots one short scene at a time. After the shooting is complete, the film is edited, in principle simply by splicing successive scenes together. In practice, this process—including the integration of the sound track—is highly complex and requires the services of many skilled and highly paid professionals. The quality of their work can have a major effect on the finished product.

Television production uses multiple cameras, and editing can be accomplished by switching from camera to camera as the action proceeds. Prior to the invention of video recording, this was the *only* method of production and editing. After the development of video recording, it became possible to perform the editing after the camera work was complete. Television practice then moved in the direction of film, although it has retained some of its multiple-camera techniques.

The design of television studios followed the evolution of production practices. Early television studios were equipped with elaborate control rooms located high above the studio floor and overlooking it through glass windows. Here the production personnel directed the action and performed visual and audio editing. With the passage of time, it became obvious that it was not necessary for these personnel to have a view of the studio, and today it is not uncommon to see scenery stacked up against the control room windows, blocking the view of the studio. In short, studio techniques have become simpler and editing more complex since the early days of television.

Studio and Field Cameras

The monochrome television era in the United States lasted from 1946 to 1964, and during this period studio and field cameras used the image orthicon tube almost exclusively. The vidicon tube was sometimes used in live cameras after its introduction in 1952, but its limited sensitivity restricted it to industrial and low-cost closed-circuit applications. For broadcasting, the image orthicon was the overwhelming choice in spite of its high cost ($1,200), short life (less than a thousand hours), and complexity.

RCA had a near monopoly of the manufacture of the image orthicon, and for many years it dominated the U.S. market for cameras. Aided by its close relationship with the RCA Tube Division and benefiting from the user's advice it received from NBC, the cameras produced by the RCA Broadcast

Equipment Division in Camden, New Jersey, were the standard of the industry. GE and Dumont produced cameras, but they were based largely on copies of the RCA design, and they enjoyed only a small market share.

The RCA TK-10 and TK-11 Cameras The first postwar television camera was the RCA TK-10. It was a fine example of industrial design, and it performed its intended functions admirably. A complete camera "chain" consisted of the camera itself, a control panel, a "master monitor" for observing both the picture and signal waveform, and a power supply. The zoom lens had not yet been perfected, and the camera contained a turret on which four lenses could be mounted. For field use, the monitor, controls, and power supply were compressed into a single case. In this configuration it was called the TK-30.

The camera had no automatic circuitry, and it required rather constant adjustment. As a result, each camera chain required two people, a camera operator to aim and focus the camera and a technician to keep the electronic circuitry in adjustment.

The TK-10 and its field counterpart, the TK-30, were replaced in 1951 by updated versions, the TK-11 (Figure 4–12) and TK-31. They contained no

Courtesy RCA Corporation.

■ **Figure 4–12** The RCA TK-11 television camera.

radical changes, but there were numerous smaller ones that greatly improved the cameras' performance and reliability. (The most visible difference was the addition of aluminum carrying handles on the camera. These were immediately copied by GE.)

The Marconi Mark IV Camera RCA's cameras were never seriously challenged by U.S. competitors, but stiff competition from foreign companies began in 1957 when the British company Marconi Communications Systems, Ltd., introduced the Mark IV image orthicon camera. It was a superb product and was the apogee of monochrome television camera design.

Its most visible feature was the use of a larger image orthicon tube—$4\frac{1}{2}$ inches versus 3 inches in previous cameras. The larger tube produced a sharper and less noisy picture, but its superiority did not end there. It incorporated numerous advanced features and was an outstanding example of product excellence. It soon became the leader in the world market outside the United States, and by the late 1950s it had begun to penetrate the U.S. market. RCA's leadership in television cameras quickly began to erode.

The RCA TK-12 and TK-60 Cameras The Marconi challenge to RCA's position in cameras could not have come at a worse time for the company. It was already reeling from the disastrous defeat by Ampex in videotape recording, and to be threatened in a product line where its leadership had been taken for granted was an especially serious blow to its prestige and its pocketbook.

With the benefit of hindsight, it might have been better for RCA to have done nothing in response. The market for monochrome cameras virtually disappeared five years later, and the cost of designing and putting a totally new product into production could have been better spent elsewhere (or returned to the stockholders). But this was not evident in 1958, and RCA decided that it had to meet the Marconi competition with its own $4\frac{1}{2}$-inch camera.

The engineering program that was begun to design the camera was a near disaster. Many of the best engineers in RCA's Broadcast Equipment Division were engaged in a desperate attempt to catch Ampex in videotape recording, and the average level of competence of the engineers assigned to the camera program was not as high as it otherwise might have been. The outcome of their efforts was the TK-12 camera, introduced in 1959. It had many good features, and with the help of RCA's reputation and strong sales organization, it sold well at first. Unfortunately, it suffered from a host of design problems that adversely affected its stability and reliability.

RCA elected to institute a massive redesign and recall program. It returned one of its top design engineers, Norman Hobson, from the tape

Courtesy RCA Corporation, with permission.

■ **Figure 4–13**   The RCA TK-60 camera.

program to manage the redesign. Eventually the problems were solved, and the rebuilt TK-12s were returned to their owners. A new camera design, the TK-60 (Figure 4–13) emerged late in 1962. It was a fine product, but very few were sold, as the market for monochrome cameras collapsed at the beginning of the color boom.

Film Reproduction Equipment

In the early days of television before videotape recording was developed, photographic film was the only available medium for storing picture information. This gave the reproduction of film over television systems a particular importance. A high percentage of nonnetwork programming and some network programming originated with film. Equally importantly, most commercials, the stations' source of revenue, were recorded on film.

Motion picture theaters used 35mm film almost exclusively, but its high cost and the stringent safety regulations that governed its use (a relic of the early motion picture days when highly flammable cellulose nitrate film stock was used) caused most stations to use 16mm. The networks used

35mm film for high-revenue shows, but the quality improvement was only marginal because of the bandwidth and scanning line limitations of the television system.

A basic problem in the reproduction of film in a television system is the incompatibility of the standard film rate of twenty-four frames per second and the television frame rate. In Europe, where the standard television rate is twenty-five frames per second (synchronous with Europe's fifty-hertz power systems), the problem was solved very easily simply by speeding up the film to twenty-five frames. In the United States, this was not possible because the difference in speeds was too great.

This incompatibility was an important factor in the choice of film re-production systems. In Europe, *flying spot scanners* are widely used. The raster from a very bright kinescope is focused on the film, the light passing through the film is focused on a phototube, and an electrical signal that is inversely proportional to the density of the film at that point is generated. Flying spot scanners produce very high quality pictures, and they have the additional advantage for color of not requiring registration of three color images. But the conversion from twenty-four to thirty frames is so difficult for scanners that *storage tube* systems are almost universally used in the United States.

Film Projectors   In the storage tube system, the image of a film frame is focused onto the photosensitive surface of the tube in the film camera. The shutter on the projector allows light to pass through the film for a period, after which the image on the tube is scanned off at the television rate. The exposure of the tube to light and the scanning need not be simultaneous, since the tube is able to store the information. It is this property that makes the conversion of frame rates possible.

One widely used film projector was the RCA TP-6. It used a "3/2 pull-down," meaning that the alternate pulldown intervals were 3/60 and 2/60 of a second. The combination of this uneven pulldown rate with the storage capability of the tube accomplished the frame rate conversion.

Film Cameras   RCA dominated the market for film cameras as well as for studio cameras during the monochrome television era. Its first film camera, the TK-20, used the same iconoscope tube employed by pre-war live cameras (but which was superseded for live pickup after the war by the image orthicon). The iconoscope was a difficult tube to operate, and the quality of the picture left much to be desired. Nevertheless, it was the best tube available until 1952, when RCA achieved another breakthrough in television pickup tubes with the introduction of the vidicon.

Vidicons had many advantages for film pickup over the iconoscope. Vidicon film cameras had a better signal-to-noise ratio, were free of shading and other defects, and were smaller, cheaper, and easier to operate. Their advantages were so overwhelming that they quickly superseded iconoscope cameras for film reproduction.

Television Transmitter Systems

The principal elements of a television transmitter system are shown in Figure 4–14. The video and audio signals are received from the transmitter and are connected to the input to the transmitter, where they modulate the visual and audio carriers. The carriers are then combined, or *multiplexed*, and carried up the tower to the antenna in a single transmission line. The effective radiated power (ERP) of a station is equal to the product of the transmitter power and the antenna gain (see illustration, bottom of next page) minus the transmission line losses.

VHF Transmitters and Antennas

Transmitters The first television transmitter to come on the market after World War II was the RCA TT-5A (Figure 4–15), which was rated at 5 kilowatts of visual power. Wartime research did not produce a satisfactory tube for VHF television, and the TT-5A used the infamous 8D21 (Figure 4–16), a tetrode in which *all* of its elements—anode, screen grid, control

---

OPERATION OF THE VIDICON TUBE

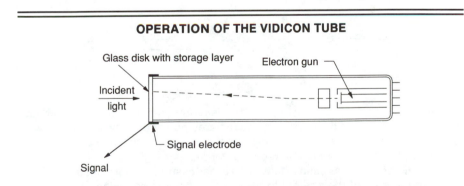

Unlike the iconoscope and the image orthicon, which are photoemissive, the vidicon is photoconductive. It has a layer of photoconductive material that varies in resistivity with the amount of light falling on it. The layer is deposited on a transparent conducting substrate, and when an image is focused on it, its resistivity varies from point to point depending on the brightness of the image. The layer is then scanned with an electron beam, and the current through the layer to the tube's output varies with the brightness.

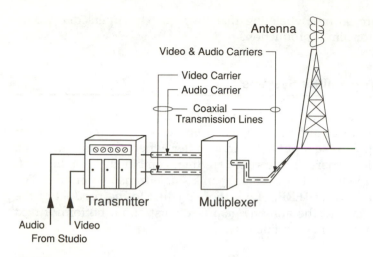

■ **Figure 4–14** A television transmitter system.

ANTENNA GAIN

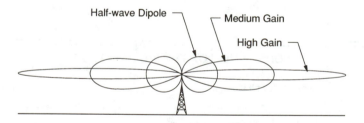

The power gain of a television antenna is defined as the ratio of its radiated power density toward the horizon to the radiated power density from a half-wave dipole. The drawing shows the radiation patterns of a half-wave dipole, a medium-gain antenna, and a high-gain antenna.

The gain of antennas used in television broadcasting is approximately proportional to the ratio of their height to the wavelength. Typical gains are 4 to 10 for low-band and 6 to 12 for high-band VHF antennas. The gains of UHF antennas are much higher, ranging up to 50.

The use of the half-wave dipole as the reference was based on the desire of engineers to be able to construct it in real life. The early work in microwave and satellite technology was done by scientists, and they preferred to use the isotropic radiator as their reference. This is an imaginary device that radiates equally in all directions.

Courtesy RCA Corporation.

■ **Figure 4–15**   The RCA TT-5A television transmitter. The 5-kilowatt TT-5A was the first television transmitter to come on the market at the end of the war.

grid, and filament terminals—were water cooled. The connections to these terminals were tiny, easily plugged up, and prone to leak. Engineers found that the leakage problem could be solved by placing sanitary napkins under the water connections, and station auditors were sometimes puzzled to find purchase orders for these products from all-male engineering departments.

Electron tube technology developed rapidly after the war, and both RCA and GE were quick to take advantage of newer tubes as they became available. At the end of 1952, RCA was offering an extensive line of VHF transmitters, all using air-cooled tubes. GE's line was not quite as extensive, but it was generally competitive.

The range of power levels in the manufacturers' product lines was dictated by the FCC's restrictions on effective radiated power. In low-band VHF, with an ERP limit of 100 kilowatts, transmitters with rated visual powers of 2 kilowatts to 25 kilowatts were offered. In high band, with a limit of 316 kilowatts, the range of transmitter powers was 2 kilowatts to 50 kilowatts.

Courtesy RCA Corporation, with permission.

■ **Figure 4–16**   The RCA 8D21 power amplifier tube. The many connections for cooling water are shown at the bottom of the tube. These are in addition to the electrical connections.

After the introduction of higher power air-cooled tubes, the changes in transmitter design were evolutionary with no major breakthroughs. Reliability and stability were improved, performance was upgraded to meet the more demanding requirements of color, solid-state circuitry was incorporated in the lower power stages, and provisions for remote control were added. The performance of modern transmitters is excellent, and they are remarkably trouble free.

Antennas   VHF television antennas must meet a variety of requirements that had not been encountered in antennas for other services. One of the more difficult, particularly for low-band channels, is providing sufficient bandwidth. The 6-MHz channel bandwidth is 10 percent of the carrier frequency on channel 2, and broadbanding an antenna to this extent is a challenging engineering problem. In addition, antennas must be mechanically rugged, be able to withstand severe winds and icing conditions, be capable of being installed by nontechnical riggers, and, of course, have the desired gain and radiation pattern.

RCA's first offering for VHF was the *superturnstile*, or batwing, antenna (Figure 4–17) in which the dipoles are large, flat structures rather than rods. Combined with suitable circuitry to couple the antenna to the transmission line, this construction provided the necessary bandwidth.

The superturnstile antenna was a very satisfactory design that was copied by GE and other manufacturers and is still in use today. For high-band, high-gain antennas, the design had the problem of complex wiring harnesses needed to feed the many layers of dipoles. To simplify the me-

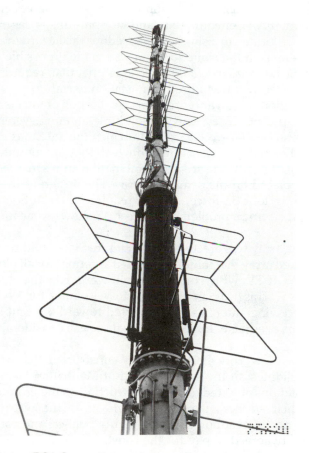

Courtesy RCA Corporation.

■ **Figure 4–17** A superturnstile antenna. Two sets of flat plate dipoles are mounted at right angles to each other around the supporting pole. When fed in phase quadrature, they produce an approximately circular pattern in the horizontal plane. The gain of the antenna is approximately equal to the number of layers. For example, an antenna with twelve layers of dipoles will have a gain of about 12.

chanical construction, RCA developed the *traveling wave* antenna in which a vertical cylindrical tube is the radiating element. Currents are induced on the surface of the cylinder through vertical slots coupled to the feed line in its center. GE also had an alternative offering, the *helical antenna*, although it was more commonly used for UHF antennas.

Multiantenna Systems Many forces motivated all the television stations in a city to locate their antennas in close proximity. Often a single mountaintop or tall building provided an ideal antenna site. In such cases, FCC rules forbid a single broadcaster to monopolize the site. The Federal Aviation Administration (FAA) encouraged, and in some instances required, all tall towers to be located in a single area to reduce the hazard to air navigation. Placing antennas in the same location also made it possible to point television receiving antennas in a fixed direction rather than requiring them to be rotated as the receiver was tuned from station to station.

In some cases, the solution was an antenna farm, a collection of towers in the same general area. In other cases, it was a multiple-antenna system. These were of two general types—*stacked arrays* in which the antennas are placed on top of each other and *candelabra arrays* in which they are mounted side by side on a single platform. There are also combination arrays in which two or more stacks are placed on the same platform. The design of these arrays was an engineering tour de force, since it required the solution of complex mechanical and electrical problems. Figure 4–18 shows some multiantenna arrays.

The first multiantenna array was erected on the Empire State Building in New York. It was a stacked array that initially included antennas for all New York's stations except WOR-TV, which built its own tower across the Hudson River in New Jersey. This was an unsatisfactory location for WOR because all the city's receiving antennas were pointed toward the Empire State Building, and it was eventually forced to move there at a considerable expense.

During the 1970s, the New York Port Authority constructed the huge World Trade Center Buildings on the southern end of Manhattan Island. New York's television stations objected to their construction on the grounds that reflections from their surfaces would cause ghosts in the pictures received in New York's northern suburbs. After lengthy political and court battles, the Port Authority agreed to pay for the construction and installation of antennas for all the New York stations on the roof of one of the buildings. RCA was the prime contractor for this project, and it designed and built two stacks of antennas, which are now a prominent feature of the city's skyline.

The first candelabra array was built for stations WFAA and KRLD in Dallas/Fort Worth. It was soon followed by a three-station array for WMAR-TV, WBAL-TV, and WJZ-TV in Baltimore. In each case, the antennas were mounted on the corner of a triangular platform, which in turn was mounted

Courtesy RCA Corporation.

■ **Figure 4–18** Multiple-antenna arrays on the Empire State Building (*top*) and in Baltimore (*bottom*).

on a tall tower. Since then, a number of multiantenna arrays have been built, including arrays on Mount Sutro in San Francisco and on the John Hancock Building in Chicago.

The design and construction of multiantenna systems is very costly, and a theologically inclined participant in the Empire State Building antenna program discovered a passage in the Holy Scriptures that was frequently quoted: "Which of you, intending to build a tower, sitteth not down first and counteth the cost, whether he have sufficient to finish it?" (Luke 14:28).

UHF Transmitters and Antennas

Transmitters Wartime research produced a tube that was ideally suited for UHF transmitters. It was the *klystron* tube, invented by the Varian brothers. It had a very high power gain and was ultimately capable of power outputs up to 60 kilowatts. Its performance was satisfactory, and klystron transmitters were simple and reliable.

GE was quick to recognize the advantages of the klystron, and its earliest UHF transmitters consisted of a 100-watt driver stage (which could also be used as a low-power transmitter) followed by a 30-kilowatt klystron amplifier.

RCA was not as wise. Its Broadcast Equipment Division allowed its engineering judgment to be overshadowed by internal political considerations. There was pressure to use nothing but RCA tubes, and RCA did not manufacture klystrons. The outcome was good politics but bad engineering. RCA's first transmitters used conventional water-cooled "gridded" tubes and were rated at 2 kilowatts and 10 kilowatts. This put its salespeople in the uncomfortable position of trying to persuade customers that 10 kilowatts from RCA were equal to 30 kilowatts from GE.

RCA's prestige in the broadcasting industry was so great that it sold a surprising number of transmitters with gridded tubes. It eventually developed models with power ratings up to 50 kilowatts, but the gridded tube was in an unequal contest with the klystron. Gridded tube transmitters were more expensive, more complex, and less reliable. The klystron was a better technical choice for transmitters in this frequency range, and RCA was forced to swallow its pride and turn to the klystron.

RCA's first klystron transmitter, the 30-kilowatt TTU-30A (Figure 4–19), was introduced in late 1964. Fortunately for the company, this coincided with an upsurge in the number of new stations, and it was able to recapture a respectable market share (although many customers of both GE and RCA later went bankrupt because of UHF's difficulties).

The power ratings of UHF transmitters offered on the market were established by the maximum power permitted by the FCC. The initial limit

OPERATION OF THE KLYSTRON TUBE

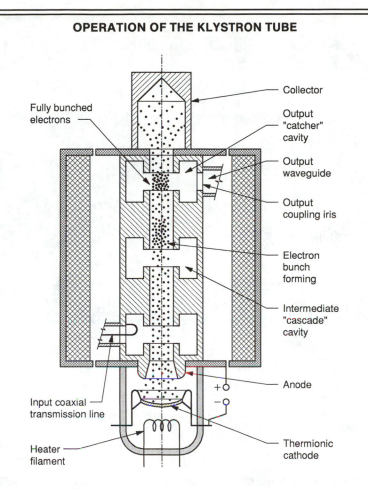

The input signal from the driver stage induces a radio frequency electromagnetic field in the first cavity. The alternating field alternately slows and accelerates the electrons in the beam and causes them to collect in bunches. The bunches induce a much stronger field in the second and third cavities. The energy extracted from the field in the third cavity by a coupling device is the transmitter output.

was 1,000 kilowatts ERP. This could be achieved with a 30-kilowatt transmitter and an antenna with a gain of 40. When the limit was raised to 5,000 kilowatts, it could be achieved by affluent stations that were able to afford two 60-kilowatt transmitters operating in parallel and feeding an antenna with a gain of 50.

Courtesy RCA Corporation.

■ **Figure 4–19** The RCA TTU-30A UHF television transmitter. The 30-kilowatt TTU-30A resulted from RCA's belated entry into the manufacture of transmitters using klystron tubes.

Antennas RCA's initial UHF antenna offerings were more felicitous than its transmitters. Its highly skilled antenna engineering department developed the *pylon* antenna, which for many years was the industry standard. Like the VHF traveling wave antenna, it consists of a slotted cylinder, which is the radiating element. Currents are induced on the surface of the cylinder by coupling the slots to a conductor running up its center. It differs from the traveling wave antenna in that the current on this conductor forms a *standing wave*.

As transmitter powers increased, RCA offered the V-Z antenna with greater power-handling capability as an alternative to the pylon. This antenna had a vertical five-sided supporting structure with zigzag conductors spaced a few inches from each face as the radiating elements.

GE's approach to UHF antennas was entirely different. It offered the *helical* antenna, which consisted of a cylindrical supporting structure surrounded by conductors wrapped around it in right- and left-handed helixes.

The horizontal components of the radiation from these helixes add, but the vertical components cancel each other, thus producing a horizontally polarized wave.

UHF antennas usually have very high gains, typically 40 to 50, and this results in a thin, pancake-shaped beam. The horizon, as seen from the top of a very tall tower, is a fraction of a degree below the horizontal, and if the beam were exactly horizontal, most of the energy would miss the earth and go out into space. It also would miss the areas between the tower and the horizon. To solve these problems, two techniques, *beam tilt* and *null fill* (Figure 4–20), are used.

■ MONOCHROME RECEIVERS

Television receivers are now commonplace in nearly every home, and they appear to be very simple devices compared with the intricate equipment found at television stations. Their simplicity is deceptive, however, because it results from years of costly and intensive engineering work. In the early years of television, one RCA engineer calculated that the amount of engineering per tube devoted to the design and production of receivers was ten times as great as that for television cameras.

This engineering effort has been accomplished by hundreds of engineers and scientists working diligently and anonymously in industrial laboratories and plants throughout the world. The objectives have been high performance, ease of operation, reliability, stability, ease of maintenance, and, above all, low cost.

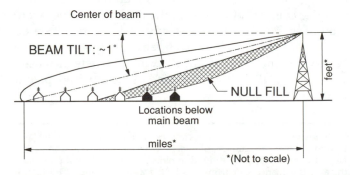

■ **Figure 4-20** Beam tilt and null fill. For high-gain antennas mounted on high towers, it is necessary to tilt the beam downward slightly so that it is aimed at the horizon for distant coverage. For close-in coverage, it is necessary to fill in the nulls in the pattern below the horizontal. Only a few degrees or fractions of a degree are involved in these adjustments, and the antenna measuring apparatus must be very precisely calibrated.

The television industry, spurred by vigorous competition, has succeeded brilliantly in achieving these objectives. The first monochrome receivers with ten-inch picture tubes sold for about $385. In 1988, a portable set with a similar size picture tube can be purchased for $100 or less. In the meantime, the Consumer Price Index has risen at least five times, so in constant dollars the customer is getting at least nineteen times as much for his or her money. This amounts to a compounded productivity growth rate of nearly 8 percent per year for forty years, a record unmatched by few if any industries. The Japanese have been responsible for some of this, but U.S. industry and its wholesale and retail distribution systems have made large contributions.

RCA's Early Leadership

RCA was the world leader in pioneering the development of monochrome receivers and their key component, the picture tube. Development work was under way prior to World War II, and RCA announced plans to offer receivers for sale to the public twice, once in 1938 and again in 1940. In both cases, its plans were stymied by standards disputes and the FCC.

At the end of the war, RCA's engineering and production facilities were put into high gear, and in 1946 it introduced the 630TS, which is sometimes described as the Model T of the television industry. (The 6 was the model number, 30 was the number of tubes, and TS meant television and sound.) The receivers were manufactured in RCA's Camden plant using picture tubes produced in its Lancaster plant.

The 630TS clearly put RCA in the lead among television receiver manufacturers, and RCA surprised the industry in 1947 when, motivated by what it perceived as enlightened self-interest, it made its manufacturing drawings available for free to its competitors. Most of the other companies were cynical about RCA's motives, but RCA was probably sincere when it stated that the industry was bigger than any single company and that helping its growth would help everyone, including itself. RCA immediately increased the market for its picture tubes, NBC's audience, and its income from patent licensing and technical aid.[13]

---

[13] RCA also hoped that this action would defuse the charges of violations of antitrust laws that continued to swirl around it. It was not successful in achieving this objective. Competing manufacturers and the Department of Justice continued to object to RCA's patent pool practices, and Philco, Zenith, and the Justice Department all filed antitrust suits. The Justice Department's civil suit contained strong implications of criminal charges against Sarnoff to follow, and in 1958 he decided to settle. RCA pleaded nolo contendere, paid a fine of $100,000, and abandoned the pool principle of patent licensing, agreeing to license individual patents instead. Its monochrome patents were put in the public domain, and U.S. manufacturers were allowed to use the color patents without payment of royalties. It was widely believed that Sarnoff was saved from personal criminal charges by the intervention of President Eisenhower. (See Bilby, *The General*.)

Automated Versus Old-Fashioned Hand Wiring

Initial technical advances in receivers were directed toward cost reduction and increasing the size of picture tubes, first to twelve, fifteen, and sixteen inches and later to nineteen and twenty-one inches. Solid-state circuitry and automated assembly were introduced cautiously. Zenith deliberately moved slowly in these areas as a marketing strategy. While other manufacturers were using automated assembly of circuit boards, Zenith vigorously advertised "old-fashioned hand wiring," including photographs of the maze of wires in its sets, which it touted as a virtue. This marketing program had no rational basis, but it succeeded brilliantly—aided by the fact that Zenith sets performed well and had an outstanding reputation for quality and reliability. Eventually the cost of hand wiring became excessive, the cost pressures from the Japanese increased, and Zenith was forced to adopt more modern designs and manufacturing methods. Today, all monochrome sets are based on highly automated assembly and all-solid-state design except for the picture tube.

Zenith Achieves First Place

RCAs's market share dwindled as its competitors proliferated. By 1953, Zenith and Philco had emerged as the strongest competitors, although GE, Sylvania, Dumont, and even a flamboyant new manufacturer, "Mad Man Muntz," achieved significant market shares. Philco's position eroded rapidly during the mid-1950s, and for a number of years Zenith was in first place, with RCA a very close second and other manufacturers far behind.

The Japanese Invasion Begins

The first indications of the Japanese invasion of the U.S. television market occurred in the early 1960s. Japanese manufacturers had already made inroads in the audio and radio markets, and monochrome television was their next target. Their market share increased rapidly, and by the mid-1970s most monochrome sets sold in the United States were imported from Japan or other offshore sources.

The Growth and Decline of the Monochrome Receiver Market

The number of monochrome sets sold in the United States rose rapidly and reached a peak in the mid-1950s as saturation of the market approached. Somewhat surprisingly, sales rose again in the mid-1960s coincidentally with the growth of the market for color receivers. Most of the sales

Deluxe television receiver

Courtesy RCA Corporation.

■ **Figure 4–21** (*Above*) A 1939 RCA monochrome receiver. (*Facing page*) The postwar Dumont monochrome receiver.

Courtesy RCA Corporation.

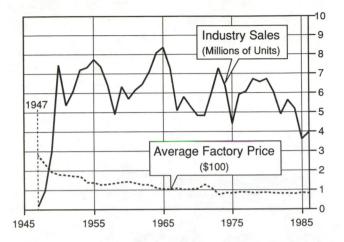

■ **Figure 4–22** Industry sales and average unit prices of monochrome receivers. The sales and unit price figures are at the factory sales level for receivers manufactured domestically and at the landed cost for imported receivers.

in this era were for small portables or table models purchased as second sets. The trend to smaller sets continued, and by 1970 the production of large console models had virtually ceased.

With dwindling unit volume, the disappearance of higher priced consoles from the market, and the takeover by the Japanese, the size of the market for monochrome receivers has declined steadily. Figure 4–22 shows the decline of industry sales at the U.S. manufacturers' level (or landed cost for imported products). But this graph tells only part of the story. With the rise of mass-marketing organizations and a reduction in the number of levels of distribution, the markup from manufacturer's price to retail selling price has declined drastically. In the early 1950s, this ratio was typically 2.5 to 1. Currently, it is typically 1.5 to 1. The declines in unit prices at the retail level have, therefore, been even greater.

■ THE DEMISE OF MONOCHROME TELEVISION

The diminishing market for small table model and portable receivers is all that is left of the once-strong monochrome television broadcast industry. Television broadcasting is now virtually 100 percent in color, and the only remaining uses of monochrome television are for industrial and a few closed-circuit applications. Monochrome technology played an indispensable role in the development of the television industry, however, and it provided a sound technical basis for the color television systems now used throughout the world.

5

■ ■ COLOR TELEVISION

On September 26, 1949, a veritable Who's Who of the television manufacturing and broadcasting industries gathered in the imposing auditorium of the Interstate Commerce Building in Washington for an FCC rule-making proceeding, a hearing convened to choose transmission standards for color television broadcasting. It pitted the powerful CBS network against its arch rival, RCA, and most of the rest of the television industry. Like a Wagner opera, it alternated between moments of high drama and hours of boring testimony. It dragged on for more than a year, and it culminated in one of the FCC's worst decisions. The CBS field sequential system was chosen, notwithstanding the fact that field sequential color broadcasts could not be received by any of the more than ten million monochrome sets then in use.

While the hearing was in progress, an equally intense effort was under way in the nation's television laboratories. This was devoted to the development of a compatible color system—that is, one in which color broadcasts could be received (in monochrome, of course) on black-and-white sets. An uneasy alliance of RCA and its major competitors, spurred by the competition from the CBS system, was successful in this effort, and the result was the National Television System Committee (NTSC) color system, a refined version of the system originally proposed by RCA.

To almost no one's surprise, field sequential color was a commercial failure, and on December 17, 1953, the FCC reversed itself and approved the NTSC system. Today color broadcasting using the NTSC system or its European variant, Phase Alternating Lines (PAL), has almost completely replaced monochrome. Although a somewhat better system could now be developed owing to thirty years of advances in basic technology, NTSC and PAL have withstood the test of time and provide highly satisfactory vehicles for the transmission of color. This chapter is the history of color television's evolution from a laboratory concept to a multibillion-dollar business.

■ COLOR TELEVISION IS CONCEIVED

The potential attractiveness of color television, as well as its technical problems, were recognized in television's earliest years. As early as 1928, Baird demonstrated a crude color system using a Nipkow disk with three sets of holes, one for each primary color (see illustration on next page). The techni-

THE PRIMARY COLORS

Colorimetric analysis is usually based on the use of primary colors. Most, although not all, of the colors in nature can be produced by mixtures of three properly chosen primaries.

Subtractive primaries are the most familiar, since human experience prior to development of color television was almost totally limited to subtractive systems. These primaries are often described as red, yellow, and blue. A more accurate description is magenta, yellow, and cyan:

Primary	Reflects or Transmits	Absorbs
Magenta	Red and blue	Green
Yellow	Red and green	Blue
Cyan	Blue and green	Red

For example, red can be produced by mixing magenta and yellow, the former absorbing green and the latter absorbing blue from the white illuminant.

The additive primaries used in color television are red, green, and blue. A mixture of red and blue lights produces magenta. Red and green produce yellow (it is not intuitively obvious, but the hue produced by red and green spotlights shining simultaneously on a white wall is yellow), and blue and green produce cyan. Pure (monochromatic) or nearly pure spectral colors are the only hues that cannot be reproduced by a proper combination of these primaries.

cal difficulties were enormous, however, and twenty-five years elapsed before a practical color system evolved from monochrome technology.

The evolution required the solution of two difficult problems: (1) colorimetry, or the production of colors that are a reasonable facsimile of those in the original scene, and (2) the transmission of both color and brightness information in the same channel.

Colorimetry

Television colorimetry is basically different from that used in photography and painting or in the perception of color in real life. All of the latter are *subtractive* systems in which the picture or scene is illuminated by an external source, such as sunlight, which includes components of all colors. The desired hue is produced by subtracting the unwanted color components by absorption. Color television produces colors directly on the picture tube. It is an *additive* system in which hues are determined by adding the primary color components in the proper proportions.

Color Transmission

Color television's other problem, the simultaneous transmission of monochrome and color information in a channel of limited bandwidth, is fundamental. In the simplest case, a color television picture would contain three times the information of monochrome—that is, a complete frame for each of the three primary colors. This could be achieved by increasing the bandwidth, thus increasing the demand for already scarce spectrum space; by reducing the number of pictures transmitted per second, thus increasing the flicker; or by reducing the duration and number of scanning lines, thus reducing the sharpness of the picture.

The dilemma facing the television technical community was the selection of the optimum trade-off among these three conflicting evils. Fortunately, it was resolved with very little compromise in picture quality and no increase in bandwidth by cleverly taking advantage of certain limitations of the human eye and the high degree of duplicate information in the monochrome television signal waveform.

■ THE FIELD SEQUENTIAL SYSTEM

CBS first proposed the field sequential color system to the FCC in 1940. It had been developed at CBS Laboratories under the direction of its head, Peter Goldmark. He was the system's leading protagonist, and he attempted to sell it to the FCC and the industry with missionary zeal.

The Technology of the Field Sequential System

In the field sequential system (see illustration on next page), red, green, and blue television fields are displayed in sequence, and the retentivity of the eye merges them into a single color picture. This system brought the bandwidth–picture quality dilemma into sharp focus. If flicker and picture sharpness were to be maintained at the level of monochrome television, a field sequential signal would require three times the bandwidth of monochrome. CBS, however, proposed compromises that reduced the additional bandwidth requirement at the expense of flicker and picture sharpness.

In 1940, CBS tentatively settled on a trade-off that divided the cost rather evenly among bandwidth, flicker, and picture sharpness. As compared with monochrome standards, the bandwidth was increased from 4 MHz to 5 MHz, the number of frames (pictures) per second was reduced from 30 to 20, and the number of scanning lines per frame was reduced from 525 to 343.

The use of disks caused RCA to describe the CBS system pejoratively as

THE FIELD SEQUENTIAL COLOR SYSTEM. (*TOP*) OPERATION OF FIELD SEQUENTIAL COLOR SYSTEM. (*BOTTOM*) COLOR TRANSMISSION SEQUENCE IN FIELD SEQUENTIAL SYSTEMS.

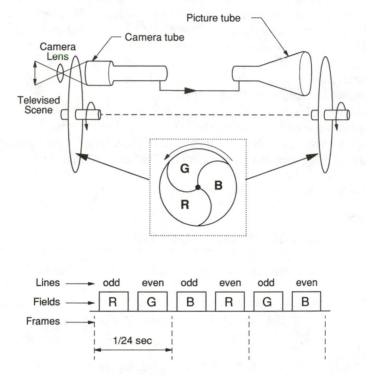

The basic operation of the field sequential color system is shown in this drawing. Light from the scene passes through a rotating disk on which red, green, and blue color filters are mounted. Thus the camera tube is exposed in sequence to the red, green, and blue color components of the scene. A disk at the receiver, similarly equipped with color filters, rotates in synchronism so that light from the kinescope passes through the red filter, for example, while the camera tube is being exposed to red light from the scene.

This drawing shows the color sequence in successive fields of the color signal as proposed by CBS in the 1949 hearing. Note that only two colors are included in each frame; for example, frame 1 has red odd lines and green even lines. Six fields, or one-eighth second, was required to scan all lines in all three colors. This caused fast-moving white objects to exhibit color breakup—that is, to appear as a series of colored images.

"mechanical," a bad word in an electronic age. This criticism was not totally fair. The CBS system was not inherently mechanical. Mechanical components were used because a key component of an all-electronic system, the tricolor picture tube, had not yet been developed. In contrast, RCA imposed an all-electronic requirement on itself for its system, and this gave CBS a head start.

CBS Petitions the FCC

In spite of the manifest limitations of the field sequential system, in 1940 CBS urged that the FCC establish it as a standard for color along with its approval of a monochrome standard. As noted in Chapter 4, the Commission refused to approve it for commercial broadcasting but did authorize continued field-testing under an experimental license.

During World War II, Peter Goldmark and his engineers continued to work on the field sequential system, and at the end of the war, CBS resumed its battle at the FCC. The essence of CBS's postwar strategy was speed. Its top management recognized that the pressure on the FCC for the approval of monochrome standards was so great as to be irresistible. But field sequential color signals were incompatible with monochrome—that is, they could not be received on monochrome sets. If monochrome standards were approved before color and a great many monochrome receivers were sold to the public, their incompatibility with field sequential color would form a major barrier to its acceptance. In 1946, therefore, CBS petitioned the FCC to approve monochrome and field sequential color standards simultaneously, with monochrome in VHF and color in UHF. With simultaneous introduction to the public, CBS believed its color system would ultimately win over monochrome in spite of its higher cost, its technical limitations, and the unknown problems of UHF.

The standards proposed by CBS in 1946 differed somewhat from its prewar system. The number of frames per second was increased from 20 to 24, and the number of scanning lines was increased from 343 to 525. These changes resulted in smaller compromises in flicker and picture sharpness but required more bandwidth—double that of monochrome. Since the amount of spectrum space available at UHF was much greater than at VHF, the added bandwidth requirement was not considered to be a serious problem.

The CBS petition was sufficiently persuasive to cause the FCC to convene a hearing on December 9, 1946, to consider it. Its high points were a demonstration by CBS of its system and a rebuttal demonstration by RCA of the "simultaneous" system it was developing.

The CBS demonstration went exceedingly well. It was held in Nyack,

New York, forty miles north of the Chrysler Building where the CBS transmitter was located. FCC chairman Charles Denny was reported by Goldmark to have been "mesmerized," not only by the quality of the pictures but also by the beauty of the model, Patty Painter. She and Denny exchanged a few words, whereupon "Denny's face again lit up. He said something gracious in reply on how wonderful she looked. The rest of the show went on and was soon over. Everyone looked pleased as they filed out of the suite. I thought we were in. So did Stanton [the president of CBS]."[1]

The RCA demonstration was a disaster. George Brown, destined to play a leading role in the development of the RCA/NTSC color system, reported:

> The receiver used three picture tubes and a lens to project the picture onto a frosted glass screen. The pictures were fuzzy with color fringes due to poor registration. Fortunately the pictures were very dim so it was difficult to see the color fringes or the lack of detail in the picture. . . .
>
> It makes me shudder even now when I contemplate that the FCC in its lack of wisdom might have adopted the RCA proposal.[2]

Given the success of the CBS demonstration and relying on the enthusiasm of Stanton and Goldmark, Paley made the costly decision described in Chapter 4 to surrender its VHF licenses in four major cities, retaining only New York. It was felt that this was necessary to prove the sincerity of CBS's advocacy of color and UHF. At the same time, he urged the CBS affiliates to do likewise.

The appraisal of Dennys' reaction to the CBS demonstration by Stanton and Goldmark is a textbook example of the wish being the mother of the thought. While Denny was friendly and gracious, his only words of compliment were to the model. He was far too shrewd to disclose his thoughts about the CBS system at that time. Both Stanton and Goldmark, however, mistook his friendliness and courtesy for assent.

The FCC Acts—Unfavorably

In view of the realities of the situation, the FCC's decision to reject the RCA system came as no surprise. In fact, RCA did not expect its approval. An evaluation of the CBS system was not as clear-cut, but the Commission decided that there were still too many uncertainties, both in the color system and in UHF, to justify its approval. On January 30, it issued an Order

---

[1] Peter C. Goldmark, *Maverick Inventor: My Turbulent Years at CBS* (New York: Saturday Review Press, 1973).

[2] George H. Brown, *and part of which I was* (Princeton, N.J.: Cupar, 1982).

denying the CBS petition to establish commercial color standards but permitting continued experimental broadcasting.

Goldmark was so certain of his position that he was convinced that Denny must have been influenced by ulterior motives. In his own mind, his suspicions were confirmed when Denny accepted a position as general counsel of NBC during the next year. In fact, there is no credible evidence that Denny's decision (or that of the other commissioners) was anything more than an objective judgment.

CBS Tries Again

CBS, egged on by Goldmark, would not give up. In 1949, it proposed a third version of the sequential system. It retained the 24-picture-per-second rate of its 1946 proposal, but it reduced the number of scanning lines to 405 and the bandwidth of the broadcast channel to 6 MHz, the same as that of monochrome. As compared with monochrome, it involved a small increase in flicker and a fairly major reduction in potential picture sharpness.

It was a skillful compromise. Given the intense competition for spectrum space, it is unlikely that the FCC would ever have approved a color system that required more bandwidth than monochrome. The 24-picture-per-second rate was lower than monochrome's 30-per-second rate, but it was the motion picture standard and very close to the European television standard of 25 pictures per second. As for sharpness, the comparatively crude television cameras used at the time did not generate much signal information at the upper end of the video channel. Consequently, the reduction in picture sharpness produced by the apparatus then available was not great. Looking to the future, however, it put a ceiling on the improvement in sharpness that could result from equipment development.

CBS also played its political and public relations cards skillfully. Goldmark was a masterful showman, and he made impressive demonstrations emphasizing the brilliant colors the system could produce but covering up its deficiencies. At the same time, CBS portrayed itself to the public, Congress, and the FCC as the good guy, fighting to bring the benefits of color to the public but being opposed by the greedy members of the television industry led by RCA.

The FCC Reopens the Issue

The lobbying was effective, and the FCC had a new chairman, Wayne Coy, who was sympathetic to CBS. On July 11, 1949, the FCC voted to reopen the color question by enlarging the issues in the television channel allocation hearing (the freeze) to include the establishment of color standards. It ordered all who wished to propose a color system to submit engineering descriptions by August 25. The hearing on the color issue was scheduled to begin on September 26, only a month later.

CBS's motives in pursuing field sequential color so vigorously and at such great cost are not entirely clear. The stated reason was that color, even in its field sequential embodiment, was a superior broadcasting medium, and it adoption would benefit the public. A more likely explanation is simply that CBS hoped to profit from the adoption of its system by license income and receiver manufacturing. A cynical explanation is that RCA and NBC had a commanding lead in monochrome technology, and CBS wished to gain time to catch up by causing a bureaucratic delay in the approval of monochrome standards and channel assignments. An even more cynical explanation is that it was a display of corporate pride and executive ego— CBS versus RCA and Paley versus Sarnoff. These explanations are not mutually exclusive, and there is probably some truth in all of them. One fact is clear: The enthusiasm and salesmanship of Peter Goldmark was a key factor in the decisions of CBS's top management. And history shows that, once established, a corporate strategy tends to achieve its own momentum, which is often difficult to reverse.

■ THE DOT SEQUENTIAL SYSTEM

The FCC decision to address the issue of color standards during the freeze hearing put RCA in a very difficult position. It simply was not ready to propose a system that was both practical and compatible. Most of its initial experimental effort had been directed toward a *simultaneous* system in which three color signals were simultaneously and continuously generated and transmitted. It required a three-tube camera, three full-bandwidth transmission channels, and a three-kinescope receiver. It was not a practical system.

The need for three times the monochrome bandwidth was an impossible barrier. Three-tube cameras were practical (most color cameras used today have three tubes), but the three-kinescope receivers were impossibly large and expensive for consumer use. In addition to these problems, an apparatus for simultaneous color had not been developed to the point where it could produce satisfactory picture quality, a weakness that had been clearly displayed in the 1946 demonstrations to the FCC.

Nevertheless, research in simultaneous color systems resulted in some technical progress. The most important was probably the development of a concept by Alda Bedford called "mixed highs." This was based on the discovery that the human eye is not sensitive to color in fine detail, the portion of a picture that requires the transmission of higher frequency components. Bedford proposed that these components be separated from the three color signals, mixed, and then added to the *green* signal. The bandwidth of the *red* and *blue* signals could then be reduced substantially, thus reducing the total spectrum space required for transmission. Although the use of mixed highs did not, in itself, make simultaneous color practical, it was an important principle, and it played a key role in the final development of standards for broadcast color television.

The work on simultaneous color also emphasized the necessity for developing a practical tricolor tube. Research on this component was successfully completed under forced draft a few years later.

It was apparent, however, that the simultaneous system, even with improvements, would never be approved by the FCC. It was necessary to develop a compatible system—that is, one in which the color signal could be transmitted in a standard monochrome channel and displayed on a monochrome receiver. In addition, color receivers would have to produce monochrome pictures from monochrome transmissions (reverse compatibility).

Work on the compatible color system that would evolve into the NTSC system began at RCA Laboratories in the spring of 1949. George Brown gives credit for the basic idea to Clarence Hansell, then director of research for RCA Communications. It involved rapid sequential sampling of the three color signals and combining the sampled waveforms into a single composite signal. The brief samples appeared as dots on the screen, hence it came to be known as the *dot sequential* system (see illustration on next page).

The requirement in the FCC's July 11 notice that organizations wishing to propose a color television system must file an engineering description by August 25 underscored the difficulty of RCA's position. This was only seven weeks later, and its engineering department was not ready. Its law department filed a response, but the commission returned it with the notice that it contained "no supporting engineering statement . . . as required."[3]

With their backs to the wall, RCA's senior technical executives, Elmer Engstrom and George Brown, worked frantically with their staffs during the next few days to compose a technical response, which they filed on September 6. It was only a paper study, but in a few weeks they had developed the essence of the system that was eventually adopted.

■ THE COLOR HEARING

The Commission

The FCC hearing began, as scheduled, on September 26, 1949, in the Interstate Commerce Commission auditorium. As is often the case in our nation's capital, the grandeur of the surroundings contrasted with the

---

[3] Letter: FCC to RCA, August 29, 1949.

THE DOT SEQUENTIAL COLOR SYSTEM

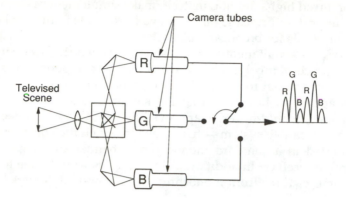

This drawing shows the principle of the dot sequential system. Red, green, and blue color signals are produced continuously and simultaneously. These signals are then sampled in sequence at a rapid rate, nominally 3.6 MHz. The output of the sampling process is a series of pulses, each having an amplitude proportional to the amplitude of the corresponding color signal at that point in the picture. This signal produces a series of tiny (approximately 0.03 inch wide) colored dots on a tricolor kinescope. These are perceived by the eye as a single color with a hue determined by the relative amplitude of the red, green, and blue pulses at that point.

mediocrity of those in positions of authority. The average competence of the commissioners who convened to decide this matter was probably not the lowest in the history of that body, but it was far from being the highest. George Brown, who became RCA's principal technical witness, described its members this way:

> Wayne Coy, the Chairman, had come from the *Washington Post* newspaper organization. He was reasonably intelligent, very opinionated, heavily biased in favor of CBS, antagonistic to industry, and short tempered. At times, a sense of humor showed through his shell of importance.
>
> Paul Walker was oldest by birthdays and in service on the Commission. He was recognized as an expert on regulatory matters concerning telephone and telegraph companies and had shown little interest in television matters. He was remarkable for his common sense.
>
> Rosel Hyde was a quiet gentleman who seemed always well informed and capable. He had come through the ranks of the FCC staff to his position as commissioner and stayed on as political parties changed by the simple expedient of declaring himself neither a Republican nor a Democrat but simply as an independent.
>
> George Sterling had been the Chief Engineer of the FCC before his appoint-

ment as a commissioner and was well known in amateur radio circles as the author of a handbook for radio amateurs.

E.M. Webster had a naval career behind him. He was a sound unflappable person who never had much to say.

Robert Jones, a lawyer, was loud and bumptious and less than bright. He was prone to asking stupid questions and not listening to answers that were not to his liking.

Frieda Hennock, another lawyer, was quite conscious that she was in the vanguard of the women's liberation movement, given to screaming at witnesses, not listening very well and completely at a loss as to how to deal with technical matters.

The Chief Counsel of the Commission was Harry Plotkin, an affable person with his feet on the ground whose intelligence and good sense exceeded that of the commissioners, individually or collectively. He never showed bias toward or against any witness or position.[4]

On the whole, these are accurate descriptions, although they came from a less than objective source. Brown could have added that Commissioners Jones and Hennock, as well as Coy and the chief of the FCC Laboratory Division, E.W. Chapin, were all biased in favor of CBS. RCA was playing against a loaded deck.

The Contestants

Three companies—CBS, RCA, and a small entrepreneurial company from San Francisco, Color Television Incorporated (CTI)—filed descriptions of proposed color television systems prior to the commission's August 25, 1949, deadline. CBS and RCA, of course, proposed the field sequential and dot sequential systems. CTI proposed a compatible system in which the red, green, and blue color components were transmitted in sequence on successive lines. Hence it became known as the *line sequential* system.

The CTI system had fundamental weaknesses that were so detrimental that it never received serious consideration. The commission was required to give it its day in court, but as a practical matter, the only contenders were CBS and RCA.[5] These two entered the hearing with different objectives and complementary strengths and weaknesses.

---

[4] Brown, *and part of which I was.*

[5] My consulting firm, McIntosh & Inglis, was retained by CTI for the limited purpose of assisting in the preparation of its technical testimony. This gave me a ringside seat at the hearing. As the hearing progressed, it became apparent to both McIntosh and me that the CTI system was fundamentally flawed. Although a technical evaluation of the system was beyond the scope of our assignment, McIntosh decided to advise the financial supports of CTI of our views. They had no technical background and were relying on the self-serving advice of the system's promoters and CTI's legal and patent counsel. The latter, all of whom were collecting very substantial salaries or fees, were outraged by McIntosh's action, and our retainer was canceled immediately. CTI pressed its case for a time, but it eventually faded from the contest.

The compatibility problem made expeditious approval of the field sequential system even more critical to CBS than in 1946. There were four million monochrome receivers in use at the end of 1949, and this increased to eleven million at the end of 1950. It was CBS's last chance, and its management urged the commission not only to make a favorable decision but also to make it quickly.

In addition to the advantages of a sympathetic FCC and its role as an underdog, CBS entered the hearing with the enormous advantage of a system that was the culmination of more than ten years of engineering development. Although its potential was limited and it suffered the near-fatal disadvantage of incompatibility, the field sequential system had been perfected to the point where it could be demonstrated very effectively under controlled conditions, particularly under the direction of Peter Goldmark.

RCA was well aware of the advantages of the CBS system. It had built a number of field sequential cameras under contract to CBS, including some of those used in the CBS demonstrations. A special closed-circuit model was built under license from CBS and installed in a satellite built by RCA for weather observation.

RCA had been caught napping by the sudden introduction of color standards as an issue during the freeze, and its system was not completely developed, even on paper. Realistically, RCA could not expect FCC approval until the performance and compatibility of its system was demonstrated. RCA's objective, therefore, was to buy time by blocking approval of the CBS system. To do so, it had to persuade the Commission that its system had sufficient potential to be worth waiting for.

Offsetting CBS's advantage of a well-developed system was the limitation of its technical resources. While this is not inevitably fatal (as Ampex was to demonstrate later with its invention of the first practical videotape recorder), it was a serious handicap. Its effect was exacerbated by the almost complete lack of industry support. Motivated by a sincere belief that the CBS system was impractical, nearly all manufacturers opposed it. This enhanced CBS's stature as an underdog and may actually have helped its case with the FCC, but it deprived CBS of the badly needed technical assistance of the remainder of the industry.

RCA's strengths and weaknesses were the reverse of CBS's. At the beginning of the hearing, its system barely existed on paper. A key component in the system, the tricolor tube, had not even been invented. Many months were to elapse before RCA could put together a satisfactory demonstration. In addition, RCA confirmed the Commission's perception of it as a powerful and arrogant monopoly by its deportment at the hearing.

The RCA system did have the crucial advantage of compatibility. Sarnoff's dedication to the success of the RCA system also was a plus, as it meant that the company's vast television technical resources were dedicated to the color system's development. Brilliantly led by Elmer Engstrom and George Brown, RCA's engineers and scientists performed miracles during the succeeding months.

Adversity makes strange bedfellows, and RCA's competitors, who hated RCA and Sarnoff with a passion, found themselves working on improvements and enhancements to the RCA system. Some of these were incorporated in the NTSC system that was ultimately approved by the FCC.

The Contestants' Leadership

As RCA and CBS mounted all-out corporate efforts at the hearing, they operated under command structures not unlike those of an army at war. At the top were the commanders-in-chief—David Sarnoff for RCA and William S. Paley for CBS. Next were the theater commanders—RCA's executive vice president, Elmer Engstrom, and CBS's president, Frank Stanton. Finally, there were the field commanders—George H. Brown for RCA and Peter Goldmark for CBS—the executives who led their troops to battle on the front lines.

Most of the RCA and CBS executives were old rivals, and they had been in competition for many years. Never, however, had they confronted each other as directly and publicly as in the color hearing.

Sarnoff and Paley Sarnoff's early years, his rise in the Marconi organization, his function in the founding and growth of RCA, his faith in new technologies, and his pioneering role in the development of radio and television broadcasting have been described in previous chapters. The development of color television was the crown jewel of his career. He had unbounded faith in the ability of RCA engineers and scientists to deliver new inventions on demand. During the color hearing, he made confident predictions of important future breakthroughs. When asked how he could be so sure his engineers would achieve them, he responded, not facetiously, "Because I told them to."[6]

He had equal faith in the ultimate commercial success of color television, even during the ten dark years that followed the FCC approval of the NTSC system. He set his sights on a long-range goal that he pursued doggedly and vigorously in spite of its adverse effect on short-range profits. (It is doubtful that he could have followed this course in the 1980s era of corporate takeovers).

His belief in color was rewarded in 1964 when color television finally took off in the marketplace with astonishing speed. Once more he had confounded the doubters, and RCA began an era of unprecedented prosperity

---

[6] FCC hearing transcript, May 3 and 4, 1950.

(although it was unfortunately dimmed by the company's mounting losses from its ill-fated effort to enter the computer business). Color television was his last great success, and he enjoyed it fully.

Paley's extraordinary skills as a broadcaster, as well as his weaknesses in technical matters, were described in Chapter 2. The consequences of these weaknesses were soon to come to the forefront.

Elmer Engstrom and George Brown In the color wars, Elmer Engstrom (Figure 5–1) was RCA's Omar Bradley. Like Bradley, he was a solid professional and a strong leader who was highly respected by his peers and subordinates. Also like Bradley, his personality was sober and dignified. He was RCA's senior technical executive at the time of the color hearing, and the performance of RCA's technical community under his leadership in developing a totally new color television system in less than a year was truly extraordinary.

After graduating from the University of Minnesota in 1923 with a degree

Courtesy RCA Corporation.

■ **Figure 5–1** Elmer Engstrom.

in electrical engineering, Engstrom joined the radio department of GE. He transferred with the department to RCA in 1930, and he remained there for the remainder of his career. His electrical engineering degree was the extent of his formal education, although he was often addressed as "Dr. Engstrom" in his later years.

Engstrom rose steadily in RCA's engineering hierarchy, and during the 1930s he became active in television research, bringing to it the then unique concept of systems engineering. In 1942, he was made head of RCA Laboratories, and it was in this role that he made the contributions to color television technology described in this chapter. In 1955, he became a senior executive vice president with overall staff responsibility for all of RCA's scientific and engineering programs. He enjoyed the full confidence of David Sarnoff, which, coupled with his technical and administrative talents, made him extremely effective in this role.

In 1961, he succeeded John Burns as RCA's president. Burns's mismanagement had made a shambles of much of the company, and Engstrom's straightforward and logical style was effective in restoring order. He also was lucky. The breakthrough in the color marketplace occurred in 1964, and this led to several years of great prosperity. Engstrom did not receive full credit from the outside world for his performance as president. Robert Sarnoff was hovering in the wings waiting to succeed his father as chief executive officer (CEO) and was expected to bring in his own top management team. As a result, Engstrom was often described as an interim president. This description does not do justice to his solid accomplishments.

Although Engstrom's personality was generally serious, he was not without a sense of humor. He was an aggressive teetotaler, and for many years the consumption of alcohol was sternly forbidden at the annual RCA Laboratories banquet honoring employees with twenty-five years or more of service. The guests expected this, and it was their practice to hold private cocktail parties prior to the banquet so that they could be properly fortified for a dry evening. Upon the urging of his staff, Engstrom finally relented and approved the serving of cocktails at the banquet. Unfortunately, not all of the guests were informed, and many of the customary prebanquet parties were held as usual. The banquet became pretty boisterous, and the wife of one of the distinguished scientists, well into her cups, swooped up to Engstrom, put her arms around him, gave him a resounding kiss, and said, "Is it all right if I call you Shorty?" (Only Engstrom's close friends called him this.) The crowd fell silent, but Engstrom handled the situation with humor and dignity. He disengaged himself from her clutches, made an admiring remark about her gown, and escorted her to her embarrassed husband with a light comment.[7]

---

[7] This incident was related to me by George Brown, who observed it.

If Engstrom was RCA's Omar Bradley, Brown (Figure 5–2) was its George Patton. Like Patton, he was a brilliant practitioner of his profession. He had the ability to motivate and inspire his subordinates. Also like Patton, his uninhibited expression of his views—which often included a low opinion of RCA's top management, FCC commissioners, and others in positions of authority—sometimes got him in trouble. During the color hearings, Engstrom gave him the dual assignment of supervising the RCA's color research and appearing before the commission as RCA's chief technical witness. His performance at RCA Laboratories was superb. As a witness, his technical expertise was most impressive, but his effectiveness was diminished by his lack of respect for some of the commissioners.

Brown received a Ph.D. in electrical engineering from the University of Wisconsin in 1933. He had become interested in radio as a boy, and this interest continued throughout his college career. His doctoral thesis was on

Courtesy RCA Corporation.

■ **Figure 5–2**   George H. Brown.

the subject of radio broadcast antennas. After receiving his degree, he joined RCA's research department, and except for a brief hiatus in 1937 when he became an independent consultant in partnership with Paul Godley, he remained with RCA for the remainder of his career.

Brown's outstanding accomplishment during the prewar years, the investigation leading to the publication of his authoritative paper on directional antennas in 1936, was described in a Chapter 2. From 1937 to 1939, Brown worked on a number of television projects, including a custom-built antenna for CBS on the Chrysler Building in New York, the development of the turnstile antenna, and the construction of a vestigial sideband filter for television transmitters.

From 1939 through the war years, his efforts were diverted from television to radio heating, and his small group worked on an amazing variety of applications, from hardening engine blocks to pasteurizing milk. Perhaps his most important radio-heating experiments were directed toward drying penicillin as part of the production process.

Soon after the war, he returned to television. In 1949, he headed a group that developed and field-tested the offset carrier concept (described in Chapter 4), which permits two television stations to operate on the same channel with reduced mileage separation. As director of the RCA systems laboratory, he was directly in charge of RCA's color television research during the color hearing.

Brown's brilliant work in color television was the high point of his career. Never constrained by false modesty, he described himself as the man "who is . . . acknowledged by many to be the architect of the [color television] system in use today throughout the world."[8] Color television did not have a single architect, but Brown was surely among the leaders. Brown was eventually rewarded for his performance by being promoted to executive vice president and elected to the RCA board of directors in 1965.

After Robert Sarnoff became president in 1966, Brown's relationship with him and his chief of staff, Chase Morsey, became steadily more acerbic. Finally, in 1972, he resigned as a member of the board and corporate officer, whereupon "within a few months my systolic pressure returned to the low value which I had enjoyed during the Engstrom presidency."[9] He retired in 1973.

Brown's personal reputation was controversial as the result of his Patton-like tongue. He was a skilled raconteur, but his stories suffered from a tendency to be variations on a single theme—the follies of other members of the human race. He made no effort to disguise his contempt for those whom he judged to be his intellectual inferiors, particularly those in high places.

[8] Brown, *and part of which I was.*
[9] Ibid.

His professional reputation, however, was undiminished by his personal foibles. He received numerous honors during his career, both in the United States and abroad, and he is universally recognized as one of the greats of television technology.

Frank Stanton and Peter Goldmark   Frank Stanton received a Ph.D. in psychology from Ohio State University in 1935. His doctoral thesis was on the subject of radio audiences' preferences and tastes, and he was brought into CBS by its president, Paul Kesten, to direct its research activities in this area. Stanton quickly displayed broader capabilities, and in 1942 Kesten urged Paley to appoint Stanton as Kesten's successor because of his failing health. Paley wrote, "[Kesten] described Stanton in glowing terms: '. . . capable, conscientious, hard-working, energetic; a man of integrity and good taste.' Paul and I agreed that he seemed to have all the qualifications the job of president required."[10] After Paley went off to war, Kesten promoted Stanton to positions of increasing responsibility. When Paley returned in 1946, he appointed Stanton president and chief operating officer as Kesten's health problems became critical.

This was the beginning of a remarkable one-on-one management relationship that lasted for twenty-seven years until Stanton's retirement in 1973. A one-on-one organizational structure requires a special relationship between the two individuals, and it often breaks down because of their inability to work together harmoniously. It was particularly difficult with Paley, who was a demanding boss. After Stanton's retirement, the president's office was occupied by a succession of executives, none of whom satisfied Paley. The Paley–Stanton team, however, was extraordinarily successful, and it guided the CBS network to leadership in both radio and television.

One reason for their success was the amicable division of responsibility. Paley devoted most of his attention to the critical function of programming, where his judgments were without peer. Stanton concentrated on the other business functions, including day-to-day operations, affiliate relations, and government relations. He was an extremely effective executive in all these areas. His appearances before congressional and FCC committees were marked by careful preparation and a firm but dignified and respectful demeanor. He became the principal spokesman in Washington for CBS—and often for the entire industry—and he was able to accomplish this without offending Paley's ego.

Stanton's weakness, like Paley's, was his lack of a technical background or an intuitive understanding of technology. This made him an easy victim of Goldmark's salesmanship. Goldmark often persuaded Stanton to sell his ideas to Paley, and this led to a number of costly mistakes.

---

[10] William S. Paley, *As It Happened* (Garden City, N.Y.: Doubleday, 1979).

Stanton was not anxious to retire, but Paley insisted that he follow the corporate policy of retirement at age sixty-five. Paley made an exception to this policy only for himself. Stanton retired in 1973 on his sixty-fifth birthday.

Peter Goldmark's autobiography includes a self-appraisal that is unusually objective and generally accurate:

> As I look back, I think my contributions were, somewhat ironically, not so much in the invention itself or in innovation (a word I would prefer because it means putting an invention to work), but in its gadfly impact on industry. The development of the long-playing record impelled the recording industry including RCA, the giant of the communications business, to change for the better its historic pattern of record production. My work in color television resulted, I think, in bringing color to the public a decade faster than it might otherwise have come, though not exactly in the form I intended. Finally, electronic-video recording [EVR], though it ended up without the auspices of CBS, fired up the video-cassette business into the potential multimillion-dollar industry whose fruits we are beginning to enjoy today.[11]

Goldmark's analysis of the effect of his work on the recording industry and on color television is correct, but his appraisal of the impact of EVR on the videocassette industry is an exaggeration. Much of the subsequent success of magnetic tape videocassette recorders resulted from their ability to make home recordings. EVR, which used film as its recording medium, did not have this capability. Like the videodisk that came later and shared this weakness, it was a commercial failure.

This appraisal also fails to describe the enormous cost of his color and EVR programs to CBS. His role as a gadfly did not come cheaply.

Goldmark was born in Budapest, Hungary, where his father was a prosperous businessman. He was brought up in an upper-middle-class home with a heritage of culture (he was the grandnephew of the composer Karl Goldmark). His family moved to Vienna after World War I because of the unsettled political conditions in Hungary. Denied admission to the Vienna Technical College because its quota of Hungarians was filled, he went to Germany. He first attended the Technische Hochschule in Charlottenberg and then the Physical Institute in Berlin, where he received a Ph.D. in physics.

He became interested in television while still in school, and he carried out experiments with the Nipkow disk system during the 1920s. As a result of this experience, he was hired by the British manufacturer Pye to establish a television department in 1932. This job did not last long because Pye, under economic pressure from the depression, closed the department. Out of a

---

[11] Goldmark, *Maverick Inventor*.

job, he decided to go to the United States, where most of the television action was. He applied for a position at many companies, including RCA, but was turned down at all of them. (Twenty years later, Sarnoff only half jokingly told Paley that both companies could have saved a lot of money if RCA had hired Goldmark.) Finally, on January 1, 1936, through contacts arranged by H.V. Kaltenborn, the CBS correspondent in Vienna, he was hired by Paul Kesten, CBS's president, to carry out research in television.

He became dissatisfied with black-and-white television after seeing the movie *Gone with the Wind* in color, and he persuaded Kesten that CBS (and the television industry) should skip monochrome and proceed immediately to color. As described earlier, this advice was economically disastrous, but he persisted and continued to receive the support of CBS's top management until its field sequential system was approved by the FCC.

Goldmark was a prolific innovator, and his CBS activities were not limited to color television. In 1945, he began the development of the long-playing record, which eventually won out over RCA's 45-rpm player. This was the only one of Goldmark's major programs that led to a commercial success. After the field sequential system finally lost the color television battle, he developed the unsuccessful EVR recording system.

Goldmark also persuaded CBS to purchase the Hytron Radio and Electronics Corporation, a manufacturer of tubes, transistors, and receivers, in 1951. He was probably sincere in his belief that this was good business for CBS, but there was also a self-serving motive. By putting CBS in the manufacturing business, the importance of its technical research was greatly enhanced, and Goldmark constantly pressed for the establishment of a major central research facility that would rival RCA's and that he would lead.

The Hytron venture was a costly failure for CBS. Goldmark and Paley gave quite different reasons for the failure in their memoirs. Goldmark attributed it to bad strategic decisions at the highest levels of the company, particularly by Paley. Paley attributed it to poor management at the operating level, including the inability of the company's engineering department to design a high-quality set. Goldmark had assured Paley that Hytron had a first-rate engineering department, but it did not. Paley called Goldmark on this years later, and he reported the following dialogue:

> "Well, Peter, I think you misled us, you know, about their [Hytron] engineers and the quality of them because after we bought the company, it seemed to me they had a second rate engineering department." And what was his answer?
>
> "Well, Mr. Paley, I was interested in color and I wanted to do everything I possibly could to keep us in the race." The answer was incredible to me, but not as incredible as it had been trying to run that company.[12]

---

[12] Paley, *As it happened.*

Much to Paley's relief, Goldmark reached retirement age in 1972. He founded his own company, the Goldmark Communications Company, which he later sold to Warner Communications. He was killed in an automobile accident in 1976.

As a person, Goldmark was warm and friendly. He was a talented amateur musician and frequently played the cello in first-rate chamber groups. He spoke English fluently but with enough accent to sound delightfully foreign. He courted the approval of the academic establishment, and the announcement of every one of his new products or systems included an optimistic forecast of its role in education.

He was a talented engineer, and he was able to recruit and lead a small group of engineers of outstanding ability. His greatest ability, however, was selling technical ideas to nontechnical executives and government officials. He was a master showman, and his demonstrations of audio and television systems were meticulously staged to emphasize their strengths and hide their weaknesses. Unfortunately, this talent was also a weakness. It was so great that he could and did sell bad ideas, not only to others but also to himself.

There was one blind spot in his personality: a hatred for RCA that bordered on paranoia. George Brown's memoir includes an appraisal of Goldmark that, on the whole, is favorable. Goldmark does not even mention Brown (or any other RCA engineer except Zworykin) in his. Like Edwin Armstrong, he sometimes seemed to be motivated more by a desire to beat RCA than by sound judgment as to what would be best for CBS.

Notwithstanding his foibles and failings, Goldmark had an enormous influence on the development of television technology. He was a most effective gadfly, and the industry benefited greatly from his efforts. He deserves to be numbered among a very select group of its technical leaders.

The Hearing—Round One

A week before the hearing convened, the FCC published an astonishingly unrealistic schedule for the proceedings. The first day was to be devoted to organizational matters and testimony from industry representatives other than the contestants. One day was allotted to each of the three system proponents; this was to include both direct testimony and cross-examination. If this schedule had been followed, the oral portion of the hearing would have been completed in one week. The reality was quite different. Industry witnesses were still on the stand at the end of the first week, and the testimony of the contestants took many months.

The Initial Industry Testimony   The RCA and other industry witnesses had the difficult and unenviable task of convincing the Commission of two negatives and a leap of faith:

1. The CBS system should never be adopted because of its incompatibility and inherent limitations.
2. The Commission should not set any standards at the time but should specify system requirements that would be necessary for approval, including compatibility.
3. The development of a satisfactory compatible color system was only a matter of time.

To make matters more difficult for the industry, its motives were suspect because defeat of the CBS system was in its immediate self-interest.

Persuading the FCC to follow these recommendations, given its predisposition in favor of CBS, would have required a superb presentation of an excellent and well-prepared case. Unfortunately, the industry stumbled badly on both counts.

The testimony of its various representatives ranged from unctuous generalities and pontifications to feisty pronouncements. The Commission was not impressed with this testimony, and most of it was counterproductive. The clumsy approach of the industry witnesses continued throughout the course of the hearing, further enhancing the image of CBS as the sole advocate of the public interest.

The substance of the industry's presentation was no better, partly because its many members seemed unable to agree on a coherent strategy. At no time was incisive, tangible, and persuasive evidence produced to demonstrate the impracticality of CBS's incompatible color. Industry prepared a poor case and presented it badly.

Industry's testimony occupied the first four days of the hearing. Much of this time was devoted to badgering by Commissioners Jones and Hennock, whose questions were as bad as the answers. When the hearing recessed at the end of the week, very little testimony on the record would have been helpful to a fair-minded Commission, let alone persuasive to a biased one.

The Initial CBS Testimony   CBS presented its initial direct case during the second week of the hearing. Goldmark was the principal witness, and he performed superbly. He was aided by a Commission that was largely on his side, and the only criticism came from Commissioner Hennock, who took him mildly to task because she did not believe he was sufficiently enthusiastic. His testimony was relatively brief, and it ended with a recommendation that the CBS system be adopted forthwith.

The Initial RCA Testimony   Now it was RCA's turn. Engstrom was the initial witness, and he presented a description of the dot sequential system that had been frantically and hastily prepared by George Brown. The Commission felt that Engstrom's explanation was inadequate, and the RCA high command decided to put Brown himself on the stand. This was the beginning of an eight-month period during which Engstrom and Brown divided their time between testifying and guiding RCA's technical efforts.

At this point, the RCA system existed mainly on paper, and no presentation could have been persuasive by itelf. The FCC could not have been expected to approve a new system without extensive demonstrations and field tests. RCA's case was further weakened by the deportment of Engstrom and Brown on the witness stand, which did not match the quality of their leadership in the laboratory. Engstrom was courteous enough, but he came across as cold and aloof. Brown could not resist the temptation to respond to less-than-intelligent questions with semifacetious answers. Both attitudes strengthened the perception of RCA as a powerful and arrogant company intent only on maintaining its dominance.

Not unexpectedly, the Commission was not persuaded by Brown's testimony. Round one of the hearing ended with a recess until October 10, when round two began with the first semipublic demonstration of the RCA system.

The Hearing—Round Two

The RCA Demonstration The audience for the RCA demonstration consisted of members of the FCC and its staff, representatives of the other hearing participants, and the press. It was an eagerly awaited event because many expected to see the unveiling of a practical compatible system by the company that was the undisputed leade in television technology.

The demonstration was held in the ballroom on the top floor of the Washington Hotel. It was an unseasonably hot day, and RCA did not open the doors until precisely 10:00 A.M., the scheduled starting time. The audience was forced to wait, jammed together in the sweltering foyer, until the appointed time. When the doors were opened and the audience poured in, the reason for the delay became obvious.

Engstrom was standing on the stage in shirtsleeves, his tie askew, his armpits bathed in perspiration, and a haggard look on his face. He had been up all night trying to get the demonstration in some sort of decent shape. He needed every possible minute, which had resulted in the delay in admitting the audience.

Nevertheless, he showed extraordinary coolness under fire. He knew the demonstration would be a disaster, and he had to face a hostile FCC and a jubilant CBS—to say nothing of an unhappy General Sarnoff. He calmly put on his coat, straightened his tie, and proceeded with the demonstration as though it were going beautifully.

There were six huge receivers (six feet high, six feet deep, and thirty inches wide) on the stage displaying sixteen-inch pictures. Four of them were triniscopes with three monochrome kinescopes, one for each primary color. The images from the kinescopes were combined and color added by crossed-dichroic mirrors. The other two receivers used only two of the primaries, red and green, and the picture colors ranged from red to orange to yellow to green, with an especially ghastly hue for flesh tones. The purpose of demonstrating the two-color receivers was never made clear.

The pictures on the triniscopes were only a little better. The hues were more or less random, with very little similarity to real life and very little consistency from receiver to receiver. Flesh tones were particularly odd, and one could see red, purple, green, and orange faces simultaneously. To make matters worse, one could not tell whether the transmission was color or monochrome, and frequent calls to the transmitter were necessary.

The problem of receiver matching was somewhat alleviated by unplanned events. In order to provide the maximum possible picture brightness, the voltage on the picture tubes was pushed to the limit. In the hot, humid atmosphere of the ballroom, the limit was exceeded on two of the receivers, and there were loud bangs as first one and then the other failed. With only two of the three-color receivers operating, the matching problem became a little more manageable.

The demonstration proceeded to its dismal conclusion. Goldmark, normally a very polite person, could not hide his glee, and he sat in the front row grinning like the Cheshire cat. The pro-CBS FCC contingent was equally pleased, and the contingent left the demonstration completely convinced that its initial judgment had been correct. More objective observers left the demonstration shaking their heads. It seemed incredible that RCA could have failed so badly.

Goldmark pronounced his judgment on the RCA system at the hearing the following day:

> **Commissioner Hennock:**
> "I had asked you a question earlier, Dr. Goldmark, and you said you would rather answer it after your testimony. Would you like to answer it now?"
> **Goldmark:**
> "Would you mind repeating the question?"
> **Hennock:**
> "I asked how long it would take to make the field tests on the RCA system with regard to propagation and apparatus."
> **Goldmark:**
> "Under the conditions, I don't think there should be a field test on the RCA system at all. I am serious. I don't think that the RCA system should be field tested because I don't think the field tests will improve the system fundamentally."
> **Hennock:**
> "Do you mean to say that nothing can improve the RCA system?"

Goldmark:
"No, nothing, I think."
Hennock:
"And then you advocate that they drop the system now?"
Goldmark:
"I certainly do."[13]

The reality, of course, was quite different. RCA's poor showing did not result from fundamental defects in its system but rather from the fact that the concept was only a few months old. Considering the complexity of the problem and of the apparatus that had to be designed, built, and tested, the wonder was not that the demonstration was so bad but that it could be done at all.

Goldmark's testimony threw Sarnoff into a rage. He circulated a memo within RCA that stated in part: "The [Goldmark testimony] is the most unprofessional and ruthless statement I have ever seen made by anyone about a competitor. I have every confidence that the scientists and engineers of RCA will answer this baseless charge by improvements which I have already seen since the first demonstrations and which will be made during the coming months."[14] Some improvements did come rapidly, and during the next ten days RCA was able to put on much more respectable demonstrations for members of Congress and others.

The CBS Demonstration The CBS demonstration held soon afterward displayed a high degree of professional showmanship. The system had been technically perfected to the extent possible within its basic limitations, and Goldmark was able to concentrate on programming and cosmetics. The room was darkened, and the viewers are placed far enough from the sets so that the reduced picture sharpness was not visible. Brilliant colors were used in the staging because experience had taught that they were more impressive to nontechnical audiences than pastels. And Goldmark had the vast production resources of CBS available.

There was a great deal of oohing and aahing over the color quality, and this time the pro-CBS faction of the FCC left the demonstration with the firm conviction that the CBS system should be adopted immediately. Had it not suffered from the enormous disadvantage of incompatibility, it probably would have been.

The Comparative Demonstrations Round two of the hearing ended with comparative demonstrations of the RCA, CBS, and CTI systems at FCC Laboratories in Laurel, Maryland, on November 21 and 22. They proved

[13] Brown, *and part of which I was.*
[14] Ibid.

very little. The RCA system had been greatly improved since the disastrous September demonstration, but its performance still left much to be desired. It continued to suffer from the use of a huge triniscope in the receiver. CBS put on its usual professional production. As for CTI, its system simply was not working.

Perhaps the most noteworthy feature of these demonstrations was the introduction of the "Chapin converter." This was a receiver circuit developed by the staff of E.W. Chapin, chief of FCC Laboratories, that would detect the presence of a field sequential signal and convert the receiver's scanning rates automatically. Chapin had given up all pretense of impartiality, and his circuit was touted by FCC chairman Coy as the answer to the compatibility problem.

The End of Rounds One and Two

At the conclusion of the comparative demonstrations, the FCC recessed the hearing until February 27, 1950, with another comparative demonstration scheduled for February 23. CBS had clearly won the first two rounds, with the second very nearly a knockout. But RCA was still on its feet, and the recess gave it a breather during which it could concentrate on improving its system.

RCA had to solve two problems: basic improvements in the method of transmitting brightness and color information in the same channel and the development of a satisfactory tricolor tube. The solutions were complex, and they had to be found quickly. The pressure on RCA's technical staff was enormous. Vacations were canceled, and seven-day weeks were common. A few employees were unable to maintain the pace, and one even suffered a nervous breakdown. For most of RCA's engineers and scientists, however, this was a time of high excitement and achievement, and many of them looked back on this period as the climax of their careers.

With significant supportive assistance from other manufacturers, the problems were solved. The RCA dot sequential system evolved into the NTSC system, named for the ad hoc National Television System Committee that was formed to develop it. RCA, goaded by an important CBS innovation, developed a practical tricolor tube. Unfortunately, although RCA and the industry moved forward with remarkable speed, it was not fast enough to deter the Commission as it moved toward its predetermined conclusion.

Formation of the NTSC for Color Standards

RCA's competitors in the television manufacturing industry, stunned by their inability to influence the FCC in the first two rounds of the hearing, decided to form an uneasy alliance with their hated rival to develop and demonstrate a workable compatible system. In January 1950, shortly after the hearing recessed, Dr. W.R.G. Baker of GE, representing the Radio Television Manufacturers Association (RTMA), proposed that a second NTSC be formed to advise the Commission on the matter of color standards as the first had concerning monochrome in 1940.

Coy was highly displeased with the formation of this Committee, since its purpose was to develop a system in competition with the one the Commission favored. There was little he could do about it, however, and he reluctantly agreed that the Committee could testify at the hearing.

The RTMA authorized the formation of the Committee on January 17 with Dr. Baker as its chairman and David B. Smith of Philco and Donald Fink, editor of *Electronics* magazine, as vice chairmen. For the next nine months, the NTSC was relatively quiescent, and it was not until November, after the FCC had issued its preliminary decision favoring CBS, that it took an active role in developing and defining a set of color standards.

The Hearing—Round Three

Round three of the hearing, which began in February 1950, was a dreary charade. Chairman Coy and several other members of the FCC, together with its engineering staff, had long since made up their minds in favor of the CBS system. Nothing that RCA or the rest of the industry could do would change them.

In January, RCA eliminated much of the color instability that had plagued its early demonstrations by introducing the "burst," a short train of sine waves behind the horizontal sync pulses. In March, it demonstrated a successful tricolor picture tube that eliminated the need for the huge and costly triniscopes (although CBS rightfully pointed out that would also eliminate the mechanical disk in its system). The FCC virtually ignored these major breakthroughs. The introduction of the Chapin converter ended any semblance of impartiality, and the hearings were directed at obtaining evidence to support the preordained result.

Opponents of the CBS system did not help their cause by the manner in which they presented their case. In an ill-advised attempt at humor, Dumont's T.T. Goldsmith ridiculed the CBS system by bringing a sixteen-inch receiver equipped with a huge five-foot color wheel onstage. Coy was infuriated by this stunt.

Although Brown and his engineers were performing miracles in the laboratory, he became increasingly sarcastic in his testimony. In a desperate attempt to aid its cause, RCA retained Clark Clifford, who was alternating between influential cabinet posts and a lucrative law practice. His role was to use his prestige and political clout to influence the FCC in favor of RCA.

On the day of his first appearance at the hearing, he seated himself prominently in the front row by the aisle. As Coy filed in with the other commissioners, Clifford gave him a friendly swat on the rear, saying, "Hi, Wayne!" Coy turned around and glared at him.

The quality of RCA's presentation was redeemed somewhat by Sarnoff's testimony. Although tainted by touches of his irrepressible ego, it was generally superb. He was on the stand for sixteen hours on May 3 and 4, and he presented RCA's case ebulliently, ably, and convincingly. He was particularly gallant to Commissioner Hennock, who was obviously flattered by the attention from the great General Sarnoff. It was too late for RCA, however, as CBS had won its case for all practical purposes.

In the meantime, CBS, with the hearing going its way, could afford to take the high road and present a statesmanlike case. Frank Stanton spoke for its top management. He was dignified, businesslike, and persuasive. He answered the criticisms of the CBS system and attacked RCA's system. He pointed out that the current AT&T intercity video distribution system could not pass RCA color signals on its narrow-band L-1 cable.[15]

The most important issue addressed by Stanton was receiver manufacture. Every other manufacturer had advised the FCC that it had no plans to offer field sequential color receivers in the marketplace (which strengthened the conspiracy theory as the explanation for the industry's opposition to the CBS system). Stanton committed the Hytron division of CBS to the manufacture of these receivers if the CBS system were adopted. This commitment could have been and nearly was disastrous to the company.

The hearing wound down to a weary conclusion on May 26, 1950. The prognosis of the industry as to the Commission's decision was mixed. On the one hand, the Commission's predilection for CBS was painfully apparent. On the other, many could not bring themselves to believe that it could make a decision for CBS in the face of mounting evidence that a practical compatible system would soon be developed. The latter group clearly underestimated the power of bureaucratic prejudice.

The FCC Miscarries

With the benefit of hindsight, it is difficult to understand how the FCC could possibly have found for the CBS system. As a matter of fact, it was difficult to understand at the time. But on September 1, 1950, the Commission issued its First Report on Color Television Issues. It shook the industry

---

[15] The L-1 cable had a bandwidth of 2.7 MHz. It was marginally adequate for monochrome and even more marginal for CBS color, but it was unusable for NTSC color. RCA engineers developed a circuit to adapt it for color, but it was never used because AT&T upgraded its system to wideband L-3 cable by the time it was needed.

by indicating its intention to adopt the CBS system, but industry was given an escape. The commission indicated its willingness to consider other systems under two conditions: (1) The superiority of the alternate system had to be demonstrated within three months, and (2) a sufficient number of receiver manufacturers had to agree before September 29 that they would manufacture receivers with "bracket standards"—that is, receivers that could receive either monochrome or field sequential color signals.

The stated purpose of the second condition was to prevent a further increase in the compatibility problem as a result of the continuing growth in the number of monochrome-only receivers in the hands of the public—even though the FCC had assigned very little weight to this problem in its Report. In fact, however, it was an attempt to blackmail the industry into building receivers for the CBS system. To its credit, the industry refused to be coerced, and, except for CBS, not a single major manufacturer agreed to build bracket standard receivers.

Surprisingly, Commissioner Hennock, in a rare display of good sense, suggested that the final decision be delayed until June 30, 1951, with no conditions on the manufacturers. RCA supported her position and petitioned the Commission to that effect. It did no good. To the continuing amazement of industry leaders, the FCC persisted in its opinion. On October 10, 1950, it denied the RCA petition and approved the CBS system. November 20 was designated as the date on which commercial color broadcasting could commence.

It was a terrible blow to an incredulous General Sarnoff. He issued the following public statement:

> We regard this decision as scientifically unsound and against the public interest. . . . The hundreds of millions of dollars the present set owners would have to pay to obtain a degraded picture with an incompatible system reduces today's order to an absurdity. Regardless of what anyone else may be called upon to do, RCA will continue to advance the bedrock principles on which the sound future of color television can be built and will be built.[16]

Sarnoff was as good as his word. Notwithstanding the enormous drain of color research on RCA's resources, both human and financial, he ordered that it continue undiminished. This required great confidence and no small amount of courage. He was criticized by some members of the RCA board and the financial community for making expenditures on a program that could not pay off for many years at best. It is doubtful that a company with a more orderly management structure would have carried it off. It required an entrepreneurial individual with enormous foresight and determination to accomplish it. Sarnoff had both these qualities to an exceptional degree.

---

[16]Eugene Lyons, *David Sarnoff* (New York: Harper & Row, 1966).

Against the advice of many of his colleagues, Sarnoff insisted on appealing the decision to the courts. The appeal included many persuasive facts, not the least of which was that there were more than ten million incompatible monochrome sets in the hands of the public. But the courts refused to overturn the decision. Following long-established precedent, they refused to overrule the judgment of an administrative body in the area of its supposed expertise. On November 15, the District Court of Appeals issued a temporary restraining order to give it time to study the record. On December 22, it affirmed the decision but continued the restraining order pending review by the Supreme Court. On May 28, 1951, the Supreme Court upheld the FCC and authorized the start of commercial color broadcasting on June 25. The CBS system became the approved mode of color television in the United States.

■ CBS COLOR IS STILLBORN

The more objective members of the CBS management could hardly have been overjoyed by the final victory of its system at the FCC hearing and in the courts. The issue would now be decided in the real world of the marketplace, and the criteria for success were quite different there. Indeed, there were whisperings within the organization that it had been a Pyrrhic victory.

The difficulties facing CBS were awesome. During the twenty-month period from the beginning of the hearing to the final Supreme Court decision, the number of incompatible monochrome sets in use had more than quadrupled, from three million to thirteen million. Even assuming that color sets were available, the public could scarcely be expected to buy them unless there was a reasonable amount of color programming. But CBS, locked in a bitter battle with NBC for market share, could not afford to weaken its position by broadcasting a significant amount of color programming that could be received only by a tiny fraction of the public. NBC, of course, never gave broadcasting CBS color a second thought.

CBS introduced its system to the public with a gala broadcast on June 25. It included an impressive array of CBS's top stars, including Arthur Godfrey, Ed Sullivan, and Faye Emerson. CBS management was represented by Paley and Stanton, and Chairman Coy spoke for the FCC. The only flaw in the program was that there was virtually no audience, as there were probably fewer than one hundred color sets in the hands of the general public. Later, CBS began regular color broadcasts from 4 to 5 P.M., a time slot chosen because it was a low point in the size of the normal viewing audience. The interruption of its regular programming, even at this time, had an adverse effect on the carryover audience into prime time. Coy took to the hustings in an effort to promote the system, but the problems of incompatibility were too great to be solved by bureaucratic intervention.

To add to CBS's problems, Stanton had committed it during the hearing to the manufacture of field sequential color sets. This would have been enormously costly, and, given the chicken-and-egg relationship of programming and receivers and the lack of support from broadcasters, it would have been financially ruinous. CBS was facing both an economic and a public relations disaster.

To CBS's relief, the National Production Authority appeared as a deus ex machina. The Korean War was under way, and this organization had the authority to control the production of civilian goods using strategic materials. On November 20, 1951, it issued Order M-90, which forbade the production of color (but not monochrome) receivers. CBS, with an ostentatious display of good citizenship, gracefully (and gratefully) acceded.

Order M-90 fit CBS's self-interest so neatly that there was a widespread belief (never confirmed) that its management had lobbied for the order. The potential production of color sets was so small that it could not have had a significant impact on the demand for critical materials. In the meantime, the much larger production of monochrome receivers continued.

It mattered little whether the Order resulted from brilliant lobbying or whether it was an extraordinary stroke of good fortune. CBS was off the hook. It ceased its costly color programming and scrapped its equally costly plans for manufacturing color receivers. For the long run, it stated that the war-produced delays made the compatibility problem insoluble, thus saving face in an embarrassing situation. And it did this cloaked in an aura of patriotism and good citizenship.

Order M-90 effectively marked the end of CBS field sequential color, but it did not really die; in fact, it had never lived—it was stillborn. The rest of the industry was equally pleased with the Order. It was now able to proceed with the development of compatible color without the distraction of an inferior but politically powerful competing system.

■ THE NTSC SYSTEM

The NTSC, which had been formed in January 1950, was relatively quiescent for most of the year, limiting its activity to relatively minor and uncontroversial testimony at the hearing. It was not until November, after the FCC had found in favor of CBS, that the Committee began to address the development of an alternate compatible color standard in depth. On November 20, Baker, the Committee chairman, appointed a subcommittee for this purpose. Its chairman was D.B. Smith of Philco, also a vice chairman of the parent committee, and it included members from Dumont, GE, Hazeltine, Sylvania, and RCA.

RCA's competitors were ambivalent about this issue. On the one hand, they vehemently opposed the CBS system, which they perceived as having a disastrous effect on the industry. On the other, they were not anxious to have the system proposed and developed by their largest and most powerful competitor adopted—with all the resulting implications of prestige, patent position, and technical leadership. This ambivalence carried over into the work of the NTSC. On the technical level, its members generally worked well together and were motivated primarily by a sincere desire to develop the best possible system. On the public relations level, however, there was a constant battle. RCA claimed that the resulting NTSC system was its system with minor modifications. Its competitors claimed that it was the outcome of an all-industry effort. The truth was somewhere in between but was closer to the RCA version.

The system had its roots in the RCA proposal, it retained most of the essential features of the RCA system, and without RCA's determination and continued research—particularly its invention of the tricolor kinescope—it could never have come into being. Others, however, particularly Hazeltine, made important contributions, and a better system resulted.

Smith's subcommittee worked at its task for several months and issued its report early in 1951. It was carefully worded to avoid giving RCA credit, and it proposed a color system having the two features that were the essence of the system that was finally adopted. The first was the use of a burst to synchronize the colors at the cameras and receivers. The second was the transmission of brightness information by a compatible monochrome signal and color by a high-frequency subcarrier. The report concluded with the statement that further work and field-testing would be required to establish numerical standards.

The report was approved by the parent Committee, which then undertook the task of overseeing the development and field-testing of the system. Engstrom was added to the roster of vice chairmen, and eleven panels were established, each with a special area to study. Twenty-nine companies were represented on the panels, which set to work in a remarkable example of industry cooperation.

Burst and Subcarrier

The use of the burst, an addition to the color signal providing a solid synchronism between the colors produced by the cameras and those reproduced at the receiver, removed a major source of the color instability experienced in the original RCA demonstration. This was clearly an RCA development, its inventor being Al Bedford of RCA Laboratories. It was first demonstrated to the FCC and the press on January 17, 1950, and it worked better than even the RCA engineers had expected. For the first time, it was possible to carry out a demonstration without constant adjustment of the equipment. This development was fundamental to the success of the NTSC system, and it is still in use today.

BURST AND COLOR SUBCARRIER

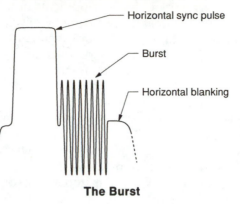

Horizontal sync pulse

Burst

Horizontal blanking

The Burst

In the first RCA demonstration, synchronization of the colors at the camera and receiver was accomplished by cyclically varying the position of the trailing edge of the horizontal sync pulse. The synchronization was easily upset by a small amount of noise in the signal, resulting in color instability. This method was superseded by adding a burst, a reference train of eight cycles of a sine wave at the subcarrier frequency. It provided an extremely stable mechanism for synchronization that was relatively invulnerable to noise.

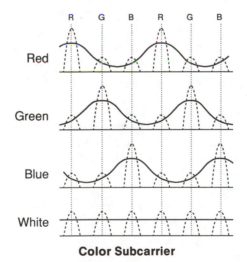

Color Subcarrier

The dotted curves show the waveforms that result from sequentially sampling red, green, blue, and white areas of the picture. The higher harmonics in these waveforms are removed when the signal is passed through a band-limiting circuit. The resulting signal consists of two components, a sine wave subcarrier superimposed on a monochrome brightness component. The phase of the subcarrier is determined by the hue of the picture; its amplitude is determined by the purity or saturation of the color. The subcarrier vanishes for a white area in the picture. After the subcarrier concept was developed, sampling was abandoned as a technique for generating color signals.

While RCA was carrying out its developments in the glare of publicity, Hazeltine, an industrial laboratory, was quietly developing the subcarrier concept under the leadership of its vice president, Arthur V. Loughren; its chief engineer, Charles V. Hirsch; and a senior engineer, Bernard D. Loughlin. This concept represented a fundamental change not in the system itself but in the method of analyzing the signal resulting from the sampling process in the dot sequential system.

To a nonengineer, the choice of method of analysis may seem trivial, but it is really very basic. Medieval astronomers, for example, described the motions of the sun and planets by a model in which the earth was the center, yielding to the then current theological concept of the universe. The model was not incorrect, but it was hopelessly complex, and our view of the solar system was marvelously simplified by putting the sun at its center.

As a result of the subcarrier concept, Hazeltine noted that the process of combining the brightness and color information on a single signal could be described not as a sampling process but as the addition of a high-frequency sine wave color subcarrier to the brightness or monochrome signal. Since the engineering profession had years of experience in working with sine waves and monochrome television, all the powerful analytic tools that had been developed could immediately be brought to bear. This change in viewpoint was a tremendously important breakthrough, and it is the only fundamental difference between the original concept of the RCA system and that of the NTSC system that was finally approved.

With the benefits of this concept, Hazeltine was able to suggest significant improvements in the RCA system. Among these was the principle of constant luminance in which the brightness signal is formed from red, green, and blue signal components proportioned in accordance with the sensitivity of the human eye. Another was bypassed luminance, which consisted of handling the brightness component separately from the chroma signal in much of the signal-generating and receiver circuitry. This concept caused sampling to be abandoned as a technique for generating the color component signals.

Hazeltine disclosed its work to RCA engineers in May 1950. The latter immediately recognized its significance, and it became the basis of much of the future technical development of color.

Procedures and Field Tests

The dedication, cooperation, and competence of the eleven panels established to set numerical standards and specify field test procedures for the NTSC system were outstanding. Information was exchanged freely among them, and their rate of progress was amazing.

On November 26, 1951, the first phase of their task was complete, and the NTSC released a document, "NTSC Color Field Test Specifications." This was actually a set of tentative system specifications contingent on the outcome of the tests.

Obtaining FCC authority to field-test a non-CBS system was not easy, but limited hours of testing were finally permitted, mostly in the middle of the night. Coy, stubborn to the end, refused permission for FCC engineers to participate in these tests except on a carefully controlled basis. He resigned on February 21, 1952, and the next chairman, Rosel Hyde, encouraged full participation of the FCC staff.

Field tests continued throughout 1952, resulting in further improvements in the system. One of the more important improvements was the transformation of the red, green, and blue signal components to I (orange-red) and Q (blue-green) signals. The purpose was to take advantage of the characteristics of the human eye, which is relatively insensitive to fine detail in the blue-green region. Thus the Q signal can be transmitted in a relatively narrow band without loss of picture quality.

On February 2, 1953, the NTSC published the Revised NTSC Color Field Test Specifications, which included all the improvements developed during the past year. Some final phases of the tests continued, however, and it was during this period that George Brown surreptitiously painted a banana in a test scene blue, completely frustrating the technical crew as it tried to adjust the system to make it yellow.

By May 1953, everyone was satisfied, and the NTSC panels formally approved the system as described in a lengthy document. Peter Goldmark, who by this time had given up on his system and had participated in the final work of the NTSC, asked and was granted permission to second the motion for approval.

FCC Approval

On July 21, 1953, the NTSC approved the system submitted by its panels and petitioned the FCC to adopt these standards in lieu of the previously approved CBS system. In the meantime, RCA had jumped the gun and on June 25 had filed a petition urging adoption of the RCA Color System, which was described as "operating on the color standards proposed by the NTSC." This infuriated RCA's competitors, particularly Philco, and the fragile truce that had been in effect during the NTSC's deliberations came to an end. Philco filed a counterpetition aimed primarily at minimizing the importance of RCA's contribution to the standards. This internecine warfare was not what the industry needed to sell its system to the FCC.

By this time, however, the FCC had very little choice. Faced with the complete failure of the CBS system to be accepted by broadcasters and receiver manufacturers, the resignation of most of its pro-CBS members, and a system that, officially at least, was the product of the entire industry, its decision was inevitable. It requested an official demonstration, the demonstration was successful, and on December 17, 1953, it approved the NTSC system, with commercial broadcasting authorized to begin on January 22, 1954.

■ THE GROWTH OF COLOR TELEVISION

A Decade of Stagnation

During the autumn of 1954, the marketing managers of RCA's Broadcast Equipment Division held a series of meetings to establish the division's sales budget for the following year. The meetings were critical. The great surge in equipment sales that had followed the ending of the freeze was nearly over, and a major new market was required if the division were to continue on its highly profitable course.

By common consent, that market was color. The RCA system had been approved for nearly a year, RCA had the products, and the broadcasters had the money. All the ingredients seemed to be in place for a major growth in the sales of color equipment.

Furthermore, RCA had unmatched sales coverage of the broadcast station market. It had enjoyed more than 60 percent of the monochrome television market, and it had the most complete line of color equipment. The top management and chief engineers of nearly all the stations were known personally to members of the sales organization. In many cases, RCA salespeople were privy to the capital budgets of their customers. No sales department ever knew its customers better.

Armed with this information, the marketing management made a station-by-station review of the broadcast industry. Forecasts were made of the anticipated purchases of color equipment by each station. A contingency factor that seemed to be extremely conservative was then applied to the total to establish the official sales budget.

As one might expect from such a systematic procedure administered by an experienced marketing organization, the budget proved to be an accurate forecast. It had only one small flaw—the timing was off. The sales forecast for 1955 did not materialize until 1965, ten years later.

The problems of RCA's Broadcast Equipment Division were a symptom of the situation in the industry—color television simply was not taking off. For the years 1953 to 1958, color receiver sales were so small and so dominated by RCA that industry sales were not made public. Beginning in 1959, when industry sales were estimated at 90,000 sets, the Commerce Department and the EIA began publishing industry statistics. Although the annual sales volume grew steadily for the next five years, it reached only 747,000 sets in 1963. During this same period, the annual sales of monochrome receivers exceeded 6 million.

This decade was a bleak one for RCA. No one except Sarnoff was certain that there was a light at the end of the color tunnel, and many were equally sure that there was no light at all. The basic problem was the lack of both programs and receivers and the question of which should be developed first. Sarnoff controlled both NBC and RCA's Receiver Division, and he insisted that both continue to invest large sums in color's future in spite of the serious negative effects on their bottom lines.

For nearly ten years, NBC was the only major network offering a substantial amount of color programming, which it did at considerable additional cost and at the expense of monochrome picture quality. CBS had no desire to help its competitor, and ABC could not afford to do so. Without Sarnoff's vision and determination, this would not have happened at NBC.

Another problem of color's early years was the relatively primitive quality and high cost of the available equipment for both the studio and the home. Color cameras were large and expensive, and they produced pictures of marginal quality, both in color and monochrome. Videotape recorders, introduced by Ampex in 1956 (see Chapter 6), became an essential element in stations' technical operations, but the initial models worked only for monochrome signals. Color receivers were expensive (although RCA was selling them at a loss) and produced inferior monochrome pictures. While none of these problems was insoluble and there were steady year-by-year improvements in both studio equipment and receivers, it would be many years before the performance of color equipment would catch up with the potential of the basic technology. In the meantime, RCA's costly color product design programs dictated by Sarnoff were bringing no return.

Sarnoff had to pay a price for his determination. The bankers on RCA's board were restless. The General had promised great prosperity after the RCA system was adopted by the FCC, and it was not being achieved. In an effort to bring more profit orientation to top management, they replaced the able Frank Folsom as president with John Burns, a senior partner of the Booz, Allen consulting firm, in 1957.[17] Burns brought in a strange collection of staff executives, and RCA's top management soon became enmeshed in a labyrinth of politics that was even more complex than the usual fictional portrayal of an executive suite. RCA's business went from bad to worse, and Burns did not endear himself to Sarnoff with his negative attitude

---

[17] Burns also had done consulting work for IBM, and when offered the RCA position, he asked T.J. Watson, Jr., IBM's CEO, whether he had any objections. Watson responded that he most certainly did. RCA had become a competitor of IBM in the computer business, and Burns had become privy to many of IBM's closely held secrets. Burns accepted RCA's offer in spite of Watson's opposition. This infuriated Watson, and IBM, an awesome competitor for everyone in the business, made a special point of opposing RCA in the marketplace.

toward color. In 1961, Sarnoff found a pretext for replacing Burns with Engstrom but not until great harm had been done to the company.

Color Television Explodes

In 1964, the industry suddenly and rather unexpectedly reached a critical mass (a term popularized by the atom bomb and signifying a confluence of events that leads to an explosive breakthrough), and the increase in television receiver sales that RCA had expected in the late 1950s finally occurred. More than six million color sets were sold in 1968, greater than the total number of monochrome sets sold and with a dollar volume three times as great. The sales history of color television receivers (Figure 5–3) illustrated a truism of market planning: Near-term sales forecasts are usually too optimistic, but long-term forecasts for a successful product are almost invariably too low.

The critical mass resulted from the cumulative effect of the slow but steady increase in the amount of color programming, enormous improvements in the reliability and performance of studio equipment and receivers, the availability of color videotape recorders beginning in 1958, and major reductions in the price of receivers.

Sarnoff's faith and determination were rewarded, as RCA was uniquely prepared to take advantage of the color boom. Engstrom's administration

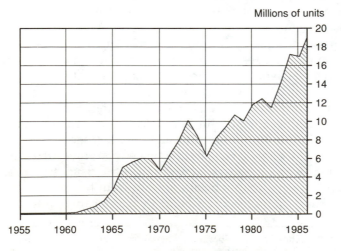

■ **Figure 5–3** Color receiver sales. The FCC approved the NTSC color system in December 1953. Color receiver sales were insignificant for ten years until a confluence of favorable events caused a breakthrough in 1964.

included RCA's halcyon years. Its picture tube and receiver manufacturing plants were operating at capacity, and the company's sales and profits reached new heights. Sarnoff gloried in the spectacular success of color television for which he, more than any other individual, was responsible. It was a marvelous climax to his career.

■ INTERNATIONAL STANDARDS

European television engineers watched the struggles to establish U.S. color television standards closely and carried out a limited amount of color research in their laboratories. In 1955, two years after the NTSC system was approved by the FCC, they convened a meeting of the Comité Consultatif International de Radio Communication (CCIR), an international standards body, in Brussels to consider the question of television broadcast standards for Europe. U.S. manufacturers, led by RCA and aided by the Department of State, made a serious effort to sell NTSC to Europe and the rest of the world. In hindsight, it is clear that the technical, commercial, and nationalistic barriers against the establishment of NTSC color as a universal international standard were so great as to be insurmountable.

From a technical standpoint, there was an opportunity to take advantage of U.S. operating experience with NTSC and correct some of its perceived faults with an improved system. From a commercial standpoint, European manufacturers were anxious to adopt a different system in order to minimize the head start in patent licensing and export markets that their U.S. counterparts enjoyed. Finally, there was the matter of national pride, a motive that was shared by engineers, manufacturers, and heads of state (particularly Charles de Gaulle).

Periodic CCIR meetings continued, and in 1961 RCA made a major effort to persuade the Europeans. It assembled an NTSC demonstration van that toured Europe, including the Soviet Union. Unfortunately, the best available NTSC color apparatus was still far from being perfected, and by today's standards the picture quality was not particularly impressive. European engineers who came to the United States to observe on-air color broadcasting often saw pictures of marginal quality or worse. The joke was that NTSC meant "Never Twice the Same Color."

This relatively poor performance of NTSC systems supported the contentions of European engineers who wanted to develop a different system for reasons of professional pride, and it provided a rationale for those who wanted Europe to adopt a different standard for commercial and political reasons. Two different systems, Phase Alternating Lines (PAL) and Sequentiel Colour avec Memoire (SECAM) were developed and proposed by European engineers.

PAL

PAL was developed by Telefunken, a German manufacturer, and it differs very little from NTSC. The phase of the color subcarrier is reversed on alternate scanning lines so that color errors produced by imperfections in the transmission system are in opposite directions on adjacent lines. The eye then tends to average out these errors and to perceive colors that are approximately correct.

The differences between PAL and NTSC equipment are not great, and PAL signals can be easily converted to NTSC and vice versa. In the end, U.S. manufacturers, recognizing the hopelessness of the NTSC battle in Europe, supported PAL. To save face, PAL and NTSC were described as being members of the same generic family, quadrature amplitude modulation (QUAM).

SECAM

SECAM was another matter. It was proposed by the French partly out of chauvinism and partly to minimize color errors produced by inferior inter-city microwave transmission systems and early television tape recorders. As microwave systems and tape recorders have been improved, this advantage of SECAM has become irrelevant, and it is left with a number of disadvantages. SECAM receivers are more complex, and SECAM signals are more difficult to convert to PAL or NTSC. It is not readily adaptable to special effects manipulation such as fades and lap dissolves. Finally, in the absence of signal distortions, it does not provide the picture quality of either NTSC or PAL.

The World Is Divided

The national proponents of PAL and SECAM fought furiously for their systems in a series of CCIR conferences in the early and middle 1960s, but it was impossible for them to reach an agreement on a single system. All the major Western European countries except France opted for PAL. France insisted on SECAM for itself and, on political grounds, persuaded the Soviet Union and its Eastern European satellites to follow. The Central and North American countries followed the United States with NTSC, but South America was divided between PAL and NTSC. Japan, anxious to exploit the U.S. market and accustomed to paying substantial patent royalties, adopted NTSC long before the European countries had made their decision.

■ FROM SYSTEMS TO PRODUCTS

From 1949 to 1953, FCC approval of the *system* for transmitting color signals from camera to receiver was the main focus of public attention. In parallel with this and continuing long after the conclusion of the hearing to the present day, industrial laboratories throughout the world have been hard at work developing the *products* required to make the system work.

As was the case with radio and monochrome television, engineers engaged in the development, design, and production of products worked in relative obscurity. They did not enjoy the glamour and publicity accorded the scientists and engineers responsible for the scientific breakthroughs. But as has been repeatedly noted, excellence in product design and manufacture can be more important to the success of an industry than scientific breakthroughs, and the history of product development is an integral part of this narrative.

A bewildering variety of products for the generation, recording, and editing of color signals is now available in the marketplace. Many of them are highly specialized, such as devices for producing the elaborate special effects used in advanced editing systems. But only three of them are basic to the production and display of television signals: cameras, video recorders, and receivers. The development and manufacture of these products was a technical *tour de force* that has been equaled in few other industries. The remainder of this chapter is devoted to the story of receivers and cameras. The history of video recorders, for both the studio and the home, is described in subsequent chapters.

■ THE TRICOLOR TUBE AND COLOR RECEIVERS

The Tricolor Tube

According to George Brown, "The development of a successful color kinescope in a six-month period [was] a triumph seldom equalled in industrial research."[18] This is not an exaggeration. Without it, RCA would have been in a hopeless situation, since it is an essential component of a practical compatible color system. Some of RCA's most brilliant engineers and scientists rose to the challenge, and in an amazingly short period of time developed a demonstrable tricolor kinescope, or triniscope.

The tube was based on the shadow mask principle (Figure 5–4), which, with variations and improvements, is used nearly universally today. The

---

[18] Brown, *and part of which I was.*

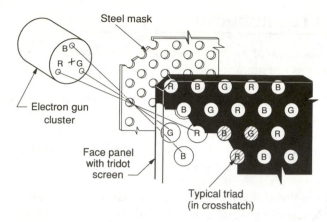

Steel mask

B

R +G

Electron gun
cluster

Face panel
with tridot
screen

Typical triad
(in crosshatch)

■ **Figure 5–4** The shadow mask principle. Shadow mask tubes are produced in a wide variety of sizes and configurations, but all of them are based on the same basic principle. Red, green, and blue phosphor dots are deposited in triads on the inside surface of the kinescope faceplate. The aperture, or shadow mask, a perforated metal plate with one perforation for each triad, is mounted just behind the faceplate. A cluster of three electron guns, located in the neck of the kinescope, are positioned so that their beams converge at the mask. Passing through the perforations in the mask, the beam from each gun impinges only on the dots of a single color because of the slight off-axis angle at which it goes through the perforation.

principle was first suggested in an impractical form by Germany's W. Flechsig in 1938. An equally impractical variation of Flechsig's concept was proposed by RCA's Alfred N. Goldsmith in 1947. Simultaneously, however, A.C. Schroeder, also of RCA, developed the format that would become the basis for the first practical tricolor tube.

With the beginning of the color hearing in September 1949, Engstrom put a program for the development of a tricolor tube in high gear under the direction of Edward W. Herold. He was given a virtual blank check on the corporation's resources, both financial and technical. He also was given a seemingly impossible schedule: feasibility demonstrated in three months and a working model in six.

As described earlier, the following months were exciting ones at RCA Laboratories. Engineers and scientists approached the problem with enormous enthusiasm, inspired not only by the technical challenge but also by Goldmark's confident statements that their task was impossible. Sixteen-hour days and seven-day weeks were common.

Dr. Albert Rose, who had gained distinction earlier as the developer of the image orthicon, kicked off the program by preparing a report titled "A Survey of Proposals for Single Tubes for Color Television." Five different approaches were investigated in depth, and in early February 1950, Harold B. Law, a key member of the team, concluded that Schroeder's shadow mask was the most promising. Development work was then concentrated on tubes using this principle, and two working prototype receivers with shadow mask tubes were produced in the incredibly short time of two months. They were first demonstrated to the FCC on March 23, 1950, and to the press on March 29.

The picture quality of the prototypes was surprisingly good, and, although neither CBS nor the FCC would admit it at the time, the demonstrations sounded the death knell for the CBS system. All the other basic technical problems of compatible color had been solved or had predictable solutions, but it could never have been practical without a satisfactory single-tube receiver. The successful demonstration of working prototypes made the ultimate approval of a compatible system inevitable.

There was, however, an enormous gap between the prototype and a reliable commercial product that could be manufactured economically. RCA was almost on the right track, but years of hard work lay ahead before it would have a manufacturable consumer product.

The most serious problems resulted from the manner in which the phosphor dots were deposited. In order to make the tube components interchangeable, the dots were printed with great precision on a flat glass plate by a silk-screen process. The plate was then mounted inside the kinescope bulb. Although the goal of interchangeability—any plate could be mounted in any bulb—was met, this construction caused new problems. It seriously limited the tube's size and performance and increased its cost. The diagonal of the original prototypes was only twelve inches, and the brightness was so low that they had to be viewed in darkened rooms.

To RCA's great embarrassment, the solution came from the Hytron Division of CBS, and its embarrassment was not mitigated by the fact that the solution was based on a photographic method of locating and depositing the phosphor dots that had been developed by Harold Law at RCA Laboratories in 1948. RCA had rejected Law's technique because it led to noninterchangeable tube components, but it turned out to be the key to the economical production of large-size kinescopes.

Norman Fyler and W.E. Rowe of CBS, aided by Russell Law (no relation to Harold), who had left RCA for CBS, successfully built a fifteen-inch prototype using Harold Law's photographic method in 1953. It was gleefully demonstrated to Sarnoff just prior to the FCC approval of the NTSC system.

If CBS's purpose was to needle Sarnoff, it was eminently successful. Sarnoff returned from the demonstration in a fury, called together his staff, and gave them a dressing down that was severe even by his standards. Soon afterward, Frank Folsom, RCA's president, convened a meeting of the top corporate executives. Shaken and grim, he said, "Gentlemen, we must catch and surpass CBS. If we do not, I can assure you that I shan't be here next year, and I will probably be joined by others."

This led to another crash program with no limit on the availability of financial and technical resources. It was coordinated by Charles Jolliffe, RCA's senior technical executive. The problem involved production processes more than basic technology, and the program headquarters were at RCA's tube plant in Lancaster, Pennsylvania, where the program was supported by the technical expertise of RCA Laboratories.

Although RCA had the basic patents on the process, it found it necessary to swallow its pride and obtain a patent license from CBS as well. According to George Brown, the fee was less than one million dollars.[19] According to Peter Goldmark, who gave CBS and himself almost total credit for the development of the shadow mask tube, the CBS legal team was able "to extract a formidable income . . . through patent infringement suits."[20] This was the only tangible return CBS received for its efforts because it had neither the will nor the resources to put the tube into production.

Bringing the tube from prototype to production proved to be as difficult as developing the prototype. The development of manufacturing processes for the tube was one of the major achievements of twentieth-century industrial technology. RCA's technical and production staffs performed superbly, and when the market breakthrough came in 1964, RCA alone was able to produce satisfactory tubes in quantity.

The initial tube designs suffered from numerous performance deficiencies. Among these were marginal brightness and contrast ratio (the ratio of the brightness of the highlights to that of the shadows), color fringing due to the failure of the three beams to converge over the entire screen, and muddy colors due to poor color purity. The latter two defects could be minimized by careful adjustment of complex external circuitry, but at best the performance of early tubes was significantly inferior to that of current models. The steady improvement in tube design that occurred in color television's first decade was one of the factors that led to the breakthrough in color receiver sales in 1964.

Design improvements did not end in 1964, however. From 1962 to 1982, typical screen highlight brightness was increased sixfold, from 10 foot lamberts to 60 foot lamberts. Some of this increase resulted from a compromise. In most tubes, the colorimetry of the phosphors deviates from the NTSC primary standards in order to increase the brightness. The public seems to prefer this, even though it results in some loss of color fidelity. (See the discussion of high-definition television in Chapter 10.)

The contrast ratio has been improved, not only because of brighter highlights but also because of blacker shadows. The phosphors used for the dots reflect a significant amount of the ambient light that strikes the screen so that early tubes appeared whitish, even when the set was turned off.

---

[19] Brown, *and part of which I was.*
[20] Goldmark, *Maverick Inventor.*

Since the black areas in a television picture can never be any darker than the tube appears when it is turned off, the reflected light limited the contrast ratio. In later designs, the amount of reflected light has been reduced by using glass in the faceplate, which absorbs a fraction of the light passing through it. This reduces the brightness, but the reflected light must go through the glass twice, and the contrast ratio is increased. The reflectivity of the array also can be reduced by surrounding each dot with a deposit of black graphite. As a result of these techniques, the picture tubes on modern sets appear much blacker when turned off, and the blacks in the picture are darker.

Major improvements also have been made in the performance of tubes with respect to fringing and color purity, and modern sets operate for long periods without the need to readjust them. In common with all consumer electronic products, the price of color tubes has steadily decreased in terms of constant dollars as a result of competition, design improvements, and more advanced manufacturing techniques.

Tube types using the shadow mask principle have proliferated since 1964. In 1986, Zenith announced its flat tension mask (FTM) tube. The faceplate is flat with no curvature, and this causes fewer problems with reflected ambient light. The shadow mask is a sheet of thin foil sealed under tension into the tube just behind the faceplate. It is distorted less by heat than is the conventional shadow mask and hence can handle higher beam currents and produce a brighter picture.

Tubes based on principles competitive with the shadow mask also have been developed. An example is the *Lawrence tube,* or *chromatron,* in which an electron beam from a single gun is deflected electrically by a comblike structure of vertical wires behind the faceplate so that it strikes red, green, and blue phosphor stripes in sequence. The chromatron was not successful commercially, but Sony, which obtained a license to produce it, used some of its concepts to develop an important shadow mask variant, the *trinitron.*

In the trinitron configuration, the phosphors are deposited in vertical stripes rather than dot triads, and the shadow mask is perforated with slots rather than holes. This construction produces a somewhat brighter picture than does the standard shadow mask. It was introduced in 1968 and was first used in smaller sets. Its application has now been extended to all standard sizes.

Color Receivers

The first color television receivers (Figure 5–5, top) had small screens, performed rather poorly, and were high priced. The first RCA model, the CT100, was introduced in 1954. It had a twelve-inch picture tube built according to the original RCA design with the phosphor dots deposited on

Courtesy RCA Corporation.

Courtesy Sony Corporation.

■ **Figure 5–5** (*Top*) An early RCA color receiver. The 21-inch model shown in the photograph was experimental; the first commercial model introduced in 1954 employed a twelve-inch tube. (*Bottom*) A modern Sony color receiver.

an internal glass plate. The tube's brightness was marginal, and the retail price of the receiver was $1,000 (nearly $4,000 in 1989 dollars).

The CT100 had a short product life—from March 25 to December 6, 1954—and fewer than ten thousand were produced. It was replaced by a new design using a twenty-one-inch tube with the dots deposited on the inside of the faceplate. Its performance was superior to that of the CT100, but it was far short of that of modern receivers.

The picture tube is the key component of a color receiver, and the constant year-by-year improvement of the performance of these tubes was reflected in the picture quality of receivers. In addition, constant incremental improvements were made in the receiver elements outside the tube. The availability of transistors provided new opportunities for receiver designers, and they were soon incorporated in sets, resulting in a significant reduction in size, power consumption, and reliability. The availability of very large scale integrated (VLSI) circuits made it possible to incorporate digital signal processing, a function that can result in picture quality that is better than the camera output.

In summary, as the result of more than thirty years of steady engineering progress, modern television receivers (Figure 5–5, bottom) are low in cost, perform well, and are reliable. If and when high-definition television (HDTV) becomes an important medium, the technology developed for NTSC and PAL receivers will be available as the basis for the design of HDTV sets.

■ COLOR CAMERAS

The Field Sequential Camera

Although the field sequential camera is not used for NTSC color, it played an important role in the development of color television. It was an effective vehicle for the promotion of color in its early years, and it provided a test bed for basic research and development in color. In addition, the threat of its competition gave RCA and the rest of the indsustry the incentive to expedite the development of a compatible system.

In principle, the field sequential camera was simplicity itself. It used a single image orthicon pickup tube and a rotating tricolor disk located between its front face and the lens. The disk had carefully shaped color filters that exposed the tube to one color at a time. It rotated synchronously with the television scanning rate so that the signal from successive fields in the camera output resulted from the exposure of the tube to the red, green, and blue components of the scene. With selected tubes, controlled conditions, and the skilled operation of Goldmark and his staff, it was capable of producing excellent pictures, judged by the quality standards of the day. It was much smaller and lighter than the bulky three-tube cameras required for compatible systems, and there was no problem of registration (that is, of precisely superimposing three images). It continued to be used in the space program into the 1980s.

The field sequential camera had its problems, however. Among these were low sensitivity (since each color was exposed only one-third of the time) and field-to-field image retention. In addition, it might not have worked as well with less skilled operators in day-to-day operation.

Goldmark did not give up on the sequential camera for broadcast use after the NTSC system was approved. He developed a device he called a "chromacoder," which converted field sequential signals to the NTSC format. His idea was to use sequential cameras in the studio with a chromacoder at the switcher output to convert the signal to NTSC for broadcasting. The sequential cameras would be smaller, cheaper, and lighter than the bulky three-tube cameras then in use, and they would be free from registration problems as well. The concept may have been sound, but the conversion process (which was accomplished by viewing three kinescopes, one for each color, with image orthicon tubes) degraded the picture quality to an unsatisfactory level.

Here is how Paley described a demonstration comparing the RCA three-tube camera with the CBS camera and chromacoder:

> Peter Goldmark, who headed the project, attended as an observer. Beads of perspiration dribbled down his face as he stood there. We watched in tense silence for fifteen minutes. When the program ended, there was a deadly pause before anyone would venture an opinion. I knew exactly what I thought. I stood up and said, "Gentlemen, I'll be glad to speak first. I think the RCA camera has us beat. It has better quality." I looked around and saw a general nodding of heads. No one spoke. So I walked out and that was the end of that CBS project.[21]

Paley went on to say that this was the awful moment when he finally realized that he had been the victim of bad advice from Goldmark and Stanton. CBS's persistence in pushing field sequential color had been enormously costly to the company, and it had little to show for it.

Three-Tube Image Orthicon Cameras

The first commercial studio cameras for NTSC color used three image orthicon tubes and evolved from laboratory models constructed to support research in simultaneous color systems. This camera, designated the TK-40, weighed nearly three hundred pounds and required a special cradle head for mounting on a studio tripod. The three color components of the scene

[21] Paley, *As It Happened.*

were separated and imaged on the tube by means of dichroic mirrors and a complex optical system. Transistors were not yet available to handle most of the electronic functions in the camera, which used vacuum tubes.

The TK-40, which was essentially an upgraded laboratory device, was superseded by the TK-41 in 1955. Although the basic principles of the two cameras were similar, the TK-41 was designed for studio and field operation and not for laboratory tests. Its performance was better, and it was easier to maintain and operate.

The TK-41, however, had many drawbacks. Its weight and size were serious problems, particularly for field use. Accurate registration of the three color images to avoid fringing also was a problem. Its solution required manufacturing tolerances in the optical and deflection systems that were an order of magnitude more precise than in monochrome cameras. Its worst problem was that it did not produce very good monochrome pictures—a serious fault as long as most of the receivers in the field were black-and-white.

The poor monochrome performance resulted from operating the image orthicons over a limited contrast range in order to provide color tracking. Monochrome pictures appeared soft and washed out, and the inability of the camera to handle a wide contrast range because of signal-to-noise limitations was especially noticeable in outdoor scenes where the lighting could not be controlled. For example, coverage of baseball games with the playing field partly in bright sunlight and partly in deep shadows was particularly difficult, and NBC's initial color broadcasts of the World Series were of inferior technical quality when seen on monochrome receivers.

The inherent deficiencies of the TK-41 were part of the reason for the slow development of color. It was, however, the best that could be achieved at the time. This is evidenced by the fact that its only serious competitor was a very similar camera offered by GE.

RCA's attempt to overcome these problems was embodied in the TK-42, which was first shown at the 1964 NAB convention. The design concept of the TK-42, developed by H.N. Kozanowski, manager of RCA's camera development group, was quite different from that of the TK-41. It was a four-tube camera with a $4\frac{1}{2}$-inch image orthicon providing the luminance, or monochrome signal, and three vidicon tubes providing the color.

On paper it looked very good. The use of a single $4\frac{1}{2}$-inch image orthicon for the luminance channel made the black-and-white picture quality equivalent to that of the best monochrome cameras then in use. Since the lower resolution color signals were not involved in producing the luminance signal, the registration problem was less severe. (The operating principle of the TK-42 was sometimes compared with comic strips in which the outlines of objects are black lines with colors painted inside.) The incorporation of a high-quality zoom lens in the camera was responsive to the growing preference in the market for this type of lens. Equally importantly, it was politically desirable for the RCA Broadcast Equipment Division because it used RCA tubes.

In spite of these desirable features, the TK-42 was very nearly a technical disaster. RCA paid a heavy price for its reluctance to use non-RCA tubes in its color cameras. The TK-42 was bigger and heavier than the TK-41. The extensive use of solid-state circuitry did not result in a smaller camera, and the zoom lens increased its weight to almost four hundred pounds. Its sensitivity was limited because of light sharing between the luminance and color channels and because the vidicons, if operated in a high-sensitivity mode, exhibited serious color smearing. Compared with today's cameras, the luminance channel was noisy and produced a snowy picture, an inherent property of image orthicons.

The TK-42's mechanical, optical, and electronic systems were extraordinarily complex, a problem that was aggravated by the initial design bugs common in new products. It was difficult to maintain and adjust because of the problems of tracking the very different characteristics of vidicons and image orthicons. The quality of its monochrome picture was of diminishing importance after 1964, and its color picture, while reasonably satisfactory with skillful operation, were often pretty awful.

In summary, the TK-42 was large, heavy, expensive (about $100,000), complex, and difficult to operate and maintain. With all these problems, no company except RCA could have pulled it off. RCA's reputation and its large and able sales and service organizations, aided by the lack of competition until 1966 when deliveries of competitive Plumbicon cameras began, managed to make it a marginally successful product. A total of five hundred cameras were shipped to customers. The first three hundred, delivered in 1965 and 1966, sold easily. The next one hundred sold with more difficulty. And the last one hundred, delivered in 1968 and 1969 after the full impact of the Plumbicon was felt, required extraordinary salesmanship.

Enter the Plumbicon

Early in 1965, CBS showed color pictures on the air with technical quality that received rave reviews from New York's television critics. The pictures were produced by experimental cameras loaned to CBS by Philips Eindhoven. Their quality resulted from the use of an entirely new pickup tube, the Plumbicon, which had been developed in great secrecy in Philips's laboratories. This tube was one of the major breakthroughs in television technology, and it was an important factor in the explosive growth of color television.

Like the vidicon, the Plumbicon depended on photoconductivity for its operation, but it could operate at low light levels without the lag or smear the vidicon exhibited. The photosensitive surface was lead oxide, hence the tube's name. The tube itself generated no noise, and all the noise in the output signal originated in the tube's amplifier. Thus it produced extraordinarily noise-free pictures. Further, the lack of noise gave camera designers elbowroom for the addition of correction circuits that enhanced picture sharpness and color fidelity without excessively increasing picture noise.

The Plumbicon was an extraordinarily difficult tube to manufacture, and the manufacturing process required a great deal of black art. Its genesis was Philips's research in X-ray imaging, and it was a remarkable achievement. With all RCA's know-how in television pickup tubes, it was never able to copy the Plumbicon successfully in spite of years of effort.

Philips first demonstrated a commercial camera using Plumbicon tubes, the PC-70, at the 1965 NAB convention. It was distinguished not only by the use of Plumbicons but also by an optical system of new and unique design. The bulky assemblage of mirrors in the RCA camera was replaced by a small optical block consisting of a system of dichroic-surfaced prisms cemented together. The use of only three Plumbicon tubes, which were only slightly more than one inch in diameter, and the compact optical system made it possible to reduce the camera's size and weight to little more than those of a monochrome image orthicon camera.

The response of the industry to the PC-70 was extremely enthusiastic. RCA's other competitors, particularly GE and Marconi, immediately recognized the potential of Plumbicons and incorporated them in their cameras as soon as they could complete the design and production cycles. Both companies chose the four-tube option to minimize the registration problem. This was beneficial to RCA because it gave RCA support in the three-tube/four-tube controversy and thus helped TK-42 sales.

The success of the three-tube Plumbicon camera created a huge problem for the management of RCA's Broadcast Equipment Division, both in the marketplace and politically within the company. It had placed its reputation on the line for the four-tube concept. Further, it felt great pressure, usually unspoken but very real, to use nothing but RCA-manufactured tubes in its cameras. Charles Colledge, the division's general manager, tried to solve its problems with another four-tube camera and a revised tube complement.

The $4\frac{1}{2}$-inch image orthicon was replaced with a tube of related design, the image isocon, which produced a picture that was somewhat more noise free. The policy of using only RCA tubes was changed to allow the use of Plumbicons rather than vidicons in the chroma channels. They had greater sensitivity and were more easily matched to the image isocon. By using an external zoom lens and incorporating other improvements, the camera's size and weight were greatly reduced as compared with those of the TK-42.

Unfortunately, these improvements were not sufficient. The camera still produced a noisy picture, and it still was difficult to set up and maintain. Designated the TK-44, it was first demonstrated at the 1967 NAB convention, where it received a lukewarm reception from broadcasters. NBC's management made it clear that it would never buy one. Instead, it advised RCA that the three-tube Plumbicon was the right choice.

In the meantime, the lower levels of RCA's Broadcast Equipment Division were trying to persuade its top management that it was on the wrong course. They were encouraged and aided in this effort by NBC's management, particularly William Trevarthen, its engineering and operations vice president. In the summer of 1967, Colledge was finally convinced, and he authorized a parallel engineering program for a three-tube Plumbicon camera. It was authorized on the basis that it would be a lower priced camera for the small-station market, but this was a deception for internal political purposes. Everyone assumed that it would eventually become RCA's top of the line.

And it did. As the year 1967 proceeded, more and more engineering effort was switched from the four-tube camera to the three-tube camera. In the fall, Colledge retired and was succeeded by Barton Kreuzer. Kreuzer immediately stopped all work on the four-tube camera, and the division's engineering resources were concentrated on the three-tube product. With the benefit of hindsight, it is clear that this decision saved the division from disaster and allowed it to regain its leadership role.

Plumbicon Studio Cameras

The Philips PC-70 was an enormous success. By 1968, it had long since passed RCA's TK-42 in sales, even in the United States, and it dominated the world market. The four-tube cameras offered by GE and Marconi were modestly successful, but their sales did not compare with those of the PC-70. Philips became so confident of its camera that it virtually stopped updating it and instead concentrated on a second-generation camera of advanced design, the PC-100. This turned out to be a disastrous decision.

Philips, flushed with victory over RCA, had underestimated the enormous resources available to RCA and its determination to use them. Norman Hobson, one of RCA's top design engineers, was put in charge of RCA's three-tube program, and many of its most capable engineers were assigned to it. RCA received a tremendous amount of help from NBC, particularly from engineer Fred Himmelfarb. An engineering prototype was sent to NBC for his evaluation, and he proceeded to dissect it, circuit by circuit. A stream of memos came to RCA's Camden plant pointing out design deficiencies and making suggestions for correcting them. RCA sometimes resented Himmelfarb's suggestions, but they were effective. When shipments began in 1969, there were few field problems, and this camera, also designated the TK-44, became one of RCA's most successful products.

The TK-44 was first shown at the 1968 NAB convention. A large contingent of Philips executives from both Holland and the United States viewed the first demonstration. When they saw that it used the same basic principles as the PC-70, including the tube layout and optical system, they could scarcely contain their glee. They had humbled the mighty RCA, which had been forced to copy their design—or so they thought.

They did not at first realize that the TK-44, although similar to the PC-70 in its basic design, was a vastly superior camera. The combination of RCA and NBC know-how led to countless refinements, which, though not large individually, in total greatly improved its performance and operation. The Philips management, however, regarded the TK-44 as confirmation of their decision to concentrate their resources on the PC-100.

The consequences of this decision were exacerbated by mistakes in the marketplace. Philips chose to base its sales approach on the advantages of three tubes over four rather than Plumbicons over image orthicons and vidicons. This was the wrong emphasis. While hindsight has shown that the three-tube approach is indeed better, the comparison was not all its favor. The superiority of the Plumbicon, however, was clear-cut. By choosing three tubes over four as its primary sales weapon, Philips gave RCA the opportunity to develop a credible opposing marketing strategy. During 1968 and 1969, RCA offered the TK-42 and TK-44 cameras simultaneously, touting them as "The Best of Both Worlds" (of three- and four-tube cameras). Surprisingly it worked, and by the end of 1969 most of the TK-42s had been sold. With a great sigh of relief, RCA dropped it from its product line.

Philips's second marketing error was to terminate its agreement with Visual Electronics, headed by veteran broadcast salesman James Tharp, which had handled its distribution in the United States. In its euphoria, Philips decided to set up its own sales force. Thus, at the very time when it began to face severe competition from RCA and had the greatest need for an in-place and experienced sales force, it approached the market with a green organization.

The final blow to Philips was the failure of the PC-100. It was intended to include many advanced features, including the use of a small triax cable to replace the bulky ones required by earlier cameras. Philips found that it had attempted more than it could accomplish, and the PC-100 never became a commercial reality. As a result of these errors, Philips lost its dominant position in the marketplace almost as rapidly as it had gained it.

Emboldened by Philips's initial success with the PC-70, other European manufacturers made serious attempts to penetrate the huge U.S. market for studio cameras. In addition to Philips, Marconi and Fernseh, the latter known by the name of its corporate parent, Robert Bosch, were among the more aggressive. None of them were very successful. Manufacturers in the three largest industrial countries—England, France, and Germany—had the advantage of captive customers in their state-owned broadcasting systems, but this was often a disadvantage in selling in the U.S. market. Their

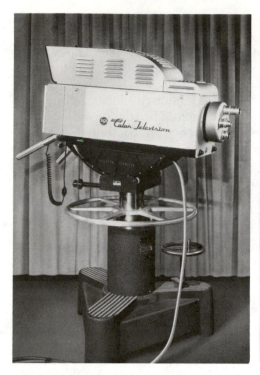

Courtesy RCA Corporation.

Courtesy RCA Corporation.

Courtesy Philips Television Systems.

Courtesy RCA Corporation.

■ **Figure 5–6**　The evolution of color cameras, 1953–1970. (*Top left*) The RCA TK-41 (1953) had three image orthicons and weighed three hundred pounds. (*Top right*) The RCA TK-42 (1964) had one image orthicon and three vidicons. It weighed four hundred pounds. (*Bottom left*) The Philips PC-70 (1965) and (*Bottom right*) the RCA TK-44 (1969) each had three Plumbicons and weighed one hundred pounds.

designs were heavily influenced, if not dictated, by the state networks, whose requirements and operating practices were often different from those in the United States. Also, European engineers, both in manufacturing and operations, tended to be more scientific but less practical than their American counterparts, and their products were more sophisticated and complex—qualities not always desired by American broadcasters.

As a result, after RCA began deliveries of the TK-44 in 1969, this product and its successors had the largest market share for studio cameras for more than ten years, until they were overwhelmed by the Japanese invasion—a story told later in this chapter.

Film Equipment

Although the Ampex magnetic video recorder was introduced soon after the approval of the NTSC color system, it was many years before magnetic recording was sufficiently perfected to serve as the primary storage medium for television programs. Its quality of color recording was marginal, and its capability for editing or making multigeneration copies was limited. These limitations, together with the huge libraries of program material on film, gave great importance and urgency to the development of equipment for the conversion of color film images to television signals.

As with monochrome television, there were two competing systems, the flying spot scanner and storage tube systems using film cameras with three vidicons or Plumbicons. The flying spot scanner was offered by most European manufacturers and by Dumont and GE in the United States. The vidicon camera was offered initially only by RCA. There was a vigorous competitive battle, and for a number of years the outcome was in doubt. In the end, film cameras using storage tubes have been almost universally used in the United States and to some extent in Europe.

As the performance of magnetic tape recorders improved, they nearly superseded film for both the production and storage of television programs. Today, film playback systems are used only occasionally by television broadcasters. They are used mainly by production houses that specialize in the high-quality conversion of film libraries to tape. The cameras are predominantly flying spot scanners or charge-coupled device (CCD) types (described later in this chapter), and they include elaborate devices for color correction, noise reduction, and reduction of wide-screen films to the 4:3 aspect ratio used by television.

Electronic News-Gathering Cameras

During the 1970s, the basic technologies used in color cameras continued to advance. Smaller Plumbicon tubes with good performance became available. Transistors were replaced with integrated circuits. Developments such as these made it possible to design truly portable color cameras that could provide live coverage of breaking news stories. This function had previously been performed by film cameras, and there was usually a delay of an hour or more while the film was being returned to the studio and developed. In combination with satellite transmission (see Chapter 8), these cameras made the live pickup of breaking news events possible. They revolutionized and greatly increased the scope of television's news departments by providing the capability for *electronic news gathering* (ENG).

One of the first and most successful of the early ENG cameras was the RCA TK-76 (Figure 5–7), introduced in 1975. It used three two-third-inch Plumbicon tubes and the same optical format as the TK-44. While its picture quality was not quite up to that of the TK-44 and other high-quality studio cameras, it produced far better pictures than the best cameras of a decade earlier. And it did so with a minimum of maintenance and adjustment.

The TK-76 was extremely popular, and more than two thousand were sold, an enormous quantity for such a product. Sadly for RCA, it was its last successful television camera, and it symbolized the end of an era.

Courtesy RCA Corporation.

■ **Figure 5–7** The RCA TK-76 ENG camera. This was one of the first and most successful ENG cameras. It was also the last RCA camera to enjoy a prominent position in the marketplace.

Charge-Coupled Device Cameras

Although the TK-76 was RCA's last commercially successful camera product, RCA made a major contribution to camera technology with the introduction of a color camera using charge-coupled devices (CCDs). The CCD is a solid-state imaging device (see box) that is an alternative to conventional pickup tubes. As compared with tubes, CCDs are smaller, use less power, have better geometric registration, and have a longer life. For broadcast use, perhaps their most important feature is an almost complete absence of lag or smear.[22] NBC made very effective use of this characteristic in the 1984 World Series by focusing the camera on the rapidly spinning balls thrown by the pitchers. The lack of smear was so complete that the stitches on the ball were visible.

OPERATION OF A CHARGE-COUPLED DEVICE

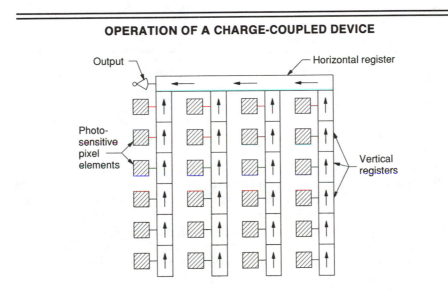

An image of the scene is focused on a rectangular array of photosensitive solid-state elements. Each element acquires a charge that is proportional to its illumination. The array is then scanned in synchronism with the television scanning rate by causing the charges to migrate into registers and thence to the output of the device.

The drawing shows a scanning format known as interline transfer (IT). An alternative format known as frame transfer (FT) also is commonly used.

---

[22] To take full advantage of this characteristic, it is necessary to add a mechanical shutter to the camera that exposes each frame for a very short period of time, perhaps one thousandths of a second.

Looking to the future, CCDs have the potential for increased sensitivity without a corresponding increase in noise. This characteristic has not been fully used to date.

The CCD was first announced by Bell Laboratories in 1970, but more than twelve years passed before the technology and manufacturing processes advanced sufficiently to make it practical for television cameras. CCDs were first used in consumer cameras offered by RCA, Toshiba, and Sony. RCA demonstrated a development model of a broadcast-quality CCD camera at the NAB and Montreux conventions in 1983, and it introduced a commercial model in 1984. NBC used this in a number of special events such as the World Series. RCA, however, did not have time to perfect the design before it withdrew from the business, and it was left to other manufacturers, primarily the Japanese but also Ampex, to make it a successful commercial product.

■ AGAIN THE JAPANESE

The Japanese began their efforts to penetrate the U.S. market for radio equipment in the 1950s. Japanese companies, notably Sony, were among the pioneers in audiotape recorders, and they offered models for both the broadcast and audiovisual markets. These markets were small and specialized, however, and they did not offer the potential of becoming the basis for a major business.

It was the advent of the transistor that first gave the Japanese the opportunity to make a major assault on the U.S. market for radio and television products (see Chapter 2). This began in the late 1950s, and since then the Japanese advance had been inexorable, albeit with some setbacks.

The markets for audio and radio products, while significant, was small in comparison with that for television, and the market for monochrome television receivers had dwindled by the time the Japanese became a viable participant. It was Japan's entry into the markets for color television equipment and video recorders that made it a major participant in the U.S. radio and television industry.

Japanese Color Receivers: Advance and Retreat

Japan's first color receiver exports began in 1968 with a small Sony transistorized portable using a trinitron picture tube. The pictures were bright and sharp, and the receivers sold well. In the early 1970s, Japanese manufacturers moved aggressively into the U.S. market, and in 1975 they exported 1,215,000 color receivers to the United States—20 percent of the market in terms of units. In 1976, exports increased to 2,834,000 units, or 35 percent of the market.

■ JAPANESE PENETRATION OF
U.S. RADIO AND TELEVISION MARKETS

Market	Initial Sales	Achieved Major or Dominant Market Share
Consumer audio	1955	1965
Radio receivers	1955	1965
Monochrome TV receivers	1959	1970
Monochrome TV cameras[1]		
Color receivers	1968	1976
Color cameras	1974	1979
Broadcast video recorders[2]	1970	1977[3]
Video recorders for the home[4]	1976	1977

[1] The market for monochrome television cameras (except low-cost industrial models for surveillance) virtually disappeared around 1968, before the Japanese were ready to address it on an international basis.

[2] Helical scan recorders; Japan was not a factor in the quadruplex recorder market.

[3] The year of approval of SMPTE Type C format, a compromise between the Ampex and Sony proposals (see Chapter 6). These companies then became the leaders in the market for broadcast video recorders.

[4] Japanese companies dominated the home recorder, or VCR, business from the beginning (see Chapter 8). The U-Matic recorder, the predecessor of the VCRs in the consumer market, was introduced in 1969.

The Japanese soon became victims of their own success, however, and further increases in their market share were inhibited by a number of factors:

- Their success encouraged competition from other eastern Asia countries and Mexico.
- Growing protectionist sentiment in the United States led to the passage of antidumping laws, quotas, and other import constraints.
- The strength of the yen versus the dollar and other currency increased the cost of Japanese products in the United States.

These developments led to a significant erosion of Japan's market share, and they encouraged the importation of kits of parts and chassis for final assembly in the United States by U.S. manufacturers or subsidiaries of Japanese companies. The results of these developments are shown in the following U.S. Commerce Department's statistics:

In spite of this erosion in market share, the Japanese remain a major force in the U.S. market for color receivers. Their market share for picture tubes is much higher, and they continue to display strong technical leadership.

■ **COLOR TELEVISION IMPORTS: FIRST TEN MONTHS OF 1987**

Country	Complete Sets	Kits/ Chassis	Total	Market Share (%)
Japan	530,500	1,002,400	1,532,900	8.2
Taiwan	2,441,700	400,100	2,841,800	15.1
Korea	1,576,700	1,231,300	2,808,000	14.9
Mexico	1,264,900	1,485,800	2,750,700	14.6
Singapore	408,400	745,900	1,154,300	6.1
Malaysia	741,900		741,900	4.0
Hong Kong	201,800		201,800	1.1
Canada	77,900		77,900	0.4
China	70,100		70,100	0.4
Total	7,313,900	4,865,500	12,179,400	64.8

Total U.S. market = 18,803,463*

* The market shares were calculated with respect to this figure for the total market. This leads to minor inaccuracies, since the total market statistic is based on EIA reports of sales to dealers, which do not coincide precisely with the imports.

Japanese Cameras

Japan's first entries in the U.S. radio and television market were consumer products. For many years, the manufacturers of broadcast station equipment felt they were reasonably immune from Japanese competition because the market was too small to attract them. This turned out to be a grave miscalculation. The Japanese *were* attracted to the business, and they were willing to make major investments to enter it. (It is said that Sony lost money in this business for thirteen years before turning a profit.) Once having made the decision, they proceeded to carry it out with their usual skill and tenacity.

Their first point of attack was the videotape recorder, a product line in which they have shown special skill. The advent of helical scan recorders gave Japanese manufacturers the opportunity to excel in tape recorders. The history of this development is detailed in Chapter 6. In the meantime, however, they were patiently developing their product plans and strategies for color cameras.

In 1979, the camera market succumbed to the Japanese. Sony, Ikegami, and other Japanese manufacturers had begun to display their cameras at the NAB conventions as early as 1976, and by 1979 their dominance over U.S. manufacturers had become evident. Although RCA made a valiant attempt by introducing computer control in the TK-47—a major advance in camera technology—neither it nor any American-made camera has successfully competed against the Japanese since 1980. Thus, they brought the era of U.S. leadership in broadcast equipment to its conclusion.

Studio Cameras The major technical contribution of the Japanese to studio cameras has been the extraordinary quality of their designs. Japanese cameras also are manufactured with high standards of quality and sold for comparatively low prices. One reason for Japanese excellence in product design is the amount of time and engineering manpower they devote to it. The smallest component shows the results of concentrated attention to detail. Figure 5–8 shows a modern studio camera from Sony.

The development of the saticon pickup tube as an alternative to the Plumbicon was another important contribution to camera technology. Like the Plumbicon, the Saticon, developed by NHK, the Japanese national network, in 1973, is a photoconductive tube that uses an arsenic–selenium compound as the photosensitive surface. The photoconductive surface was first developed by RCA, but the Japanese carried it through to a product. As compared with the Plumbicon, it has better static resolution and less flare. It has worse dynamic resolution (resolution of moving objects) and worse handling of highlights, however. It is made in a two-thirds-inch version with mixed field operation that combines magnetic focusing and electro-static deflection. This tube is particularly useful for portable cameras.

CCDs, ENG Cameras, and Camcorders The characteristics of CCD cameras, particularly their light weight, low power drain, and ability to stop fast motion, make them ideally suited for use in ENG and camcorder applications. (A camcorder combines the functions of a camera and a re-corder in a single portable unit. See Chapter 6.) Building on RCA's experi-ence with CCDs in 1984, the Japanese have developed CCD cameras for both purposes.

The characteristics of CCDs versus Plumbicons are summarized in the published specifications of two commercial ENG cameras:

■ **COMPARISON OF CCD AND PLUMBICON ENG CAMERAS**

	CCD	_Plumbicon_
Model	Sony BVP-330A	Sony BVP-5
Imagers	2/3″ CCDs	2/3″ Plumbicons
Weight	7 lb., 1 oz.	10 lb., 9 oz.
Power	10.5 w	24 w
Sensitivity	2,000 lux, F5.6	2,000 lux, F4.5
Resolution	500 lines	600 lines
Registration error		
Center	0.05%	0.1%
Corners	0.05%	0.4%
Signal-to-noise ratio	58 dB	58 dB

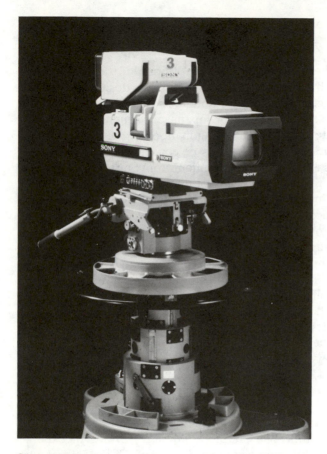

Courtesy Sony Corporation.

■ **Figure 5–8**   A modern studio camera made by Sony.

The performance of the CCD is particularly impressive in light of the fact that the Plumbicon has had the benefit of twenty more years of engineering development in commercial camera applications. Clearly, the CCD will play an important role in future color cameras.

6

■ ■ BROADCAST VIDEO RECORDING

Early in 1956, General David Sarnoff issued a stern edict from the RCA holy of holies, the executive offices on the fifty-third floor of the RCA Building in New York. RCA and NBC were to spare no effort or expense in sending a single message to the upcoming NAB convention in Chicago: Color is here! Sarnoff hoped this would persuade the doubters in the broadcast industry, especially the other television networks, to adopt a more aggressive attitude toward color programming. More than two years had elapsed since the FCC had approved the NTSC/RCA color system, but color programming was limited, and color had failed to make much progress in the market-place. Sarnoff was impatient, and RCA's bankers were critical. The pressure was on.

NBC's role was to develop a promotional program based on the dedica-tion of the new color studios of its Chicago station, WMAQ-TV. RCA's Broadcast Equipment Division was directed to demonstrate its new line of color equipment in a massive exhibit. In addition, a small army of advertis-ing and public relations executives was busily planning a series of press conferences, receptions, and other promotional events. In total, many hun-dreds of engineering and marketing specialists were engaged in planning and executing the programs.

In the meantime and without RCA's knowledge, a group of six engineers at a small company in Redwood City, California, was making plans of a different sort. The company was the Ampex Corporation, a highly re-spected manufacturer of high-quality audio recording equipment. It was preparing to demonstrate a revolutionary new product—the first practical magnetic tape recorder for video signals.

The development of a satisfactory video recorder had been an extraordi-narily difficult problem, and for many years it had frustrated the world's major television research laboratories. Now a small group of engineers at Ampex had solved the problem, and the company was ready to introduce its product to the marketplace.

Partly to maintain preconvention secrecy, the recorder was not located in the main NAB exhibit area but was installed in a dingy suite in a rear wing of the convention hotel, the Conrad Hilton. In spite of its inconspicuous location, the demonstrations of this product were a sensation.

The excitement was well justified, because this was a historic milestone in the development of the television industry. It was a massive technical breakthrough, and it changed the nature of television broadcasting forever.

As for RCA, it suffered a triple blow:

- Its efforts to promote color at the convention were completely upstaged by the Ampex demonstrations.
- The Ampex recorder could not reproduce color, and this caused a further delay in the progress of color broadcasting.
- RCA's technical community, with its vast reputation for television research, had suffered a humiliating defeat at the hands of a small company with no previous television experience.

From this modest but dramatic beginning, video recording has developed into a multibillion-dollar worldwide business. Video recorders could not be built at any price prior to the Ampex breakthrough but now are mass-produced and can be bought for a few hundred dollars. This chapter and Chapter 9 tell the story of the development of this new industry.

■ RECORDING ON FILM

Kinescope Recording

The recording of video signals on magnetic tape is now such an integral part of television that it is hard to imagine a broadcasting system without it. Prerecorded programs, reruns, instant replays, electronic news gathering, electronic editing, and network time delay all require a rapid, reliable, and high-quality method of recording and playing back video signals.

In the early days of television broadcasting, the industry was forced to record video signals by the only means available—kinescope recording, or the photographing of television images on a kinescope. In principle, kinescope recording was a satisfactory system. In practice, it suffered from so many technical and operational problems that it was only marginally successful and became obsolete as soon as practical magnetic recorders were available.

Kinescope Recording's Problems Kinescope recording's most serious technical problem was the cumulative degradation of the picture signal as it was processed and transmitted through a lengthy series of system components from television camera lens to home receiver. Each step contributed to the degradation of one or more of the basic elements that determine picture quality—signal-to-noise ratio ("snow" in a television picture and graininess in a photograph), gray scale, and sharpness. The conversion of the picture information from electrical to photographic form and then back to electrical form was particularly troublesome.

The graininess of the film, when added to the noise from the camera tube and the transmission channel, resulted in pictures that were significantly snowier than those produced by live pickup. Good gray-scale reproduction was extremely difficult to achieve because of the tendency of the camera tube and film to compress blacks and whites. The final picture was often mostly dark blacks and bright whites with only a few gray areas. (The ability to provide good gray-scale reproduction was probably the most striking advantage of early magnetic recorders.) The loss of sharpness was the least serious of the problems, but only by comparison, and it contributed to the overall degradation of the picture.

By contrast, a magnetic recording system, in principle, is simplicity itself. It has fewer components, and the picture information remains in electrical form from television camera to receiver.

In addition to marginal picture quality, kinescope recording in the United States had another annoying problem resulting from the incompatibility of the 24-frame-per-second film standard and the 30-frame-per-second rate for television. To make this conversion, it was necessary to photograph the upper and lower halves of alternate film frames from different television fields. Unless the equipment was very carefully adjusted, an annoying line known as a shutter bar would appear at the splice across the center of the picture. The requirement for recording color introduced a whole new set of problems that were never solved sufficiently to permit widespread on-air use.

Kinescope recordings (kines) had a final problem—the time required to develop the film. For network time delay, less than an hour could be allowed. This could be accomplished in monochrome with reversal film and rapid developers, although they resulted in grainier prints. Instant replays, an integral part of today's sports broadcasts, were impossible. For color film, a one-hour development time was out of the question. In short, the industry was in desperate need of a better method of video recording.

Broadcast Kinescope Recorders Uncertain that a better recording system would be developed, the engineering departments of major television organizations in the United States and Europe worked diligently to improve the quality and reliability of kinescope recording. Kinescope recording research was a marvelous opportunity for engineers who had infinite patience and a desire for long-term employment. There are more than a dozen independent variables in television and film systems, and each had to be investigated. Improvements came slowly and in small increments, and there was never a major breakthrough leading to a dramatic improvement in quality.

In spite of these difficulties, there was sufficient demand for kinescope recorders from broadcasters to encourage manufacturers to offer off-the-shelf products. RCA and General Precision Laboratories were the two leading suppliers of this equipment in the United States.

The major networks and other major users of kinescope recorders were not satisfied with off-the-shelf products, and most of them custom built their own equipment. In the end, however, obtaining satisfactory results in day-to-day operations was more art than science. This led to the development of a mystique about kinescope recording that its practitioners assiduously cultivated. It included a jargon that was unintelligible to outsiders.

Because of the importance of kinescope recording to NBC, RCA was one of the leaders in its research. Its most ambitious program was the investigation of ultraviolet-sensitive film as the recording medium. Paper studies had shown that this would result in a significant improvement in all aspects of picture quality. Achieving this improvement in practice was another matter, and the research was lengthy and tedious. Whether it would have been ultimately successful will never be known. The program, which had started in 1950, was still far from finished in 1953 when the approval of the NTSC color system brought it to a halt. This led Sarnoff to believe that monochrome systems would soon become obsolete, and all research in monochrome apparatus was deemphasized at RCA.

The emphasis in kinescope recording research then shifted to color. RCA investigated two techniques. The first and most straightforward consisted of recording a triniscope image on color film. With skillful operation, surprisingly good results could be obtained under laboratory conditions. But the film was expensive, and rapid development was impossible. The second was the use of lenticular film, monochrome film with cylindrical lenticules, or lenses, embossed on it. When viewed through a special filter, a color image could be obtained on playback. This permitted rapid film development, but unfortunately it never worked very well.

In spite of these difficulties, RCA continued to make optimistic prognoses of the future of kinescope recording. I participated in this and in 1955 addressed the Academy of Television Arts and Sciences in Hollywood on the subject of magnetic and kinescope recording. I described the problems of magnetic recording of video in gruesome detail, predicting that it would be many years before they would be solved. At the same time, I gave an upbeat forecast of the future of kinescope recording, emphasizing the point by showing an excellent and carefully selected color kine. Members of the audience, most of whom came from the film industry and saw magnetic recording as a competitor, loved it. Mercifully, this speech was forgotten when Ampex demonstrated a successful magnetic recorder a year later.

Electronic Video Recording

Long after kinescope recording has passed into the history books, the recording of television images on film had one more lease in life. The indefatigable Peter Goldmark developed a new method of photographic recording in which the film was exposed by scanning it with an electron beam. This required the film to be placed in a vacuum chamber during the recording process, but once the negative was exposed and developed, any number of positive prints could be made easily. Color signals could be recorded by exposing three film negatives simultaneously to the red, green, and blue signals and making a composite print.

Two varieties of the concept were announced in 1967. A higher quality system called broadcast electronic video recording (BEVR) was intended to produce films suitable for transmission over the air. A lower quality system called electronic video recording (EVR) would be used to produce films for distribution to the general public. Functionally, EVR was equivalent to the videodisk introduced more than ten years later. The films were mounted in cassettes and played back through a player attached to an ordinary television set. The recording process required costly professional equipment, and home recordings were not possible. Like the videodisk, the system's consumer uses were limited to prerecorded material.

The use of an electron beam rather than a kinescope image for exposing the film solved many of the problems of kinescope recordings, and Goldmark was able to demonstrate some rather impressive results. The VCR and videodisk were not yet invented, and the possibility of an inexpensive medium for the distribution of programs to the public was appealing. Although Goldmark was not in the good graces of Paley because of his key role in CBS's color television debacle and the CBS–Hytron disaster, he was ever the salesman, and once more CBS management agreed to support one of his ideas.

Neither BEVR nor EVR was a success, and in 1971, four years after its announcement, CBS wrote off the business. We will never know whether this failure was the result of bad management or a lack of patience or whether it was just a bad idea. It was probably a combination of all three. Perhaps the best commentary came from Paley himself:

> One might think that men occupying such lofty positions as do William S. Paley and Frank Stanton always learn from such mistakes [the Hytron fiasco]. We would never do such a thing again, would we? But a few years later, CBS Laboratories developed a marvelous little invention called Electronic Video Recording, or EVR. . . . The possibilities for its use indeed seemed unlimited in education, in industry, in government, in the professions and in the home. There were predictions of a new billion-dollar business in the making and CBS was in the lead with the first such system on the market.
>
> . . . . I was promised and assured that the EVR operation would not escape our strict managerial and financial control, as had Hytron. But over the next four years we encountered great problems in manufacturing here and abroad. . . . . I discovered our original marketing projections were absolutely unrealistic and overstated by a huge amount. It seemed to me that we were once again in a hopeless situation, pouring millions of dollars into an inven-

tion for which there was only a questionable existing market. Our costs relative to projected sales were far, far out of line. Then several people associated with the EVR project began to admit to me that they were wavering on their earlier estimates of the success of the video recorder. Except for a few specialized applications, we never did get to the market with EVR. The manufacturing problems seemed to overwhelm our hopes and dreams.

Finally, in 1971, after consultations with some of my associates, I had to say "Enough, we've had enough." We sold off our overseas interests and began to phase out our domestic operations, and eventually wrote the whole venture off. I began to look on Peter Goldmark, whose fame as an inventor for CBS had spread far and wide, as a thorn in my side. That year, he turned sixty-five and retired from the company.[1]

■ THE GENERAL'S BIRTHDAY PRESENT

It was September 27, 1951, and General Sarnoff was in a euphoric mood. The date was very nearly the forty-fifth anniversary of his association with the radio industry, which had begun with his employment by the Marconi Wireless Telegraph Company of America on September 30, 1906. The CBS color system had won the initial round at the FCC, but it was going nowhere commercially and it was obvious to all that a modified RCA system would be the eventual winner. What pleased the General most on this day was the renaming of RCA Laboratories in Princeton, New Jersey, to the David Sarnoff Research Center. Naming buildings, ships, and forts after philanthropists, politicians, admirals, and generals is a standard practice in public life, but it is rare indeed in industry. And so Sarnoff regarded this as a high honor.

The official ceremony took place at a gala party that served to celebrate both the rechristening and Sarnoff's forty-fifth anniversary. The affair began with an elegant luncheon. It was followed by speeches and other testimonials, including a telegram from President Truman paying tribute to the General's many accomplishments and his contributions to radio and television.

The climax was a speech by Sarnoff himself in which he challenged RCA to deliver three birthday presents for his fiftieth anniversary in radio. They were an electronic light amplifier, which would make bright large-screen television possible, an electronic refrigerator, and a television tape recorder.

His exact words concerning the final item were as follows:

> Another present I would like to ask from you also relates to television. I would like to have you invent a television picture recorder that would record the video signals of television on an inexpensive tape, just as music and speech

---

[1] William S. Paley, *As It Happened* (Garden City, N.Y.: Doubleday, 1979).

are now recorded on a phonographic disk or tape. Such recorded television pictures could be reproduced in the home, or theater, or elsewhere at any time.[2]

In making this request, Sarnoff seemed to have lost his touch as a prophet of new technologies. To this day, thirty-five years later, no one has developed a commercially satisfactory television light amplifier or electronic refrigerator. To RCA's enormous embarrassment, the first practical television tape recorder was developed by Ampex, then a tiny competitor. And the manufacture of inexpensive recorders for home use was pioneered and has been dominated by the Japanese.

It goes without saying, however, that Sarnoff's wish was the Laboratories' command. For the next few years, a large share of its vast resources was devoted to these three programs. Its task was eased somewhat by establishing objectives that were more limited than Sarnoff had in mind. He expected his presents would be the basis for commercial products that could be sold in volume and at a profit. In contrast, the laboratory's efforts were directed toward the development of working models that could be demonstrated with little or no concern as to whether they could evolve into practical products. After all, that was the job of the operating divisions!

Harry F. Olson

Responsibility for the video recording project was given to Dr. Harry F. Olson, vice president for acoustical and electromechanical research. He was a scientist of worldwide renown who had distinguished himself previously by major contributions to acoustics and audio technology. He was the holder of more than one hundred patents and a fellow of the American Physical Society, the IEEE, the Audio Engineering Society, the Society of Motion Picture and Television Engineers (SMPTE), and the Acoustical Society of America. He was fifty years old and at the peak of his career.

Sadly for RCA (and for Olson himself), he was an unfortunate choice as the leader of this program. He was the wrong man for the job, and his lack of qualifications was aggravated by the wrong motivation.

His most obvious weakness was his inexperience in television, but this need not have been fatal. He had available the experience and talents of the world's foremost television research organization, which had just distinguished itself by its brilliant work in developing a compatible color system. His more important weaknesses were the intangible ones of attitude and approach.

As a result of his age and previous achievements, he had become an elder

[2] George H. Brown, *and part of which I was* (Princeton, N.J.: Angus Cupar Press, 1982).

statesman. His manner toward young, innovative engineers was imperious and dictatorial. New ideas were often dismissed out of hand and in a manner that humiliated the proposer. The recording of video signals on magnetic tape was an extremely difficult technical problem that desperately needed fresh approaches, but Olson did not create an atmosphere in which they could flourish.

In addition, Olson failed to take the normal first step in any research program—reviewing the technical literature to learn what others had done. In part, this failure reflected the attitude that too often characterizes major research institutions: If it was not invented here, it cannot be worth much. Had Olson taken the trouble to carry out this review, the outcome might have been quite different.

The RCA Longitudinal Recorder

Olson's sole motivation was to present the General with his birthday present on schedule, and to him this meant a working demonstration model. To accomplish this, he preferred a relatively surefire, brute-force technology over a riskier one, even though the latter might have resulted in a superior product. Given this preference and drawing on the audio background in which he was comfortable, he chose longitudinal recording (pulling the tape past a fixed record and playback head), the method universally used for audio. The technical demands of video recording, however, are orders of magnitude greater than those of audio, and Olson's team was unable to develop a unit that was satisfactory, even as a laboratory demonstrator.

Olson and others who attempted to develop a satisfactory video magnetic recorder faced three basic problems:

1. *Bandwidth.* The bandwidth of a video signal is nominally 4 MHz, more than 250 times as great as the 15-kHz bandwidth of a high-fidelity audio signal. In the audio recorders of that time, each kilohertz of bandwidth required a tape speed of 1 inch per second. For example, 15-kHz recorders operated at 15 inches per second. If this same ratio were extended to video, a speed of 4,000 inches, or 333 feet, per second would be required. The wide frequency range, about eighteen octaves from low to high end, also created serious problems.

2. *Timing stability.* Slight variations in the speed of the tape as it is pulled past the recording/playback head of an audio recorder result in wow and flutter. Although one would assume intuitively that it should not be difficult to achieve a sufficiently uniform tape-to-head speed, it is a problem even for high-quality audio recorders and next to impossible for video recorders.

3. *Linearity.* Magnetic tape is inherently a nonlinear medium, and recording an audio or video signal directly on the tape results in intolerable distortion. In audio, this is handled by superimposing the recorded signal on a high-frequency bias waveform, but this technique cannot be applied to video recording because of the high frequencies required.

Olson attacked the first problem with a combination of improved recording heads and multiple recording tracks. The pole pieces on the magnetic heads were pressed into contact so that the gap between them was infinitesimally small. Five recording tracks were used—one for each color, one for the high-frequency components, and one for the sync signal (see illustration on next page). This made it possible to optimize the characteristics of each track in relation to its signal. As a result of these techniques, it was possible to slow the tape speed, first to thirty feet and later to twenty feet per second.

But even twenty feet per second was far too fast for a practical tape recorder. It required a twenty-inch reel to hold sufficient tape for fifteen minutes of playing time. The inertia of a fully loaded reel, rotating at a high speed, was enormous, and a powerful braking system had to be provided to stop it. As an additional precaution, an engineer with heavy leather gloves sometimes stood by to stop the reel in the event the brakes failed. This precaution was prudent because if there were a brake failure with no backup system, the room would soon be filled with thousands of feet of tape.

Had Olson been more sensitive to the need to develop a practical recording system, he would probably have discarded the longitudinal method at the beginning of the program. The impossibility of achieving a lower tape speed could be predicted by a paper study. But Olson's objective was primarily to produce a laboratory demonstrator, not a prototype for a commercial product, and so the program progressed.

Electronic time base correction devices are available today to correct small timing errors introduced by the tape-handling mechanism, thus solving the timing stability problem. But they were not available to Olson's engineers, who had to rely entirely on the design of the mechanical system. This problem was never really solved, and pictures from the Olson recorders usually had an unsteady quality.

Finally, there was the problem of recording the signal with satisfactory linearity. This is necessary in order to provide proper gray-scale reproduction in monochrome and to avoid hue distortions in color. Olson's group attempted to solve this problem by using a different value of DC bias for the low- and high-frequency tracks. The technique was not totally successful, and the linearity of the signals left much to be desired.

Olson's group apparently never tried and perhaps never considered what, in hindsight at least, should have been an obvious candidate for the

THE RCA VIDEO RECORDING SYSTEM RECORDING TRACKS

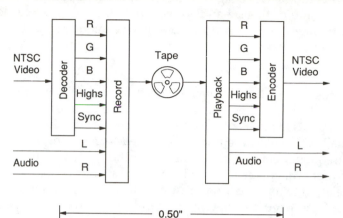

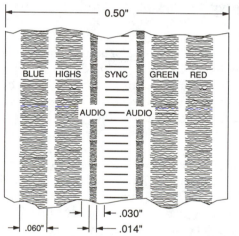

(*Top*) Block diagram. (*Bottom*) Recording tracks. The composite NTSC color signal was separated into five components: high frequencies (above 1.5 MHz), red (R), green (G), blue (B), and sync. Each was recorded on a separate track. The high-amplitude, low-frequency R, G, and B components were recorded with a fixed magnetic bias to improve the linearity. After playback, the five components were combined to reform the composite NTSC signal. Ampex avoided this complexity and achieved better results by using FM.

solution of this problem—the use of FM. The failure to do this was particularly strange, since the use of FM was extensively investigated for the audio signal. There seemed to be a blind spot in the RCA engineering hierarchy toward the use of FM for video. In a 1950 meeting with a group of RCA's top engineering executives, a major customer suggested the use of vestigial sideband FM for microwave systems as a means of spectrum conservation. At first there was a shocked silence. Finally, one of the engineers spoke up, saying, "There ain't any such thing." That was the end of the discussion. But six years later, this form of FM was a key technology in the Ampex recorder, and its simplicity and performance were among the key elements of Ampex's success.

In spite of these difficulties, Olson pushed doggedly ahead. Three years later, in the autumn of 1953, laboratory equipment had been developed that could record and play back color signals after a fashion. The tape speed at this stage of development was thirty feet per second, and a seventeen-inch reel was required to hold enough tape for four minutes of playing time. Sarnoff, always anxious for publicity, insisted on a demonstration. A color signal was transmitted from the NBC studios in New York to the Princeton laboratories, where it was recorded on a jury-rigged test system. It was then played back before a variety of guests—RCA licensees, the press, and other members of the electronics industry. The results were not good, but they were considered acceptable as a first step.

After this demonstration, work on the recorder continued. The tape speed was reduced to its ultimate limit of twenty feet per second. The timing stability also was improved, although it still fell short of the requirements for regular broadcast use. By 1955, sufficient improvements had been made to warrant another public demonstration.

The vehicle for the demonstration was the dedication of a new research building by the Minnesota Mining and Manufacturing Company in St. Paul in May 1955. A color program was transmitted from NBC to KSTP-TV in Minneapolis by AT&T landline and was then broadcast over the air for pickup at the 3M building where the recorder was installed. Here it was recorded and played back for an appreciative invited audience, which, having no standard for comparison, was generally complimentary.

At the same time, however, 3M was purusing another research program in great secrecy. Its objective was to develop a tape especially tailored to the requirements of the Ampex recorder.

NBC was now drawn into the RCA research program, and a unit was installed on its premises for field-testing and ultimately operational use. This led to an internal political battle that pitted the higher ranking NBC and RCA executives against lower level managers who had direct operational responsibility. The former, motivated by a desire to please Sarnoff, insisted it was ready for on-air use. The latter, who had to make it work, knew better. Even with constant engineering supervision, it was unable to produce pictures that would consistently meet even the relatively crude picture quality standards of the time. But no one in authority had the courage to speak up and tell this to the General.

The General's Birthday Present Is Stolen

This impasse and RCA's program to develop a broadcast-quality video recorder came to a crashing halt with Ampex's announcement at the 1956 NAB convention. The attitude of RCA Laboratories toward this technical coup was strangely relaxed. Olson did not even go to Chicago to see the Ampex unit demonstrated. George Brown (who had no direct responsibility for the tape program) seemed more relieved by the resolution of the political impasse than concerned by its effect on RCA's commercial position in the broadcast equipment market:

> Help [in resolving the impasse] arrived unexpectedly. I arrived in Chicago on the morning of April 14 in 1956 to attend the convention of the National Association of Broadcasters and, a few hours after my arrival, I learned that the Ampex Corporation had that morning disclosed a television magnetic tape recorder to the managers and owners of television stations which were affiliated with the Columbia Broadcasting System.
>
> Before the day was over, I was invited to a demonstration of this really novel device. . . . It was obvious to me that this would spell the end of the RCA . . . product . . .
>
> The following morning, April 15, . . . Robert Sarnoff, at that time president of NBC, was obliged to tell his father by telephone that some people from California had stolen one of his birthday presents.[3]

It was not a joking matter to the personnel of RCA's Broadcast Equipment division. They recognized it as a major competitive breakthrough that would end forever RCA's total dominance of the broadcast equipment business. It was a glum group indeed that returned to the Camden plant at the end of the convention.

The group's gloom was well justified. While RCA continued to enjoy many prosperous years in the manufacture and sale of broadcast equipment, the seed of its ultimate failure had been sown, and RCA would withdraw from the business thirty years later.

■ THE AMPEX BREAKTHROUGH

David and Goliath

In 1955 and 1956, while Olson and his large, prestigious team of engineers and scientists were continuing their futile pursuit of a longitudinal video recorder at RCA, a group of six young and relatively unknown

---

[3] Brown, *and part of which I was.*

engineers, some without engineering degrees and all under thirty years of age, were completing the development of a recorder based on an entirely different principle. They had pioneered a fresh and innovative approach, and it was spectacularly successful. It was a twentieth-century version of David and Goliath.

Alexander M. Poniatoff

The engineers were employed by the Ampex Corporation, a small but highly regarded manufacturer of professional audio recorders for the broadcast, motion picture, and record industries. Founded by Alexander M. Poniatoff (Figure 6–1), a Russian immigrant engineer, Ampex was an acro-

Courtesy Ampex Corporation.

■ **Figure 6–1** Alexander M. Poniatoff.

nym consisting of the founder's initials plus "ex" for excellence. Starting with six employees in a garage, Poniatoff, adhering to the standard suggested by the company's name, had built a profitable business in audio recorders that achieved a sales volume of nearly $10 million and a profit of about $400,000 in 1955.

Unlike most of the contemporary leaders of the radio and television industry, Poniatoff was a quiet and modest man who neither sought nor received wide publicity. It was not that his life lacked excitement. He was born in Kazan, a small city in southeast Russia, in 1892. He received a technical education culminating in a mechanical engineering degree from the Technical College in Karlsruhe, Germany, in 1914. He enlisted in the Russian Imperial Navy, learned to fly, and was a naval pilot from 1916 to 1918. After the collapse of the Russian Empire, he continued to fly with the White Russian Army until its defeat in 1920. He then obtained a position as assistant engineer of the Shanghai Power Company, where he remained until 1927.

Poniatoff emigrated to the United States and was employed for three years by GE's laboratories in Schenectady. In 1930, he moved to the San Francisco Bay Area, where he engaged in engineering development work for a number of organizations until 1940, when he was hired by the Dalmo Victor Company for the design of radar components. He left Dalmo Victor in 1944 to form the Ampex Corporation.

Initially, Ampex devoted its efforts to the production of radar components, but after the end of World War II, its efforts shifted to magnetic recorders, the products for which it was to become famous. Ampex's first professional magnetic tape audio recorder, introduced in 1947, revolutionized audio technology.

To the outside world, Poniatoff did not play a highly visible role in the management of Ampex. Rather, he appeared as an elder statesman who established the tone and culture of the company—high technology and high quality—but who left its active management to others. This in no way diminished the respect accorded him, both inside and outside Ampex, and he is universally regarded as one of the industry's leading pioneers.

Charles Ginsburg and His Team

Poniatoff had a keen interest in adding video recorders to the Ampex product line, and he was sufficiently impressed by this technology to decide to pursue it on a modest basis. Poniatoff's enthusiasm was not shared by all of Ampex's top management, and this skepticism was to plague the project for the next three years. Nevertheless, the company went ahead, and Charles P. Ginsburg was hired to lead the program.

The decision to hire Ginsburg was an act of faith and also of extraordinarily good judgment. His educational and professional background, while respectable, was not outstanding. His résumé might not get a second look from a high-tech company recruiter today. He was born in 1920 and entered the University of California at Berkeley in 1937 as a premed major. In 1939, he transferred to the Davis campus of the University to study veterinary medicine. He dropped out of school in 1940 and for the next seven years worked at a variety of technician jobs, including telephone system installer and radio station operator. He also continued his education by correspondence and intermittent attendance, working variously toward a degree in mathematics, physics, and electrical engineering. He returned to college full-time in 1947 and received a degree in electrical engineering from San Jose State College in 1948. He resumed his career as a radio station operator, where he remained until he joined Ampex in 1952 with the specific assignment of seeking a practical method of recording video signals on magnetic tape.

A large corporation would probably not have chosen a person with Ginsburg's background for such an important assignment. (Compare his formal qualifications with those of Olson at RCA, for example!) A smaller company like Ampex could be more flexible. Ginsburg had the right combination of managerial ability and technical competence, but most of all he had the ability to organize, guide, and motivate a small team of highly creative individuals. The selection of Ginsburg to manage this program was an outstanding choice.

Ginsburg's first task in his new assignment was to assemble his team. His selection of personnel also was excellent, and the group included precisely the right combination of talents.

The first team member to be selected was Ray M. Dolby, later to become famous in audiophile circles for his development of the Dolby noise-reduction system. While at Ampex, he was a brilliantly creative engineering student, first at San Jose State and later at Stanford.

One by one, the remaining team members were chosen. Charles E. Anderson was an outstanding electronics engineer. It was at his urging that Ampex tried FM and found it to be successful. Shelby Henderson was a skilled model maker. Alex Maxey was a highly talented engineer who was responsible for the development of the variable-position tape guide, one of the three critical technologies that made the recorder successful. While working with the team on the Ampex recorder, he conceived an entirely different recording principle, *helical scan*, which is now used universally for broadcast and consumer video recorders. Fred Pfost was another highly talented engineer who was responsible for the design of the record/playback heads. Figure 6–2 shows the six-man Ampex team.

Ampex was noted for the excellence of its mechanical design, but few suspected that it had the resources to tackle the extremely difficult problem of video recording. Even the company's top management had its doubts,

Courtesy Ampex Corporation.

■ **Figure 6–2** The Ampex recorder development team. From left to right: Charles E. Anderson, Ray Dolby, Alex Maxey, Shelby Henderson, Charles Ginsburg, and Fred Pfost. The men are gathered around a development model of the quadruplex recorder and are displaying the Emmy Award that Ampex won in 1957.

and engineering funds for the program were doled out a little at a time. Fortunately, management's skepticism did not extend to the engineering team. The engineers pursued the program with extraordinary dedication and enthusiasm for more than four years, often having to solve seemingly impossible problems.

How was it possible for a small and relatively inexperienced team to succeed while a much larger group of highly talented and experienced engineers at RCA was failing? There were several reasons.

First, the Ampex engineers were more open-minded. They were not too proud to review the literature to find out what others had done and to gain new ideas and insights. Ginsburg encouraged innovation, and he selected his team members for their creativity. Each of them made unique and major contributions to the system. Because of their youth, they were not inhibited by the "We tried that ten years ago, and it didn't work" or "There ain't such a thing" syndromes. The result was a highly exciting and creative atmosphere, not unlike that at RCA Laboratories during its development of compatible color.

Second, the Ampex engineers were pursuing a more valid objective. Their purpose was to develop a practical product that could be used in day-to-day operations by broadcast stations and networks. They wasted little time on impractical approaches, such as the longitudinal principle, that would not have resulted in a satisfactory product even if their technical problems were solved. A mere laboratory demonstration was of no interest to them.

Third, Ampex was working toward a more modest objective. Sarnoff insisted that all of RCA's efforts be directed toward a color recorder. Ampex was content to develop a monochrome machine first with the hope that it could be upgraded to color later. This turned out to be a more sound technical strategy.

Finally, the Ampex team enjoyed a substantial amount of good luck. Its critical technologies could not easily be rigorously analyzed mathematically, and their success could be determined only by experiment. A prudent engineer, estimating the odds that all of them would be successful before beginning a major development program, would have rated them as extremely low. But Ampex was lucky; all of them were successful, and the result was a revolutionary new design.

The Breakthrough[4]

In one sense, the development of the Ampex recorder was not a breakthrough. Its success was not the result of a single massive technological invention or discovery. Ginsburg recognized this in his description of the program:

> The work that led to the development of the first practical videotape recorder did not flow from a divine inspiration or a miraculous breakthrough onto the road to success.

---

[4] Much of the information in this section is taken from Charles Ginsburg, "The Birth of Video Recording" (Paper presented at the 82d Convention of the Society of Motion Picture and Television Engineers, New York, 5 October 1957).

The first videotape recorder was the product of over four years of hard—and at times, inspired—work by a team of individuals who brought their own unique skills to bear on the endless problems that confronted the development team. At times, progress was slow—and twice the project was put on the shelf.[5]

In a larger sense, however, the product was a breakthrough. Its design required a combination of three critical and untried technologies plus the solution of a host of detailed and difficult engineering problems. Two of these technologies were previously known and had been reported in the technical or patent literature, but their innovative and imaginative use and their skillful embodiment in a practical apparatus was an engineering achievement of the highest order.

The Three Critical Technologies

The three critical technologies in the Ampex recorder were developed in direct response to the three basic problems of video recording described earlier.

1. The problems of high head-to-tape speed and timing stability were solved by mounting the recording heads on the rim of a rapidly rotating headwheel. The tape is pulled past the headwheel in a direction parallel to its axis so that the head is moved past the tape rather than vice versa, as in longitudinal recorders (see illustration on next page). This makes it possible to achieve a tape-to-head speed of more than 120 feet per second while moving the tape at only 15 inches per second. It also provides timing stability for monochrome signals and, with suitable electronic correction, for color.

2. The rotating headwheel, while solving two of the three basic problems of video recording, created a new one. Minute differences between the ratios of the headwheel and the tape dimensions on record and playback—which might be caused by tape shrinkage or headwheel wear—resulted in a venetian blind effect in which vertical lines appeared as a sawtooth. This problem was solved by the use of a variable-position tape guide, which made it possible to compensate for small variations in headwheel diameter and tape stretch (see illustration on next page).

3. The problem of recording linearity was solved by the use of vestigial sideband FM. Both RCA and Ampex initially used forms of AM, but non-

---

[5] Ibid.

linear distortions led to unacceptable picture quality. Ampex was sufficiently broad-minded to try FM, and it worked.

The Development Program

The development program for the Ampex video recorder began in October 1951. Its genesis resulted from conversations between Poniatoff and two of his technical advisers, Myron Stolaroff and Walter Selsted. They urged the investigation of an approach for video recording using a rotating head described in certain patents of Marius Camaras held by the Armour Research Institute, of which Ampex was a licensee. Another rotating head

THE THREE CRITICAL QUADRUPLEX RECORDING TECHNOLOGIES

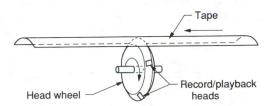

The Rotating Headwheel

Four record/playback heads are mounted on the V-shaped rim of a two-inch headwheel. The two-inch tape is partially wrapped around the rim and is pulled past it at a speed of 15 ips as the wheel rotates at 14,400 rpm. The recorded tracks are transverse to the tape.

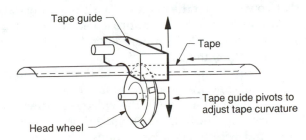

The Adjustable Tape Guide

The vacuum block tape guide holds the tape in contact with the headwheel. By pushing it upward, the stretch of the tape can be controlled, thus matching the dimensions of the tape and wheel and eliminating the venetian blind effect. It was fortunate that the Mylar tape base, which had just been developed, had the proper degree of elasticity.

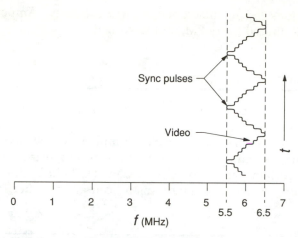

Vestigial Sideband FM

The diagram shows the frequency deviation of the carrier for low-band recording as standardized by the SMPTE. The sync tip was clamped at 5.5 MHz, and the carrier was deviated upward to 6.5 MHz at peak white. The upper sidebands were beyond the frequency range of the recorder and were attenuated, thus producing a form of vestigial sideband modulation.

technology was described in an Italian patent, and this also may have played a role in the Ampex development.

The rotating head principle demonstrated by Camaras was the basis for Ampex's early work. It used arcuate scanning, meaning that the heads were mounted on the end of a drum and the tape was pulled across this end as it rotated. Thus the recorded tracks were a series of parallel arcs across the tape. For nearly two years after the initiation of engineering development in October 1951, Ampex pursued this approach in a format described by Ginsburg: "Three heads were to be mounted on the flat surface of a drum, scanning in arcuate fashion the surface of a 2-inch wide tape. The head-to-tape speed was to be approximately 2500 ips to allow dependable recording of 2½ megacycle signals and the tape was to move at 30 ips."[6]

In spite of Poniatoff's endorsement, the video recording project did not have a high priority in the Ampex engineering program, and work proceeded slowly and with many interruptions. The team had to contend not only with difficult technical problems but also with a doubting management. Nevertheless, by October 1952, Ginsburg was able to demonstrate a picture that, though quite poor, was sufficiently promising to persuade management to continue to fund the project.

[6] Ibid.

In March 1953, a second and much improved model with four recording heads was demonstrated. In spite of the improvements, however, there were a number of major problems with no obvious solutions. This caused the Ampex management to lose whatever enthusiasm it may have had for the project, and in June 1953, it halted the project, using higher priority programs as the excuse.

These were dark days for the team, and the project probably would have died but for the optimism of a few of its members. Doggedly, they kept it going during this period with little funding and a far greater bootlegged effort. They devoted their time to studies of the project's fundamental problems and a consideration of possible solutions. The hiatus may actually have been helpful because it gave the team time to study these problems without any pressure from management.

The team's tenaciousness was rewarded, for in August 1954, Ginsburg was able to propose solutions that were sufficiently encouraging to persuade Ampex management to recommence the project. Ironically, RCA's highly publicized but unsuccessful attempt to develop a longitudinal video recorder was of great help to the Ampex team in selling its program. They pointed out that video recording must be important if RCA was devoting so much effort to it.

Somewhat grudgingly, the Ampex management authorized up to eighty man-hours to update the experimental model by incorporating the changes suggested by the studies. In a larger company, eighty man-hours would scarcely have been enough to handle the paperwork, and even in the small Ampex organization, it would have been inadequate but for a continuation of the off-hours effort for the next few months.

The most important change in the new design was to place the recording heads on the rim of a headwheel rather than on the end of a drum. (The term *drum* survived the switch to headwheels, and Ampex personnel described them as drums for years thereafter.) Four heads were used, spaced 90 degrees apart. Because of this design, the format became known as quadruplex.

The quadruplex format required the development of the second critical technology, varying the tape stretch by altering the position of the tape guide. Developd by Maxey in February 1955, it was an extremely innovative concept, and it solved one of the recorder's fundamental problems.

The third major improvement was the replacement of AM with a form of vestigial sideband FM for the signal recorded on the tape. This development was introduced in December 1954 at the urging of Charles Anderson. Once FM was introduced, the ultimate success of the project was ensured, although the team members may not have felt this degree of confidence at the time.

Vestigial sideband FM was not a new idea. In 1950, E.W. Chapin, then director of FCC Laboratories in Laurel, Maryland, had experimented with the transmission of video signals through television broadcast channels by using this form of modulation. These experiments had been carried out during the FCC freeze on new station applications, and their purpose was to determine whether FM broadcast signals would be more immune to cochannel interference than AM signals. Although AM had become too firmly entrenched as the broadcast video transmission medium to permit serious consideration of FM, Chapin's experiments showed that vestigial sideband FM was practical for video signals and that it had a number of advantages over AM. One of these, its relative insensitivity to nonlinear distortion, solved one of the fundamental problems of recording video signals on magnetic tape.

Chapin's work was probably known to RCA, but his overt favoritism of CBS in the FCC color hearing had destroyed his credibility with RCA's management. RCA ignored his work, and to the end it never gave FM serious consideration.

After the incorporation of FM into its recording system, progress was rapid at Ampex. One by one, the problems were solved. Often the solutions were not perfect, and improvements continued to be made for twenty years after Ampex first introduced a commercial product. But they were adequate for the time.

The progress of the program is demonstrated by photos showing the picture quality obtained in February 1955 and the picture demonstrated to the Ampex board of directors in March, just a month later (Figure 6–3). The March demonstration was critical because it finally convinced the Ampex management that its engineers were on the right track, and from that time on there was adequate financial support for the project. Philip L. Gundy, the general manager of the division, moved the group into isolated and much larger quarters, where intensive development work continued at an accelerated pace and in great secrecy.

With management's full support ensured, the next year was a busy one, with everyone working with extraordinary dedication and enthusiasm. Tremendous progress was made, and by February 1956 Ginsburg felt that the time for a public demonstration and announcement was approaching. But first there was a demonstration to the Ampex management. Here is Ginsburg's description of that demonstration:

> In early February of 1956, we gave a demonstration for what was originally supposed to be a very small management group but turned out to be one attended by about 30 Ampex people. For all of us on the engineering project, this was the most dramatic demonstration we were to make. The guests arrived, were seated, a few words were spoken to the effect that we would show them what we had produced and the machine was then put in the playback mode and played back a program we had recorded an hour earlier.
>
> We then announced that we would record a sequence and immediately play it back. Completely silent up to this point, the entire group rose to its feet

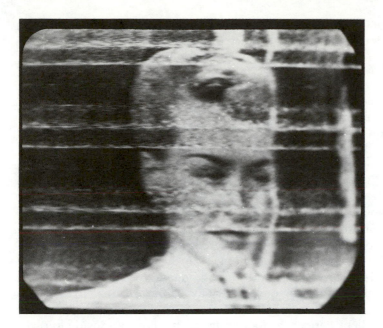

Courtesy Ampex Corporation.

Courtesy Ampex Corporation.

■ **Figure 6–3** Photographs of early videotape recordings. (*Top*) February 1955; (*Bottom*) March 1955. These photographs show the rapid progress being made in the development of the Ampex recorder in early 1955.

and shook the building with hand-clapping and shouting. The two engineers who had done more fighting between themselves than the rest of the engineering crew combined, shook hands and slapped each other on the back with tears streaming down their faces.[7]

The Ampex management realized that it was onto something big, and it wasted no time in introducing this dramatic new product to the industry. The senior technical executives of CBS and ABC, William Lodge and Frank Marx, respectively, were invited to Redwood City for private demonstrations, as were representatives of the BBC and the Canadian Broadcasting Corporation (CBC). NBC was pointedly omitted from the list of special guests.

Lodge was so impressed by the demonstration that he made an offer on behalf of CBS that Ampex could not refuse. He proposed that Ampex make its first semipublic demonstration of the recorder at a meeting of CBS affiliates in Chicago scheduled for April 14, 1956, the day before the opening of the annual NAB convention. In return, Lodge agreed to make the resources of the CBS engineering department available to Ampex to assist its engineers.

It was an excellent bargain for both parties. It gave CBS inside knowledge, early exposure, and a priority position for the availability of a product that was critical to its operations. It enhanced its position with its affiliates. And it was an opportunity to needle its hated rival, RCA. In return, Ampex received the enthusiastic assistance of the very able CBS engineering department, which suggested features and modifications to improve the performance of the machine and make it easier to integrate into an operating system.

The 1956 NAB Convention

The agreement with CBS presented the Ampex engineers with a new challenge. Only six weeks remained before the opening of the convention, and during that time they had to build a completely new demonstration model while incorporating the design improvements being made daily. Meeting this deadline required some small miracles and many hundred-hour weeks, but the task was completed on time. The machine that was delivered to Chicago made pictures that Ginsburg described as the best he had ever seen.

The private demonstration to the CBS affiliates was a sensation, and word quickly spread through the convention. To see a demonstration of the

---

[7] Ibid.

Ampex recorder became a must for all those at the convention, technical and nontechnical alike.

At the convention itself, the recorder was not demonstrated in the main exhibit area but in a small suite at the rear of an upper floor of the Conrad Hilton hotel. To reach the suite, people had to wind their way through several corridors, making the correct turn at each corner. Finding the way soon became easy, however, because at times the line of people waiting to be admitted extended nearly to the elevators.

Small groups were admitted to the demonstration, one at a time, from the head of the line. It opened with a brief speech by Gundy, who described the principle of the machine. An off-the-air recording lasting about a minute was made from CBS's Chicago station, WBBM-TV; the tape was then rewound and played back. It was a simple demonstration, but everyone left the suite knowing the he or she had seen history being made.

Ampex had humbled the mighty RCA. It had beaten the vaunted RCA Laboratories in a critical television technology, and it was now in a position to challenge RCA's leadership for broadcast equipment in the marketplace. Although it had but one product to sell, it was a major one that had no competition. Every television station, the networks, and international broadcasters were potential customers.

Product Design

Notwithstanding the enormous success of the demonstrations, much work remained before Ampex could deliver a practical recorder. There is a vast gulf between a working model that can be demonstrated in a controlled situation and a product that is manufacturable and salable. Bridging this gap takes time and money. For a product of the complexity of the Ampex recorder, product design normally requires at least two to three years, and the engineering cost is at least three times that of producing the original model. This assumes that all the basic technical problems have been solved.

Ampex was faced with an overwhelming demand for its product, and its management was desperately impatient to begin deliveries. It initiated a crash program to carry out the product design phase and bring the product to market. Miraculously, the task was completed in one year, and product deliveries began in mid-1957.

Long hours and much dedication, stimulated by the recognition that this was an opportunity that seldom comes to a manufacturing company, were the order of the day. Costs were of secondary importance, and customers were willing to accept imperfections in the product, so Ampex was able to take shortcuts in the normal engineering and production cycle. Nevertheless, it was an extraordinary achievement, nearly as remarkable as the development of the product though not as widely recognized.

The product design program was both aided and complicated by a decision to build sixteen prototype models for delivery to the networks and other major customers. The first of these was delivered to CBS in an astonishingly short time, and it was first used on the air on November 30, 1956 (the program was "Douglas Edwards and the News"). Ampex received invaluable feedback from the customers during this phase of the program as the prototypes were put into operation.

The product that emerged from this process was known as the VR-1000 (Figure 6–4). For the next two years, until 1959 when RCA began shipments of a competitive model, it was the only recorder on the market, and hundreds of the units were sold to broadcasters all over the world.

By today's standards, the VR-1000 was pretty crude. Figure 6–5 shows the picture quality typically obtained. Its softness is particularly noticeable,

Courtesy Ampex Corporation.

■ **Figure 6–4**   The Ampex VR-1000 videotape recorder. The VR-1000 mechanism and circuitry were housed in the console and the middle rack. The other racks contain test and monitoring equipment.

Courtesy Ampex Corporation.

■ **Figure 6–5** A photograph of a VR-1000 recording. Although this photograph does not do justice to the picture's quality, it was soft and noisy by today's standards. By the standards of 1956, however, the quality was remarkable.

due in large part to the limitations of the recorder. The recorder also had a number of operational problems. It could not be synchronized with other signal sources on playback, and the picture rolled every time the output of the recorder was switched to "on" from another source. Editing had to be accomplished by the clumsy process of mechanically splicing the tapes. And the recorder worked only with monochrome. Nevertheless, it was a remarkable achievement.

Happy Days in Redwood City

The success of the Ampex demonstrations was followed by an equal success in filling its order book. The VR-1000 was offered at a price of $50,000, and the demand was so great that unit sales probably would have been affected very little if the price had been $75,000 or even $100,000. Customers were clamoring for the product, not only in the United States but all over the world. In some countries, Ampex became a generic term for a video recorder. For at least three years after deliveries began in 1957,

Ampex's recorder sales were limited by its ability to produce rather than by the number of orders.

Ampex's profits from sales of the recorder, although they could have been larger with a more aggressive pricing policy, grew rapidly. In a short time, they became the major profit contributor of the entire company (Figure 6–6).

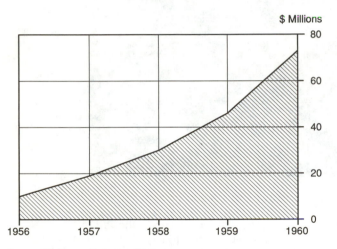

Source: Shearson-Lehman-Hutton.

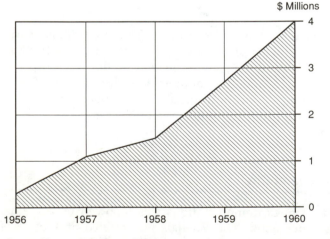

Source: Shearson-Lehman-Hutton.

■ **Figure 6–6** Ampex Corporation sales and profits, 1956–1960. For four years after the introduction of the videotape recorder in April 1956, Ampex's sales grew at a compounded annual rate of 65 percent. Profits grew at a rate of 92 percent. (*Top*) Revenues. (*Bottom*) Pre-tax profits.

Ampex's success in the marketplace did not go unnoticed by the financial community. The first hint of this appeared just before the 1956 NAB convention. It was widely believed that a number of Ampex, CBS, and ABC executives and engineers, though possibly not insiders by the Securities and Exchange Commission's definition, had knowledge that the company was about to introduce a revolutionary new product. In any case, there was a miniboom in Ampex stock during this period. After the convention, when the recorder became public knowledge, the stock price began to rise in earnest, and the rise continued as Ampex's sales and profits soared.

With sales and profits escalating, and with many of their number becoming wealthy from the rise in the price of the company's stock, Ampex executives succumbed first to euphoria, then to complacency, and finally to an overconfidence bordering on arrogance.

In 1958, the Ampex management proposed that RCA, which two years earlier had more than 60 percent of the broadcast equipment market, cease its direct marketing efforts and become an original equipment manufacturer (OEM) supplier to Ampex. RCA rejected the proposal as preposterous, but the very fact that it was made was an indication of the Ampex state of mind.

Pride leadeth to a fall, and Ampex fell into this trap. In 1961, while it was resting on its laurels and making only incremental improvements in the VR-1000 design, RCA leapfrogged it with the first all-transister video recorder. Suddenly, Ampex found itself on the defensive, and sales of its lead product plummeted. After the dust had settled, Ampex's president, George Long, and many of his staff had been let go.

Where Are They Now?

The enormous success of the Ginsburg team came not at the end but at the beginning of its members' careers. All of them could look forward to thirty years or more of active participation in the industry. What happened to them in subsequent years? Were they able to repeat their success, or was this a one-shot effort? The truth is somewhere in between.

Ginsburg, whose talents lay in innovative product development rather than the nitty-gritty of product design, remained with Ampex in a series of engineering management roles, all involved in advanced technology and future planning. His work as leader of the team that developed the Ampex recorder was widely recognized, and he received numerous awards and citations.

Anderson, Pfost, and Maxey spent most of their ensuing careers in the Ampex engineering department, where they continued to make significant contributions to recording and computer technologies. Anderson rose to a senior management position in the Ampex engineering department, which he held at his retirement in 1984. Maxey received the Poniatoff Award from the SMPTE for his invention of the helical scan format.

Dolby completed his undergraduate work at Stanford and went to Cambridge, where he received a Ph.D. in physics in 1961. For a number of years, he specialized in electronic microscopy. After holding a variety of research posts in England, India, and the United States, he founded his own company, Dolby Labs, Inc. This company developed the famed Dolby noise-reduction system, which became a standard component of the best high-fidelity audio systems.

In short, all the team members enjoyed successful careers subsequent to their invention of the VR-1000. Still, it was a hard act to follow, and, with the possible exception of Dolby's, none of their subsequent achievements quite equaled it.

■ THE AMPEX–RCA COMPETITION

A humbled group of engineers and executives from RCA's Broadcast Equipment Division returned to its Camden plant after the 1956 NAB convention. The division had suffered a humiliation because Ampex had beaten its sister division, RCA Laboratories, in the race to develop a satisfactory video recorder. But it also faced a critical and agonizing business decision: Should it take out after Ampex and develop its own recorder?

This was a particularly unpropitious time for such a decision. Both the Broadcast Equipment Division and the corporation were suffering from serious profit problems as a result of the failure of the color television market to develop. On the one hand, the company could ill afford the major engineering expenditures that would be required to design an RCA recorder, even though Ampex had shown the way. On the other hand, failure to do so could lead to even more serious consequences.

A number of grim meetings were held to decide the issue. The group favoring an aggressive course was led by the broadcast division's brilliant marketing manager, A.R. "Hoppy" Hopkins. He pointed out that the recorder market was a major business opportunity and urged that it be pursued on this basis alone. But he argued further that RCA had no choice but to proceed if it wished to maintain its leadership in the broadcast equipment business. If Ampex were left with a monopoly in video recorders, its profits would be so enormous that it could fund major programs to compete with and eventually surpass RCA in the market for cameras, transmitters, and other broadcast products. Thus, it was necessary for RCA to provide competition as a defensive measure to keep Ampex from becoming too strong.

Hopkins's judgment was correct. A number of years later, Ampex did attempt to expand its business by adding other major broadcast products.

Its efforts were not successful, however, largely because, faced with RCA's competition, it was unable to generate sufficient profits from the recorder business to support adequate engineering programs. Had RCA chosen to stay out of the recorder business, the story might have been quite different. Ironically, by this time Hopkins had had a falling out with RCA's top management and was serving Ampex as a consultant.

In spite of Hopkins's persuasiveness and the validity of his arguments, it is unlikely that the RCA corporate management would have approved the program, with its short-term adverse effect on profits, solely in the interest of the future of the Broadcast Equipment Division. (One of Hopkins's favorite axioms was that, given the choice between a "fast nickel and a slow dime," corporate management would invariably choose the former.[8]) But RCA had a broader reason for wishing to compete with Ampex, and this was the basis for its painful decision to proceed. The reason was color.

The Ampex recorder would not work with color, and Ampex had limited resources to add this capability. Further, it had little incentive to do so, as two of its most influential customers, CBS and the BBC, were aggressively uninterested in color. Color broadcasting was already having difficulty getting off the ground, and monochrome-only recorders could only make matters worse until color recorders were available. RCA's management believed, probably correctly, that it was the only company that had the will and the resources to make this happen within a reasonable period of time. With the profits of nearly every division dependent on the growth of color, RCA had no choice but to proceed.

To a large extent, the short-term profit problem was solved by diverting funds and engineering resources from other programs. The adverse effect of this policy on the design of the TK-12 monochrome camera has already been described in Chapter 4.

RCA Responds with a Color Recorder

The RCA broadcast division's engineering department, stung by Ampex's success, responded to this challenge with enthusiasm. The program was placed under the leadership of A.H. Lind, one of its ablest product engineering managers, and RCA assigned many of its best engineers to the task. Although Ampex had a two-year head start, RCA had certain strengths and advantages that were extremely useful in its efforts to catch up.

First, Ampex had grown overconfident. Ampex had promoted a mystique about its recorder to the effect that its manufacture required certain esoteric processes that could not be duplicated by others. This may have

---

[8] Personal communication.

been useful as a sales tool, but it became a problem when Ampex itself began to believe it.

Second, RCA had greater technical resources. The shallowness of Ampex's video expertise, perhaps an advantage during the invention phase because it encouraged a freewheeling and innovative experimental approach, became a disadvantage in the product improvement phase. During meetings of the SMPTE technical committee, which assumed the responsibility for establishing industry standards, it sometimes appeared that the RCA engineers understood some of the subtleties of the Ampex recorder better than Ampex.

Third, there was Ampex's huge task in gearing up for production. This was a distraction to the engineering department in its efforts to upgrade the design.

Finally, there was Ampex's lack of interest in color.

As is often the case with a technical breakthrough, the most important secret of the Ampex design was that it worked. Once this was demonstrated, there was no more concern that it might be a blind alley. Fortified by the knowledge that the quadruplex principle was feasible, the RCA team proceeded confidently with its own design, drawing on all the corporation's technical resources. The resulting design was far more than a copy because it could record color.

The most serious problem in recording color was the timing instability, or jitter, of the signal as it was played back. It was not a particularly serious problem for monochrome, but it upset the precise phase relationship of the burst and subcarrier required for correct reproduction of picture hues. RCA solved this problem with a clever circuit that made the burst-controlled subcarrier and the picture signal jitter together, thus maintaining the same relative position. The solution was rather crude, and more sophisticated methods of time base control were eventually introduced, but it worked.

In about a year, the RCA team had developed a working model (Figure 6–7) that could produce reasonably satisfactory color pictures. But before the design of a commercial product could begin, it was necessary to resolve the patent issue.

The Cross-Licensing Agreement

One school of thought at RCA held that everything unique in the Ampex machine was based on prior art and that no patent license would be required. A more conservative view was that this would be a risky course and a license should be obtained. The latter opinion prevailed, and a lengthy series of patent negotiations began.

RCA opened the negotiations by inviting a group of Ampex engineers and executives to its Camden plant for a demonstration of its model as it

Courtesy Ampex Corporation.

■ **Figure 6–7**   The first working model of the RCA quadruplex recorder. The model was examined early in 1957 by Dr. George H. Brown (*left*), then chief engineer of RCA's Commercial Electronic Products Division, and Theodore A. (Ted) Smith (*right*), executive vice president and general manager.

recorded and played back color signals. The demonstration came as something of a surprise to many at Ampex because it destroyed the mystique that its recorder was based on impenetrable secrets. Ampex's top management, however, remained sufficiently confident of RCA's inability to produce a competitive machine that it was willing to negotiate a patent and technical aid agreement that made RCA's problem easier to solve.

In addition to this confidence, the Ampex management had a positive reason for being willing to make an agreement with RCA: It needed RCA's color patents and know-how, and its recorder patents provided a bargaining chip to obtain these without cost. Also, it may have had legal advice that its own patent position was not immune to challenge. In any event, the negotiations proceeded positively in the weeks following the demonstration.

RCA proposed a cross-licensing agreement that gave Ampex the right to use all of RCA's color patents in its recorder products and RCA the right to use all of Ampex's recorder patents. This proposal became the basis for the final agreement, which also provided that RCA pay Ampex a sum of $200,000. In addition, each company was permitted to send a team of engineers to the other's plant to exchange technical information. After this period, the iron curtain of secrecy was to be drawn.

The agreement was signed in October 1957, and both companies followed its terms scrupulously. There was some reluctance on the part of the Ampex engineers to divulge their secrets, and they volunteered very little. But with patient digging, the RCA engineers were able to obtain most of the information they needed.

The First Competitive Battle

While the patent issue was being negotiated, RCA proceeded as rapidly as possible with the design of a commercial product to be known as the TRT-1A. Its engineering department was faced with a formidable task. The product not only had to meet the normal standards of the marketplace but also had to have features that would distinguish it sufficiently from the Ampex machine to enable the sales department to overcome Ampex's head start.

The RCA machine's most important feature was the ability to record color, but Ampex had closed this gap by offering a color model by the time RCA could start deliveries. (Thus RCA's corporate strategy of forcing the availability of color recorders was successful). A number of other features were added, including vertical mounting of the tape reels for easier loading and electronic delay lines to compensate for slight differences in the spacing of the heads on the headwheel. These features provided useful ammunition for the salesforce, but none of them was sufficiently fundamental to give RCA a clear-cut product superiority. Fortunately for RCA, Ampex did not take full advantage of its lead, and, except for the addition of color, its design was a nearly stationary target.

The TRT-1A (Figure 6–8) was first demonstrated at the 1958 NAB convention, but the demonstration model was barely working when the convention opened. One of its problems was an unstable circuit that required

Courtesy RCA Corporation.

■ **Figure 6–8** The TRT-1A television tape recorder. The TRT-1A was all rack-mounted as contrasted with the Ampex VR-1000, which used a combination of rack and console mounting. The closeness of the technical race between these products is indicated by the fact that vertical versus horizontal mounting for the tape reels became a major competitive issue.

constant adjustment. The test point for this circuit was in front of the recorder, but the adjustment had to be made from the back. RCA solved this problem by leaving a small crack between the equipment racks and placing the test oscilloscope in front of the crack. One of RCA's engineers, Roy Marion, then spent the convention behind the racks, peering through the crack and making adjustments of the errant circuit as required.

At the 1958 convention, RCA announced that it would begin deliveries of its commercial product one year later. This announcement initiated one of the most intense rivalries in the history of commercial products. Ampex was determined to drive RCA from the marketplace, but RCA was equally determined to achieve a market share approaching 50 percent. No quarter was asked, and none was given.

There was not enough difference between the products to produce a clear winner. Technical superiority, while hotly debated, did not decide the battle, which to a large extent was fought on nontechnical issues.

On the one hand, Ampex entered the contest with the enormous advantage of a two-year head start. It had a proven product. Many customers had already bought one or two Ampex machines, and as they expanded their tape rooms, they usually did not wish to mix brands. Without a doubt, Ampex was successful in persuading some customers that it had a unique mystique. It was able to persuade others that the RCA recorder had compatibility problems. (This claim had some foundation. The Ampex recorder was more tolerant of badly recorded tapes on playback.)

On the other hand, RCA had an entrenched position in the marketplace, a large number of loyal customers, and a more experienced and highly professional sales force under the able leadership of Ed Tracy and Dana Pratt. Given half a chance, RCA salespeople did not often lose orders, and they fought for the recorder business with skill and tenacity.

This competition became highly visible at the 1959 and 1960 conventions, where the RCA and Ampex exhibits were located across the aisle from each other. The aisle became a no-man's-land, with each company's sales staff warily watching the other's and trying to lure the competitor's customers to its exhibit.

The scorecard for the competition was the number of machines shipped each year by the two companies, and this made clear the advantage that accrued to Ampex as the result of its head start. This advantage was particularly important in the international market, where RCA had neither as strong a position nor as skilled a sales force as in the United States. Ampex retained more than 60 percent of the business in the United States and more than 80 percent internationally. Nevertheless, RCA was making significant inroads, and the stage was set for the RCA breakthrough in 1961.

The RCA Breakthrough

Charles Colledge, formerly vice president for operations at NBC, became general manager of RCA's Broadcast Equipment Division at the beginning of 1959. The division was having serious profit problems as the result of competition from Ampex and the delay in the growth of color broadcasting.

Colledge's somewhat contradictory assignments were to "get the division moving again" by reestablishing its technical leadership but at the same time to improve its short-term profitability. He correctly perceived that RCA needed a spectacular breakthrough to overcome Ampex's lead, and he was given that opportunity when the engineering department proposed the development of a radically new videotape recorder that would be totally solid-state (transistorized). Although the division could ill afford the cost, College approved the program.

Today solid-state devices and circuitry are so common that it is difficult to appreciate the magnitude of this development. A video recorder requires an extremely wide variety of circuit types, each requiring specialized active components. Early recorders used vacuum tubes almost exclusively for these functions. Little by little, transistors replaced tubes as suitable types became available. But in 1959, contemplating a total replacement of tubes with transistors required an act of faith.

The initial development and design program also was carried out under the leadership of A.H. Lind, ably assisted by one of RCA's most talented engineers, Arch Luther. It was successful, and by early 1961 a demonstrable prototype had been completed. As compared with the original Ampex development, it did not require the same degree of innovation, but it was a tour de force of complex and advanced engineering. It truly merited the description "breakthrough."

With the prototype available, Colledge had to decide whether it should be shown at the 1961 NAB convention. His staff (including myself) unanimously voted against it. The TRT-1A was selling reasonably well against Ampex, and there was a concern that showing a new product that could not be delivered for at least a year would adversely affect the sales of the older product. Colledge outvoted his staff and ordered that the new product, called the TR-22 (Figure 6–9), be shown.

There was a suspicion on the part of Colledge's staff that his decision was motivated more by a desire to impress RCA's top management than by good marketing judgment. Whatever his motive, the decision was the right one. The broadcast industry was enormously responsive, and RCA's somewhat tarnished technical reputation was refurbished.

To the staff's surprise, showing the TR-22 did not harm the sales of the TRT-1A. On the contrary, it gave them a boost. With the benefit of hindsight, the explanation is fairly obvious. One of the impediments to RCA recorder sales, a problem that was not fully recognized by the sales department, had been the concern that RCA might not be in the recorder business for keeps. The TR--22 persuaded the doubters that it was, and the TRT-1A enjoyed a surge of new orders.

Trouble in Redwood City

The reaction of the Ampex contingent at the convention was a combination of dismay and disbelief. The exhibit halls were in the Wardman Park (now the Sheraton Park) Hotel in Washington, and RCA's exhibit was so

■ **Figure 6–9** The RCA TR-22 videotape recorder. The combination of all-solid-state circuitry and excellent styling gave the TR-22 a high-tech look. The tape reels were mounted at an angle, a compromise between the vertical and horizontal mountings previously used by RCA and Ampex.

large that it was placed in a separate wing. Thus it was able to maintain an unusually high degree of secrecy during the setup of the exhibit prior to the convention's opening. As soon as the exhibits were opened to the attendees, a large group of Ampex executives and engineers rushed into the RCA display to have a look at the new product, which had been the subject of lively rumors. When they saw the TR-22 operating, they could not believe their eyes. They left the exhibit in stunned silence.

This was the beginning of a bad time for the Ampex Corporation. The broadcast equipment market was in a slump because of an economic recession, and RCA was capturing an increasing share of the video recorder business. In its euphoria, Ampex had overexpanded and established an overhead structure based on an assumption of everincreasing growth. Its sales dropped from $73 million in 1960 to $70 million in 1961. As a result, its profits fell precipitously, from a profit of $4 million to a loss of $3.9 million.

With Ampex in serious trouble, its board of directors became dissatisfied with the performance of its president, George I. Long, Jr., who had a banking background. This experience proved to be inadequate for extricating a high-technology company from a serious competitive problem, and he resigned.

The Ampex board replaced Long with William Roberts, a manufacturing executive with a fine reputation who was recruited from Bell & Howell. Roberts and Peter G. Peterson, later to become secretary of commerce under President Nixon and chairman of Lehman Bros., had been Bell & Howell's crown princes under its youthful president, Charles Percy. When Percy became chairman in 1961, Peterson was chosen over Roberts to succeed him as president. Having been passed over, Roberts was ripe for an offer from Ampex, and he accepted it enthusiastically.

Roberts initiated a program of austerity at Ampex but continued the engineering programs that were so essential to its future. As long as Ampex concentrated on high-technology products, he was a fine chief executive, and his policies eventually led to a second era of prosperity for Ampex.

In the meantime, however, it was RCA's turn to prosper. The TR-22 was an excellent product in spite of the fact that it was pushing the state of the art in the application of solid-state circuitry. With a more advanced product and a more experienced sales force, RCA's market share improved dramatically.

Now it was RCA's turn to become overconfident as it gained temporary leadership in the recorder market. RCA rested on its laurels with the TR-22. Although it had an active engineering program for incremental improvements, it did nothing to bring about a fundamental improvement in the TR-22's color performance. This was an obvious need that RCA, of all companies, should have recognized.

Ampex Strikes Back with "High Band"

At the beginning of 1965, the position of RCA's Broadcast Equipment Division was somewhat like that of Moses as he approached the Promised Land. It had struggled through ten lean years awaiting the arrival of the color boom. It had survived the traumatic setback of the Ampex video recorder. It had spent large sums for the development of an entirely new line of equipment named the New Look, the brainchild of its creative marketing vice president, John P. Taylor. Even the color of the equipment was changed to a bright blue from a drab umber gray. And, in 1964, the elements that were essential to the success of color broadcasting—technology, color receiver sales, and color programming—had reached a critical mass, and the transition from monochrome to color began to move at an accelerating pace.

Mount Pisgah for the division was the 1964 NAB convention. Its exhibit was a stunning success. The TR-22 video recorder was at the peak of its popularity, and even the monstrous TK-42 color camera was well received. Ampex demonstrated a video recorder with a new feature called "high band," but it was a cloud no bigger than a man's hand. The division appeared to have regained technical parity or leadership for all its major products, and the New Look gave the impression of a company on the move. RCA's Promised Land, a quantum leap in sales and profits, was in sight for 1965. Accordingly, it approached the 1965 NAB convention with confidence and anticipation.

But, like Moses, the division never quite reached its Promised Land. Two competitive products, the Philips PC-70 Plumbicon camera (see Chapter 5) and the Ampex VR-2000 high-band video recorder, a production model of the prototype displayed the previous year, were introduced at the convention. Their superiority over RCA's offerings was so pronounced that they destroyed RCA's leadership in these product lines. Although the market for color equipment grew so rapidly during the next few years that the division prospered in spite of itself, its success was far less than it would have been had it retained its leadership.

Ampex's success was a consequence of the determination of its engineering department to continue working on high-band recording in spite of the company's financial problems. It was a fundamental advance in recorder technology, and it resulted in an almost order-of-magnitude improvement in the quality of color pictures.

The advance was an increase in the maximum instantaneous frequency of the carrier recorded on the tape from 6.5 MHz to 10.0 MHz (see illustration on next page). This modification had two important advantages.

First, it made possible a nearly threefold increase in the total frequency deviation of the carrier, which improved the signal-to-noise ratio by the same amount (4.6 dB). Second, it moved the high-energy sidebands of the recorded signal farther out of the video base band and well away from the color subcarrier. This nearly eliminated the herringbone patterns and other spurious effects resulting from interaction between the base band and modulated signals that had plagued low-band recordings.

When Ampex exhibited the prototype at the 1964 convention, RCA had some advance information about Ampex's plans, but it was caught essentially unprepared. In an effort to counter the Ampex announcement, it

DISTRIBUTION OF SIGNAL AND SIDEBAND ENERGY

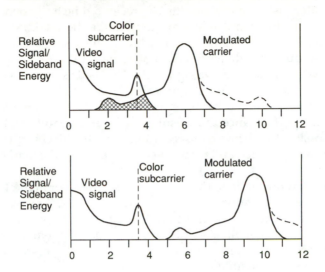

(*Top*) Low-band recording. (*Bottom*) High-band recording. These graphs show, in principle, the distribution of signal and sideband energy for recorded color signals. In both cases, the upper sidebands are attenuated because of the limited bandwidths of the recorders. In low-band recording, there is considerable frequency overlap between the lower sidebands of the modulated carrier and the video base band. This leads to herringbone patterns and other moiré effects. High-band recording moves the lower sidebands out of the spectrum of the video signal.

demonstrated a TR-22 that had been hastily modified for mid-band recording with its carrier midway between low band and high band. Although its performance was less than marginal, the industry did not yet realize the significance of high band, and the Ampex demonstration did not spoil RCA's triumph at the convention.

During the next year, RCA engineers worked frantically to design a mid-band version of the TR-22 that would perform as well as a high-band unit. The result was the TR-22HL. It was better than the experimental model shown at the convention, but it was not a completely satisfactory product. Try as they might, RCA's engineers could not design a modified TR-22 that would perform as well as a high-band recorder.

To make matters worse, RCA also announced kits for modifying TR-22s in the field for mid-band use. The kits were completely unsatisfactory, RCA was unable to keep its promises, and it lost untold credibility in the marketplace—especially with ABC, which had previously purchased a large number of TR-22s.

Finally, after losing many months of precious time and spending many equally precious dollars in an attempt to upgrade the TR-22, RCA decided that it would have to design a new machine. The result was the TR-70. It was a fine product, but deliveries did not begin until 1967.

Ampex, which had wisely designed the VR-2000 as a new machine and not as an upgrade of an older one, was able to start deliveries in quantity in 1965. By this time, the marketplace had learned of the dramatic improvement in color performance that resulted from high band, and it became nearly impossible to sell low-band recorders for use with color. The VR-2000 was an excellent product, and for two years Ampex had the market nearly to itself.

Happy days returned to Redwood City. With the color market at its flood and the clear technical superiority of its video recorder, Ampex prospered. Every low-band machine in the field, whether RCA or Ampex, was a candidate for replacement. Again, Ampex's sales soared. Its expansion, which had been so rudely interrupted in 1961, was resumed. Roberts seemed to have fully justified the board of directors' confidence in him.

Disaster at Ampex

Once again, pride would lead to a fall. As part of a corporate strategy of growth by expansion into related businesses, Roberts decided to enter the prerecorded audiotape market. The decision was a mistake, and it very nearly ruined the company.

Prerecorded tapes are part of the entertainment industry, which is no place for a management team trained in the more disciplined and orderly world of product manufacturing. The management styles and techniques for the two types of business are so different that an integrated management can seldom run both successfully.

Matters came to a head in 1971, when it became apparent that Ampex had a financial disaster on its hands. The disaster went beyond bookkeeping; it extended to the real world of cash. Ampex was unable to pay its creditors, and it was forced to seek the relief of Chapter 11 of the bankruptcy law. Its write-offs and losses in 1972 and 1973 totaled nearly $100 million—more than the company's cumulative profits to that time (Figure 6–10).

Now it was Roberts's turn to leave. He resigned and was succeeded by his executive vice president, Arthur Hausman, a solid and able professional. Slowly and methodically, Hausman took the difficult steps necessary to rescue the company. Ampex emerged from the protection of Chapter 11 a leaner and better managed company. Its sales and earnings increased rapidly, and it was able to negotiate its sale to the Signal Compa-

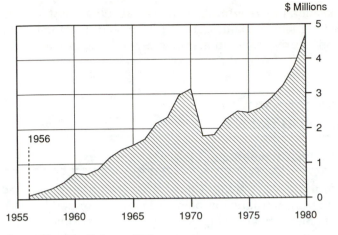

$ Millions

Source: Shearson-Lehman-Hutton.

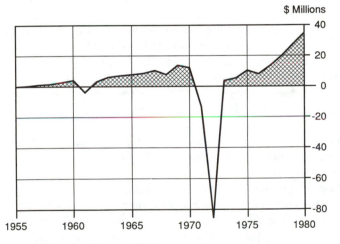

$ Millions

Source: Shearson-Lehman-Hutton.

■ **Figure 6–10** The sales (*top*) and profit (*bottom*) saga of the Ampex Corporation, 1956–1980. During those years, the Ampex Corporation enjoyed three periods of rapid growth: from 1957 to 1960, as the result of the VR-1000 videotape recorder; from 1963 to 1970, largely as the result of its high-band color video recorders; and from 1976 to 1980, as the result of its strength in video recorders and controlled expansion in related product lines. It suffered two severe reverses: in 1961, as the result of competition from RCA and a depressed market for broadcast equipment, and in 1971, as the result of the prerecorded tape debacle.

nies, Inc., from a position of strength. It became a subsidiary of that company in 1981. Allied Signal Companies, Inc., in turn sold it to the Lanesborough Corporation, a privately held manufacturer of specialty chemical products, in 1987.

■ QUADRUPLEX RECORDERS: CLIMAX AND DENOUEMENT

Final Incremental Improvements

Intensive engineering development of quadruplex recorders continued in the years following the introduction of high band by Ampex in 1965. There were no further major breakthroughs, but there were steady year-by-year design enhancements, spurred by the spirited competition between RCA and Ampex and to a lesser extent by the German company Fernseh (which has now taken the name of its parent, Robert Bosch).

The development of a suitable pole-tip material for the recording heads was a difficult and enduring problem. Specifications for this material included stringent magnetic properties, mechanical strength, and resistance to abrasion. The preferred material was alfecon, an alloy of aluminum, iron, and silicon, but ferrites were used as well. By the 1970s, the typical head life had been increased from one hundred to three hundred hours.

One of the most important improvements was the ability to synchronize the recorder playback signal with signals from live programs or other recorders. This was made possible by improved servo systems and the availability of increasingly precise solid-state time base correction devices. At first, synchronization was frame by frame, and this permitted sources to be switched without picture roll. Later, synchronization became more precise, proceeding from line by line to picture element by picture element and finally to color burst by color burst. With the achievement of color burst synchronization, recorders could be fully integrated into live productions or complex editing systems.

The Cartridge Recorder

In 1971, RCA began shipments of the TCR-100, its cartridge tape recorder. This machine permitted segments of program material up to two minutes in length to be recorded on cartridges rather than on reels. It solved a serious problem during station breaks when a series of short commercials had to be played in rapid sequence. With reel-to-reel recorders, a separate machine had to be loaded for each commercial, and extremely precise manual timing of the machine's operating controls and switching was required. The cartridge mechanism performed these operations automatically.

Although the TCR-100 did not include any dramatic breakthroughs, it had to be meticulously designed to meet rigid reliability specifications. One failure in a thousand plays was barely good enough. RCA's first machines did not meet this standard. There were no fundamental problems, but there were a host of small ones that added up to unacceptable reliability. It was necessary to recall the first eight machines shipped, make numerous design changes, and rework all the machines that were in production. This delayed shipments for another year, but the redesigned machines were a commercial success, and the product earned an Emmy for RCA in 1974.

The Demise of Quadruplex

Although the performance and features of quadruplex recorders were brought to a very high level, some serious inherent drawbacks remained. The high-pressure contact between the heads and tape continued to limit the life of both in spite of major improvements. The mechanical and electronic complexity of the machines was great, and their maintenance required highly skilled technicians. It would probably have been impossible to develop a quadruplex recorder that was sufficiently cheap and simple to be sold for home use.

For broadcast use, the greatest single problem was the sensitivity to poor maintenance. If a machine was well maintained and skillfully operated, it was capable of producing superb picture quality. But if either were allowed to slip even a little, the performance deteriorated rapidly. It was particularly difficult to keep the outputs of the four heads precisely matched; if they were not, "banding" (the division of the picture into horizontal bands) resulted. The eye is particularly sensitive to small discontinuities in any picture characteristic—brightness, contrast, noise, or colorimetry—and the tolerance in output variation from head to head was very low.

These problems made quadruplex recorders vulnerable to competition from other technologies, and they eventually lost out to helical scan recorders. Helical scan recorders began to be used by broadcasters for specialized applications as early as 1970, but they did not become a serious competitor until 1977. Their acceptance in the marketplace grew rapidly, and by 1980 the only quadruplex recorders being sold were for playback of the vast libraries of quadruplex tapes that were still available. By 1981, twenty-five years after their introduction, quadruplex recorders were no longer offered by any of the major manufacturers.

Quadruplex recording played an indispensable role in the development of television broadcasting. It was the dominant technology for more than twenty years, and during this period more than ten thousand machines with a market value in excess of $500 million were manufactured and sold by Ampex, RCA, Fernseh, and Toshiba.

■ THE BIRTH OF HELICAL SCAN

Competing Recording Formats

In spite of the success of quadruplex recorders, most major broadcast equipment manufacturers investigated other recording formats.

RCA Laboratories stubbornly refused to abandon the longitudinal format and directed its efforts toward a lower priced product that might be sold in the audiovisual educational market. Its engineers constructed a working model of such a product and turned it over to the broadcast division for product design. It did not work very well, and the broadcast division received it with less than overwhelming enthusiasm (it was irreverently known as "little slimy"). After a number of unsuccessful attempts to design a salable product, the program was terminated. Some years later, a similar attempt to design a consumer product using this format also failed.

The Japanese and Germans also experimented with the longitudinal format. Toshiba and BASF introduced longitudinal video recording (LVR) recorders for consumer use. Both brands used a continuous loop of tape mounted in a cartridge. The tape speed was 215 ips in the Toshiba machine and 160 ips in the BASF model. The cartridge for the Toshiba machine held approximately five hundred feet of tape, which gave about twenty-five seconds of playing time for one pass. To achieve a reasonable playing time for the cartridge, multiple tracks were used. At the end of each pass, the head jumped to the next track during a vertical blanking interval. Since the tracks were extremely narrow—a little over one thousandths of an inch— three hundred tracks could be recorded side by side on half-inch tape. The Toshiba machine had a total playing time of two hours, while the BASF model, with its slower tape speed, had a playing time of three hours.

While the two LVR machines were the result of skillful engineering, they never achieved much success in the marketplace. Longitudinal recording simply was not a competitive format for video except for a few highly specialized applications.

The Helical Scan Format

The format that finally prevailed was known as helical scan or slant track and was first developed by Alex Maxey of Ampex. The tape path is a helix, and the diagonal configuration of the recorder tracks gave rise to the expression *slant track* (see illustration on next page).

HELICAL SCAN RECORDING

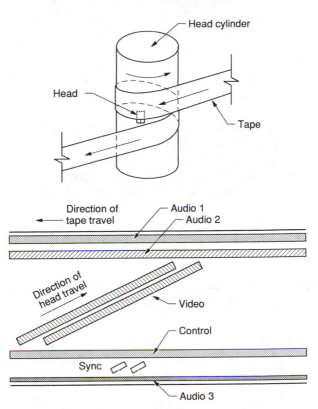

The tape is wrapped in a helix around a cylinder or drum on which the recording head is mounted. As the drum rotates, the tape is pulled past it in the opposite direction. The head-to-tape speed is the sum of the drum circumference speed and the tape speed, the former being much larger.

The resulting recording format is a series of tracks recorded at a slight angle to the edge of the tape. This angle is exaggerated in the drawing; in the SMPTE Type C format, for example, it is only about 2.5 degrees.

The helical scan format has several important advantages over quadruplex. Perhaps the most important is its ability to record a complete television field in one pass of the head across the tape. This eliminated the banding problem and the need for precise maintenance that had always plagued quadruplex recorders. The ease with which certain helical scan formats can be made to produce still-frame and slow-motion pictures also is an important operational feature.

The operating costs of helical scan recorders are lower than those of quadruplex. This results from their greater simplicity, easier maintenance, lower tape usage, and longer head life. As an example of the savings, LaVerne Pointer, engineering vice president of ABC, reported that operators are assigned to helical recorders on a one-for-three basis, whereas it was standard practice to use a one-for-one basis with quadruplex.

Finally, the helical scan format is extremely versatile with many possible variations, and it can be adapted to meet the requirements of recorders in every price range, from the finest broadcast-quality machines to low-cost consumer products.

The helical scan format also had some problems. Since a small but significant portion of the head-to-tape speed results from the motion of the tape, it had some of the timing stability problems that were so serious for longitudinal recorders. An even more difficult problem was tracking—that is, causing the head to follow precisely the long recorded track segments during playback.

Furthermore, the industry had great difficulty in establishing standards for helical scan machines. Ampex, having invented the quadruplex recorder

---

TAPE-WRAP CONFIGURATIONS

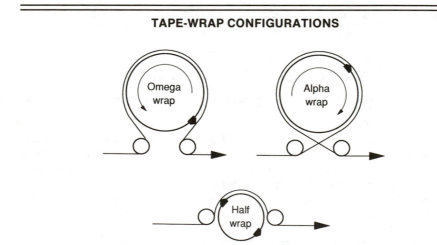

The helical scan principle is extremely versatile, and many variations are possible. Variable parameters include the number of heads, the amount of tape wrap around the drum, the drum diameter and speed, the tape width, and the tape speed. Three of the tape-wrap configurations are shown. The omega wrap and alpha wrap are full-wrap types. A complete television field is recorded with each revolution of the drum. With the half-wrap configuration, a complete field is recorded with each half revolution of the drum, and alternate fields must be recorded with different heads. Half-wrap is universally used with cassette recorders because of the ease of threading.

and having a two-year lead over RCA, its only competitor, was able to establish de facto standards for the quadruplex format. No single manufacturer enjoyed such a dominant position in helical scan, and products using a variety of standards appeared on the market.

For the broadcast market, helical scan had an additional standards problem—its incompatibility with quadruplex. Since quadruplex had a head start of many years, thousands of machines were in use, and libraries containing millions of reels of tape recorded in the quadruplex format had been established. To overcome this problem, it was not sufficient that helical scan machines be equal to or a little better than quadruplex; they had to be significantly better.

Markets for Helical Scan Recorders

In spite of these difficulties, the promise of helical scan was so great that many large companies, including Ampex, RCA, a Bosch–Philips joint venture, and a host of Japanese manufacturers, as well as some smaller companies such as International Video Corporation (IVC) initiated engineering programs as early as the late 1950s to develop products for the commercial market.

These programs were directed toward products for three different markets:

1. The broadcast market, in which the recorders would compete directly with the Ampex and RCA machines and range in price from $40,000 to $100,000
2. The audiovisual market for education and training, with a price range of $1,000 to $10,000
3. The consumer market for home recorders, in which a price under $1,000 was required

Products in the middle-range audiovisual market proved the easiest to develop. They had neither the rigorous performance requirements and quadruplex-compatibility problems of broadcast recorders nor the extraordinarily difficult cost targets of home recorders. Accordingly, they were the first to appear on the market. The evolution of these products into the VCR for home use is described in Chapter 9.

The Role of the Japanese

Helical scan gave the Japanese an opportunity to enter the broadcast equipment export market. They had a special talent for the design and manufacture of recording equipment, and there was no entrenched competition. Helical scan recorders of all types became one of their strengths, and they totally dominated the consumer market (see Chapter 9).

Early Technical Developments

The development of broadcast helical scan recorders was not as dramatic as that of quadruplex because the nature of the broadcast equipment business had changed. The technologies had become more complex, the market had grown severalfold, and the size of engineering departments and programs had grown correspondingly. An inevitable consequence was more emphasis on teamwork and less on individual effort and creativity. It is doubtful that Ginsburg and his small team could have been successful in this environment; certainly their operating mode would have had to be quite different. Thus the history of technology became one of organizations and not of colorful leaders.

Although many manufacturers dabbled in the application of the helical scan format for broadcast use, only four expended major efforts during the 1960s and early 1970s. They were Ampex and IVC in the United States, Sony in Japan, and the Philips–Bosch joint venture in Europe. All of them had to face entrenched quadruplex competition, but they faced it with different motives. For Ampex it was a defensive move; broadcast recorders were such a large part of its business that it had to be prepared to move to helical scan if it superseded quadruplex. For IVC it was an opportunity to enter the broadcast market by upgrading its audiovisual helical scan recorder. Although its broadcast model was not a commercial success, it probably stimulated Ampex to accelerate its work on helical scan. For Sony it was an opportunity to enter the broadcast recorder market on equal terms with Ampex. For Philips and Bosch it was an opportunity to establish a position of leadership for European manufacturers.

All the manufacturers had to solve the basic problems of helical scan recorders—tracking and time base instability.

The tracking problem was solved in part by careful mechanical design of the drum and the tape-handling system. Additional precision was provided by the addition of Automatic Scan Tracking (AST®). This was originally offered by Ampex to correct the inherent mistracking that occurs when the tape is played back at other than normal speed, but it had the additional advantage of improving the tracking reliability under standard speed conditions.

Time base stability was achieved by breakthroughs in solid-state technology and a dramatic decrease in the cost of memory chips. The result was the introduction of digital electronic time base correction. Digital correction met the requirements of helical scan recording because it could correct rather large timing errors with great accuracy. It was a completely successful solution, and after it was introduced in 1973, the ultimate success of helical scan and the demise of quadruplex was assured (although this was not obvious to everyone at the time). In the immediate future, however, the industry was faced with the major tasks of product design and the establishment of standard industry formats.

■ HELICAL SCAN: ANALOG RECORDING

The technology of digital recording had not progressed sufficiently in the early 1970s to be used in video recorders, and the initial helical scan products used analog recording with FM (as with quadruplex). Three standard tape widths evolved—one-inch for studio recorders and three-quarter-inch and half-inch for use in ENG and other portable applications.

One-Inch Analog Recorders

Product Development   Ampex and Sony were the pioneers in the development of high-performance one-inch analog recorders for studio use. The engineers at both companies were aided by technical advancements in other industries.

The computer industry, with its vast engineering resources, responded to the urgent need to record large amounts of information on small areas of magnetic tape or disks by achieving order-of-magnitude increases in the packing density (that is, the amount of information stored per unit area). Much of this work was directly applicable to video recording, and the television industry's engineers combined it with their own innovations to produce a dramatic reduction in tape usage. The introduction of metal tape, which used fine grains of metal alloy rather than iron oxide as the recording medium, was a particularly important advance. The standard Type C format uses one-inch tape running at 9.5 ips. By contrast, standard quadruplex machines used two-inch tape running at 15 ips. In the early years of helical scan, the reduction in tape usage was touted as an inherent property of the format, but it was more the result of basic advances in tape technology and recording head design. The effect of these advances was even more dramatic in digital recorders.

Rapid developments in solid-state technology also helped the designers of helical scan recorders. These developments made possible the design of sophisticated signal-processing circuitry and product miniaturization. For example, a Sony Type C recorder (Figure 6–11) has a width of two feet and a height of two and a half feet, compared with four by four feet for the last quadruplex console models.

In spite of the advantages of helical scan recorders and the excellent

Courtesy Sony Corporation.

■ **Figure 6–11**   A Sony Type C helical scan recorder.

products being offered, sales of helical recorders in the broadcast market were minor during the mid-1970s. The problem was twofold—the inherent incompatibility with quadruplex and incompatibility between helical scan machines produced by different manufacturers. There was an urgent need for industry standardization.

The Standard SMPTE Formats   Three helical recording formats for one-inch tape machines were candidates for standardization in the mid-1970s. The U.S. entry was the Ampex model VPR-1; the Japanese entry was the Sony BVH-1000; and the German entry was the Bosch BCN format, which had sold in significant quantity in Europe.

Although the Ampex and Sony models were incompatible, they were similar in that both used a full-wrap format (see page 346), described as non-segmented since the same recording head was used for all fields, and both used one-inch tape. They were sufficiently similar to form the basis for a common standard. The Bosch machine used a half-wrap format, described as segmented since alternate fields were recorded by different heads. It was inherently incompatible with the full-wrap formats.

The barriers to standardization were imposing. Each manufacturer had made a substantial investment in its own product. If a competitor's format were accepted as the industry standard, it would mean a costly redesign, a loss of prestige, and a head start in the market for the competitor.

Ampex pointed out that its VPR was compatible with some thirty thousand audiovisual models already in the field. In addition, its format had been officially recognized as the SMPTE Type A standard. Bosch noted that its BCN format had become the de facto standard in Europe and was being resold in the United States by a number of manufacturers, including RCA. Sony took the hardest line of all. Its broadcast equipment general manager, David MacDonald, stated, that there was no hope of standardizing until everyone agreed that Sony had the best system. Sony would cooperate with others, but would not change its format.

Enormous pressure from customers was required to break this impasse, and it finally came in the form of a white paper from CBS and ABC at the 1977 winter meeting of the SMPTE. They urged that the SMPTE break the deadlock between Sony and Ampex by establishing standards for a one-inch nonsegmented helical scan recorder. The SMPTE responded by setting up a working group with this assignment. Recognizing the acceptance of Bosch's segmented recorder in Europe and the preference of some broadcasters for this format, the SMPTE set up a second working group to establish segmented standards as well.

Faced with the pressure from two of their biggest customers, Ampex and Sony reluctantly agreed to cooperate in the establishment of a compromise standard. The SMPTE working group met regularly during the next several months and in July 1977 announced there was an agreement in principle on a standard that reconciled the differences. Both Ampex and Sony had to make significant concessions to reach an agreement, but it appeared that the resulting format resembled Sony's more than Ampex's. The details of the standard were developed during the remainder of the year, and it was officially adopted as SMPTE Type C with its publication in the *Journal of the SMPTE* in December 1977.

There was no contest for the standard for the segmented format, and SMPTE Type B was adopted, giving official sanction to the Bosch Philips recorder.

Three-Quarter-Inch and Half-Inch Recorders

While manufacturers were designing one-inch recorders for studio use, they also were upgrading three-quarter-inch and half-inch VCRs for broadcast use. Their main advantages were low cost and portability. In addition, they used cassettes rather than reels, making for easier loading. They were originally intended for use in ENG and other portable applications, but in time their performance was improved sufficiently to make them satisfactory for some studio applications.

One of the first VCRs to be used in the broadcast market was the Sony U-Matic (see Chapter 9), which used three-quarter-inch tape. It was later upgraded to the SP (superior performance) U-Matic. This format has been standardized by the SMPTE as Type E.

Sony also developed the Betacam with half-inch tape in a cassette for use in ENG systems. It had a playing time of only twenty minutes, but it was a success in the marketplace, as more than twenty-five thousand were sold.

In 1985, Panasonic, a subsidiary of Matsushita, announced the MII format. MII also used half-inch tape in a cassette, but its playing time was one hour. It was developed in cooperation with NHK, the Japanese national network, which was its first customer. Like the Betacam, it was an outgrowth of an earlier format developed for the lower priced audiovisual market. Unlike the original Betacam, it was intended for both studio and portable use.

There was considerable industry skepticism over the performance of the MII format in the studio. The industry was stunned in 1986 and 1987, therefore, when NBC ordered nine hundred MII recorders. Michael Sherlock, the president of NBC Operations and Technical Services, announced that the MII machines would replace the one-inch Type C machines in almost every case, not only for fieldwork but also in the studio. He gave cost savings as the reason, pointing out that MII was a universal format that could be used in both the studio and the field, that it used less tape, and that its maintenance and operation required fewer technicians. He estimated that the cost savings over five years would be $26 million.

NBC's purchase gave MII instant respectability as a studio-quality recorder, but its acceptance was not universal. Panasonic's own evaluation of its performance is that it is suitable in any application that does not require multigeneration copies, as in complex editing operations.

Not to be outdone by MII, Sony announced the Betacam SP recorder, an upgraded version of the original Betacam. Its cassette holds ninety minutes of tape, and its performance is satisfactory for all but the most demanding studio applications. In 1986, Sony and Ampex signed agreements licensing Ampex and others to manufacture and sell the Betacam SP.

Helical Scan
Sweeps the Broadcast Industry

With its technical problems solved and the standards issue settled by the SMPTE, helical scan rapidly superseded quadruplex in the marketplace. Quadruplex sales fell off slightly in 1977, then dropped sharply in 1978. The drop-off continued in 1979 and 1980, and in 1981 both Ampex and RCA dropped quadruplex machines from their product lines. In contrast, sales of helical scan recorders burgeoned. Every quadruplex recorder in the field became a candidate for replacement, and both Ampex and Sony prospered.

The Plight of RCA and the Demise of Its Broadcast Division

The rapid transition from quadruplex to helical scan recorders in the broadcast market created a serious problem for RCA's Broadcast Equipment Division. Its management had not anticipated that the transition would occur as soon or as rapidly. The vast libraries of tapes that had been recorded with the quadruplex format were an enormous barrier to the introduction of a new standard. It appeared that the impasse between the Ampex and Sony standards would take years to resolve. Most importantly, the division had failed to appreciate the magnitude of customer preference for helical scan machines because of their easier maintenance and other advantages. The division's management concluded, therefore, that an engineering program directed toward a broadcast helical scan recorder was not urgent.

With the benefit of hindsight, it is clear that RCA should have started such a program no later than the beginning of 1976 and preferably sooner. In fact, however, the RCA engineering department had done very little work on helical scan recorders when Neil Vander Dussen (later to become president of Sony America) became general manager at the beginning of 1977.

As it became increasingly apparent that the days of quadruplex were numbered after the SMPTE working group reached agreement in principle on standards in mid-1977, Vander Dussen was faced with the problem of getting helical scan machines designed, manufactured, and delivered to the market on a crash basis.

This problem, a difficult one at best, was aggravated by pressure from powerful members of RCA's technical community who believed that the company should leapfrog the Ampex–Sony format and design a digital recorder. The digital recorder engineering program was ahead of its time, and it was a distraction that siphoned off engineering effort that would have been better spent on the helical scan program.

The pressure from the market was so great that Vander Dussen felt compelled to have a helical scan machine in the RCA product line much sooner than its own units could be available. Accordingly, he reached an agreement with Sony in 1978 providing that Sony would supply machines to RCA that would be resold under the RCA label. It was an advantageous arrangement for both parties. It gave RCA a product that could be sold at a profit, and it gave Sony the benefit of RCA's powerful sales department in its competitive battle with Ampex.

It was not thought to be sound strategy, however, for RCA to depend on another manufacturer indefinitely for such a major product, and RCA proceeded simultaneously with its own design program. Vander Dussen's charge to the engineering department was to design a machine that was essentially a copy of the Sony product. There was no time to reinvent the wheel, and it was more prudent to play catch-up than leapfrog. After his promotion from the division in 1979, however, the product objective evolved into something quite different. The result was the TR-800, which was intended to be more advanced and to have more features than the Sony machine.

The TR-800 was a disaster. The design objectives were too ambitious, and in spite of huge expenditures, the engineers were unable to design a reliable product. Even worse from RCA's point of view, engineering funds were diverted from camera and transmitter programs, and RCA's position in these product lines deteriorated as well.

These internal problems could not have come at a worse time. The competition from the Japanese was growing by the year, and RCA, in common with its larger competitors, was experiencing increased competition from small specialty companies. RCA found it impossible to compete with these companies on small specialty items such as microphones, and eventually its own product line was limited to major products. Inevitably, this led to an erosion of its market share.

As a result of these problems, the Broadcast Equipment Division, once one of RCA's most profitable arms, sank into the red. RCA's new top management team, Thornton Bradshaw and Robert Frederick, judged the situation hopeless, and in October 1985 they decided to withdraw from the business at a cost of $81 million. It was a sad ending for an organization that had once dominated its market, had made major contributions to RCA's profits, and had provided much of the technology that was the basis for the broadcasting industry.

■ HELICAL SCAN: DIGITAL RECORDING

Why Digital Recording?

Digital technology (see Chapter 10) has had a unique appeal to the electronics engineering profession. In part, this has been an emotional response to digital's technical elegance and sophistication, but it also has been a recognition of digital's tangible benefits. Digital technology is basic to the design of computers, and it is indispensable to that industry. It is not as fundamental to radio and television, but it has important potential advantages for the transmission and recording of radio and television signals.

The most important advantage of digital signals is that they can be transmitted or recorded with virtually no loss of quality. Analog signals suffer at least a small loss of quality, even with the best transmission circuits and recorders. This characteristic of digital signals is particularly important in systems where the signal is subjected to repeated degradations; examples are long-haul communication circuits with many transmission paths in series and multigeneration recordings that are copies of copies. While the degradation of the signal in any single step may not be serious (or perhaps even noticeable), the cumulative effect of repeated degradations can seriously damage its quality or even destroy its usefulness.

The use of digital recording for video signals is particularly important in editing systems. Complex modern editing practices may require as many as ten or more generations. With digital recording, this can be accomplished without significant loss of quality. The results with analog recording would be marginal at best (see below).

ANALOG AND DIGITAL RECORDER PERFORMANCE

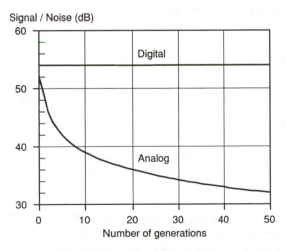

The signal-to-noise (S/N) ratios of the two recorders shown in this example are both excellent—54 dB for the digital recorder and 49 dB for the analog. (The S/N ratio for quadruplex machines typically varied from 40 to 46 dB.) The S/N ratio for the digital recorder remains constant out to as many as fifty generations, while the performance of the analog machine deteriorates rapidly.

The Technical Challenge
of Digital Recording

There are two basic problems in the development of video digital recorders. The first is the fact that our senses of vision and hearing perceive the
world in an analog fashion. Microphones and cameras produce analog
signals, and analog signals are required to activate loudspeakers and picture
tubes. Video recording, therefore, requires the use of analog-to-digital
(A/D) and digital-to-analog (D/A) converters in the input and output channels. At one time, these converters were complex and costly, but by the
mid-1960s the rapid decrease in the price of solid-state devices made them
economically practical. Their use is now routine.

The other problem is more difficult and more fundamental. It is the need
to use very high bit repetition rates, which require extremely large bandwidths. The bandwidth problem is particularly difficult in component recording, which is used in advanced editing systems. The standard bit rate
for this configuration is 228 megabits per second (Mb/s),[9] equivalent to a
bandwidth of 114 MHz. The difficulty of achieving this can be appreciated
by recalling that high-band quadruplex recording, which increased the
bandwidth of the recorded signal from 6.5 to 10.0 MHz, was considered to
be a major technical breakthrough in the mid-1960's. With a head-to-tape
speed of 30 meters per second (the tape speed is 11.2 ips, only slightly faster
than the 9.5-ips rate for Type C analog), the wavelength is 0.9 micrometers,
which is in the range of infrared light.

Digital Video Recorders

Component Recording The most important application for early
digital recording was in editing or postproduction systems, where its ability
to record repeated generations without degradation is critical. This market
strongly prefers component recording to composite because it produces a
higher technical quality in the edited program. Therefore, the first digital
recorder to enjoy significant sales, the Sony DVR-1000, used component
recording.

In component recording, the composite NTSC (or PAL) color signal is
divided into three components (see Chapter 5)—Y (brightness), Y minus
blue, and Y minus red—each of which is recorded separately. The editing is
carried out in the component mode, and the components are recombined at

---

[9] Component digital recorders use a 4-2-2 format, which means that the luminance signal,
Y, is sampled at 4 times the color subcarrier frequency, while the color components, R - Y and
B - Y, are sampled at 2 times the frequency. Each sample requires an 8-bit code, and the
resulting bit rate is 228 Mb/s.

the end to form a composite signal. The editing process requires frequent change of tapes, which are contained in cassettes to provide easy changing.

Since Sony was initially the only manufacturer of component digital recorders, its specifications were adopted by the SMPTE as the D-1 standard. Even without opposition, however, the development of the standard required eight years.

Composite Recording   Although digital recording is most beneficial in postproduction systems that require complex editing, its uniformly high quality is also advantageous for more routine applications. Accordingly, a market for digital recorders of composite signals developed. Both Ampex and Sony produced recorders for this market, and this created another standards controversy.

With the unpleasant memory of past battles in mind, the companies decided to settle the issue amicably, and in 1986 they reached agreement on a compromise standard that, in this case, more closely resembled the Ampex proposal. This standard was adopted by the SMPTE as D-2.

The bit rate employed in the D-2 format is 114.4 Mb/s, approximately equal to a bandwidth of 57 MHz. The tape speed is only 5.1 ips, less than that of Type C analog recorders. Like the D-1 component recorder, it uses cassettes.

■ HELICAL SCAN FORMAT SUMMARY

The array of standard formats for helical scan recorders is somewhat bewildering. They are designed to serve a variety of applications, and they reflect a decade of technical evolution. The formats in use in 1988 are summarized in the following table:

■ HELICAL SCAN BROADCAST RECORDER FORMATS

Format	Analog/ Digital	Tape Width	Reel/ Cassette	Component/ Composite	Remarks
A	A	1″	R	Composite	Obsolete
B	A	1″	R	Composite	Used in Europe; segmented format
C	A	1″	R	Composite	Most commonly used U.S. format
E	A	3/4″	C	Composite	U-Matic
Super Betacam	A	1/2″	C	Composite	Leading half-inch format
MII	A	1/2″	C	Composite	First studio-quality half-inch format

Format	Analog/ Digital	Tape Width	Reel/ Cassette	Component/ Composite	Remarks
D-1	D	19mm	C	Component	Postproduction format
D-2	D	19mm	C	Composite	Standard digital format
NA	D	1/2"	C	Composite	

■ THE CAMCORDER

In the early years of ENG, it was necessary for the camera and the recorder to be packaged in separate units. The recorder was usually mounted in a small van so that the camera operator was attached to it by the umbilical cord of the connecting cable. As cameras and recorders were miniaturized, it became practical to house both in a single unit that was light enough (typically less than ten pounds) for the camera operator to carry on his or her shoulder. Power consumption also was reduced sufficiently so that the unit could be operated with rechargeable batteries for periods of up to one hour or more. This freed camera operators from the constraints of the video, audio, and power cables so that they could position themselves for the most advantageous shots.

The combined camera–recorder units are now universally called camcorders. There are two types—low-cost consumer products (see Chapter 9) and professional units used by broadcasters.

One of the first professional camcorders on the market was the Hawkeye, an offering of the fading RCA broadcast division in a joint venture with Matsushita. It was one of the last hurrahs of the division, and RCA withdrew from the business before its design could be perfected. Camcorders are now offered by a number of Japanese manufacturers, and they have become standard equipment for ENG and sports crews. Figure 6–12 shows the Sony broadcast camcorder.

■ EPILOGUE

The technical history of video recording has been a succession of miracles. Repeatedly, last year's impossibility has become next year's commonplace. These advances have been made possible by an order-of-magnitude growth in technical personnel devoted to recording technology. In 1956, it was limited to the six-man Ampex team plus perhaps fifty more engineers at RCA and other companies who were pursuing unsuccessful approaches. Today recorder engineering personnel number in the hundreds in the

Courtesy Sony Corporation.

■ **Figure 6–12**   A Sony broadcast camcorder.

United States, Japan, and Europe. Their efforts have been greatly aug-
mented by technical developments in the computer and solid-state industries. Ampex, the pioneer, has not retained the leadership it enjoyed in the
quadruplex era, but it is one of the few major U.S. companies that has been
able to compete sucessfully with the Japanese.

With hundreds of able and innovative engineers devoted to developing
new recorder products, it is reasonable to expect more miracles in the years
ahead. Recorder technology will continue to progress, and products will
become even more versatile, reliable, and available at a lower cost. HDTV
(see Chapter 10) will put more demands on recorder performance, and
these will be met. But notwithstanding the enormous future progress that is
expected in recorder technology, it is doubtful that any single development
or breakthrough will have as much impact on the television industry as the
achievement of six young engineers working in a back room of the Ampex
laboratories in 1956.

7

■ ■ CABLE TELEVISION

Success usually has a thousand fathers, and cable television was no exception. A number of individuals claimed the distinction of installing the first system for bringing television signals into the home on a coaxial cable rather than over the air, but the evidence seems to award the honor to E.L. Parsons, the owner of radio station KAST in Astoria, Oregon. In the summer of 1949, he erected an antenna system to receive station KING-TV in Seattle, 125 miles away, and he distributed its signal by cable to twenty-five "subscribing neighbors."

The concept spread, and during the next two years a number of systems were established. Among the earliest were installations in Mahoney City, Lansford, Pottsville, and Honesdale, Pennsylvania, and Bellingham, Washington. (John Walson of Mahoney City is believed by some to have preceded Parsons in installing a system.) These communities were too small to support their own television stations, and the reception from nearby stations with home receiving antennas was poor, usually because of an intervening mountain range. But a high-gain antenna on a tall tower or mountaintop could receive a satisfactory signal, which was transmitted by cable to individual homes. From this modest beginning grew today's multi-billion-dollar industry.

The growth of the cable television (CATV) industry can be divided into three phases.

In the first, or mom-and-pop, phase, cable systems were used to extend the range of local television stations or to fill in holes in their normal coverage areas created by mountains or other obstructions. They were small and generally locally owned, and their function was described by the term *community antenna systems.* Subscribers typically paid an installation fee of $100 to $150 and a $3 to $5 monthly service charge. The system owner was often a local radio and television dealer, and his or her motive was as much to increase the market for receivers as to derive income from its operation.

The second, or *distant station importation*, phase began during the late 1950s. It added a totally new dimension to cable television by using microwave to transmit signals from broadcast stations to cable systems located far outside their normal service areas. (Later, microwave also was used to distribute pay-TV and other nonbroadcast services.) Cable systems could now carry broadcast stations located outside their immediate area, and it became profitable to install systems in cities that had service from a number of local stations. As a result, the term *community antenna systems* was dropped in favor of *cable television* or *CATV* during the mid-1960s. The importation of signals from distant stations also created a host of regulatory and copyright problems, some of which are still unsolved.

The third and final phase began in the late 1970s with the use of satellites to distribute programs to cable systems. It brought about an even more fundamental change in the industry because satellites made profitable the nationwide distribution of programs that did not originate on television stations. In a very short time, entrepreneurs offered dozens of program services on satellites, and the quality and variety of these programs created a widespread demand for cable service, even in cities with a large number of local broadcast stations.

The developments of the third phase vastly increased the size and scope of the cable television industry. As of January 1, 1988, it was estimated that 8,500 systems were in operation in the United States with 42,750,000 subscribers—nearly 50 percent of the country's television homes.[1] Cable television has become big business.

■ CABLE TELEVISION TECHNOLOGY

Cable System Design

Figure 7–1 shows the layout of a typical cable television system. Signals are received at the headend (Figure 7–2) from nearby broadcast stations and by microwave and satellite. The signals are processed, combined, and fed

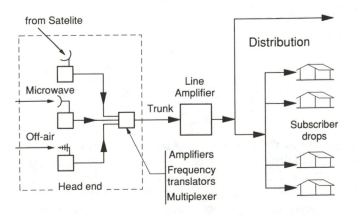

■ **Figure 7–1** Typical cable television system layout.

[1] Michael C. Taliaferro, *TV & Cable Factbook*, (Washington, D.C.: Warren Publishing Company, 1988).

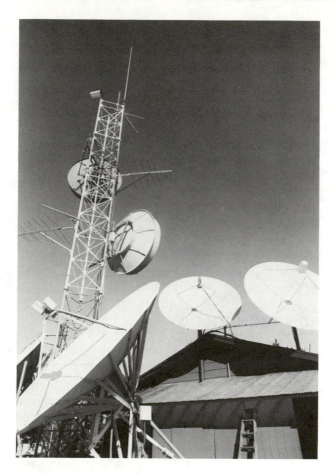

Courtesy Bend Cable Communications.

■ **Figure 7–2**   Typical cable television head end antenna system. This system includes three yagi antennas for the reception of television stations, two microwave antennas mounted on the tower, and three dishes for satellite reception.

into a coaxial cable *supertrunk*. (A coaxial cable consists of an inner conductor surrounded by an insulating material that holds it in the center of a cylindrical outer conductor.) The processing may include shifting channels; for example, a station on channel 21 may be transmitted over the system on channel 4. The supertrunk signals are fed into the distribution system *trunks* through a *splitter* and brought into homes through *subscriber drops*.

The amplifiers placed in the cable trunks must be carefully designed to compensate for the greater attenuation of the cable at higher frequencies. (The cable attenuation is typically twenty times greater at 500 MHz than at

50 MHz.) They also must have precise automatic gain control to adjust for temperature variations and other factors that may cause changes in the attenuation of the cable or the gain of the amplifiers. Determining the amplifiers' spacing requires a trade-off between noise, distortion, and crosstalk between channels. If they are too widely spaced, the noise level will be excessive. Reducing the spacing increases the number of amplifiers and thus the distortion and cross talk. In short, the design of a cable television system requires a series of difficult compromises and trade-offs.

Coaxial cable has served the industry well, but its high signal attenuation and limited bandwidth prevent cable systems from achieving optimum performance. Cable's limited bandwidth prevents the use of digital transmission to obtain the advantages of that mode (see Chapter 10). To accommodate the large number of transmission channels carried on modern systems (see the section on channel assignments later in this chapter), it is necessary to jam them together with no guard bands between. With fifty or more channels transmitted on a single cable, unequal signal levels or distortions in the trunk amplifiers often cause cross talk between channels, resulting in a herringbone pattern on the screen. To date, it has been difficult to transmit television with stereo sound over cable without seriously impairing the video signal.

FCC Technical Regulations

Since cable television is not a broadcast service, its technical regulation by the FCC is less severe. There are a few rules for cable systems, however, which are intended to ensure a degree of compatibility with broadcast transmissions, minimum standards of signal quality, and freedom from spurious radiation that could cause interference with other services. FCC Rules specify the following points:

- The channels to be used by cable television systems
- Frequency tolerance of the visual and aural carriers
- Minimum signal levels
- Uniformity of signal levels from channel to channel
- Uniformity of signal levels with time
- Maximum spurious radiation

Channel Assignment

Early community systems used the five low-band VHF channels (channels 2 to 6). Cable attenuation was lowest at these frequencies, so fewer line amplifiers were required.

As the demand for additional channels grew, systems expanded into the seven high-band channels (channels 7 to 13). The spacing between line amplifiers had to be reduced, but the system capacity was increased to twelve channels, and unmodified standard receivers could be used.

The attenuation of cable signals in the broadcast UHF channels is so great that it is more economical for systems with more than twelve program channels to use the portion of the spectrum between the VHF and UHF bands. These cannot be received directly on standard television sets, and *set-top converters*, which translate all incoming channels to a single output channel, or *cable-ready receivers*, which can tune to the special cable channels, must be used.

A standard channel configuration for cable television provides fifty-five channels (see table below) with the highest channel at 403.25 MHz, just below the broadcast UHF band. In early 1988, 7 percent of the cable systems in the United States with 18 percent of the subscribers had more than fifty-five channels, usually as a result of competitive pressures in obtaining their franchises. These systems are more costly because they use frequencies above 403 MHz and require larger diameter cable and more closely spaced trunk amplifiers.

■ **CABLE TELEVISION CHANNELS**

Channel Designation	Visual Carrier (MHz)
Low band (5 channels)	
2	55.25
3	61.25
4	67.25
5	77.25
6	83.25
Mid-band (11 channels)	
A-2	109.25
A-1	115.25
A–I	121.25–169.25
High band (7 channels)	
7–13	179.75–211.25
Superband (14 channels)	
J–W	217.25–295.75
Hyperband (18 channels)	
AA–RR	301.25–403.25

This configuration provides a total of fifty-five channels. Systems requiring more channels continue upward from 403.25 MHz in the spectrum at 6-MHz intervals.

■ THE FIRST PHASE: MOM-AND-POP COMMUNITY SYSTEMS

Industry Growth

The beginnings of the mom-and-pop phase of the cable television industry were described at the beginning of this chapter. It was a cottage industry, largely consisting of systems owned by local entrepreneurs in small towns and cities. As long as cable systems depended on off-the-air signals from nearby cities, the industry's opportunity for growth was limited to communities that met rather narrow criteria. There could be a maximum of one or two local stations. The community had to be near enough large-city television stations that their signals could be picked up with a master antenna but beyond the range of these stations for standard home antennas (either because of distance or an intervening mountain range).

The limited growth opportunity for the industry in the mom-and-pop phase was reflected in the number of subscribers (Figure 7–3). By 1957, there were about 500 systems in operation throughout the country with 350,000 subscribers, an average of only 700 per system. The rate of growth in the number of subscribers was slowing, and if cable systems had not begun the importation of distant stations, the industry would have remained small.

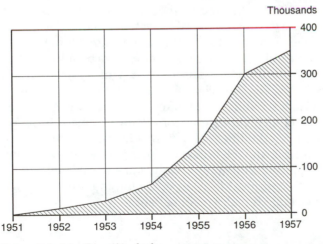

Source: Television Digest Yearbook.

■ **Figure 7–3** Cable television subscribers during the mom-and-pop phase, 1952–1957. After four years of moderately rapid growth, the annual increase in the number of subscribers dropped off in 1957 as the market for distribution of local signals approached saturation.

The Pioneers

Milton Shapp   In spite of the limited prospects for the industry, strong leadership soon developed. Both RCA and Philco began the manufacture of distribution equipment, but both found the market too small to be attractive to large companies. Instead, Jerrold Electronics, a small Philadelphia manufacturer (later to become a division of the General Instrument Corporation), became the leading supplier of cable television equipment and systems under the leadership of its aggressive and colorful president, Milton Shapp.

Shapp understood the market, and his small manufacturing plant was ideally suited for the construction of low-cost equipment. He also added a construction department that offered turnkey installations, including cable and construction. The company prospered, and Shapp retired to become governor of Pennsylvania in 1971.

Martin Malarkey and the Founding of the NCTA   Among the system operators during this phase, Martin F. Malarkey, Jr. (Figure 7–4) was perhaps the leading spokesman of the industry. He was born in Pottsville, Pennsylvania, and in 1939 at the age of twenty-one became vice president of the family radio and music store, Malarkey's, Inc. He became president of the company after the end of the war in 1946.

The company's ability to sell television sets was limited by the poor quality of reception in Pottsville, and he was one of the first to recognize the potential of increasing the market for sets by using cable to bring television signals into homes. In 1951, he founded the Trans-Video Corporation, which received a franchise to construct a cable system in Pottsville.

Malarkey was also one of the first to foresee the need for a trade association to represent the cable industry as the NAB represents broadcasters. He organized a meeting in Pottsville on January 16, 1952, which established the National Community Television Association, later to become the National Cable Television Association (NCTA). Malarkey was elected its first president. Initially, most of its members were from Pennsylvania, but it soon grew to become a thriving and influential nationwide organization. Its first convention was held in 1952, and it has become an annual event of growing size and importance. Still later, Malarkey founded one of cable television's most prestigious consulting firms, Malarkey & Taylor, in partnership with Archer Taylor, a former radio broadcast consultant.

■ THE SECOND PHASE: DISTANT STATION IMPORTATION

The first application for a microwave system for the importation of distant stations was filed by J.E. Belknap & Associates of Poplar Bluff, Missouri, on October 6, 1951. The company proposed to bring signals from broadcast

Courtesy Malarkey and Taylor.

■ **Figure 7–4** Martin F. Malarkey, Jr.

stations in St. Louis and Memphis into cable systems in Poplar Bluff and Kennett, Missouri, and Mount Vernon, Illinois. Its application specified that this would be accomplished by a microwave system operated as a common carrier.

This concept promised to change the character of the industry, as it opened up an immense new potential market for cable service. The technology required was well within the state of the art, but its development was delayed by a Pandora's box of regulatory issues.

Government Regulation

When community television systems first began operating, the FCC and local governments were somewhat puzzled as to what action, if any, they should take. Cable systems were neither fish nor fowl. They were not public utilities in the same sense as power companies, but they had many of the characteristics of a utility, such as, a large fixed investment and the need for easements to operate along and across streets. They were not anticipated by the Communications Act of 1934, and it gave little policy guidance.

The FCC at first responded by doing little or nothing. The systems were small, they did not require spectrum space, most broadcast stations did not object, and they created little controversy or pressure for action. Local municipalities assumed they had the power to grant exclusive franchises, and most of the early systems operated under these, often with rate regulation.

The Belknap application brought the issue to a head at the FCC in a manner that could not be ignored. It proposed a microwave system that occupied spectrum space, and it proposed common carrier operation. It was clearly within the jurisdiction of the FCC, and it forced the Commission to face the broader issues posed by cable television, which had become more complex with distant station importation.

In the ensuing years, the regulation of cable systems by federal, state, and local governments has gone through an incredibly complex series of changes. Administrative agencies, federal and local governments, and the courts all have been involved in an attempt to resolve the conflicting interests of broadcasters, cable system operators, local municipalities, the owners of copyrights, and the public. The issues have by no means been settled as of this writing, and it is possible only to summarize the situation as it exists today.

The Cable Act of 1984   The Cable Act of 1984 resulted from the attempts of Congress to resolve many of the thorny issues that had troubled the cable television industry. Among its more important provisions were the following:

- Municipalities and other franchising authorities could grant one or more franchises at their option for cable television systems. (The courts' treatment of this provision is discussed in the next section.)
- Franchise fees were permitted.
- The franchising authority could require that certain channels be designated for public, governmental, and educational use.
- Larger systems must make channels available for commercial use by unaffiliated persons.

- Common carriers could provide facilities but not programming.
- Ownership by television broadcasters in the same area was prohibited.
- Authority for establishing technical standards was vested with the FCC.

Franchising Authority of Municipalities   Congress thought that it settled the issue of franchising authority by passing the Cable Act, which explicitly gave municipalities the authority to grant one or several franchises. Their right to grant an *exclusive* franchise was soon challenged in the courts on First Amendment grounds, but the courts' rulings have not been uniform. An exclusive franchise in St. Paul was found to be constitutional, but one in Sacramento was not. Sacramento was forced to make a substantial financial settlement with the applicant for a second franchise on the basis that he had been unconstitutionally deprived of the profits of his proposed enterprise. As long as the constitutionality of the law is uncertain, it is questionable whether municipalities will expose themselves to the risk of an unfavorable judgment by granting an exclusive franchise.

Rate Regulation   State and local governments have limited authority to regulate rates. They are empowered to establish rates for basic service—that is, excluding premium services for which an extra charge is made—provided that fewer than three unduplicated broadcast signals serve the community.

Must-Carry Rules   When distant signal importation began, there was a fear that cable systems would bump local stations in favor of distant stations with more popular programming. The public broadcasting stations expected that they would be particularly vulnerable. To prevent bumping, the FCC adopted must-carry rules requiring that signals from most local stations be carried on cable systems.

Like the franchising authority, the must-carry rules were challenged in the courts. On December 11, 1987, they were declared unconstitutional, also on First Amendment grounds, by the U.S. Appeals Court in the District of Columbia. On January 29, 1988, the Court "clarified" its ruling by stating that an A/B switch to enable the viewer to switch the receiver from the cable system to an external antenna could be required as a means of easing the transition to a world without must-carry channels. On May 31, 1988, the Supreme Court refused to consider an appeal, and the ruling became final.

The court's ruling caused consternation. Commissioner Quello of the FCC termed it a disaster. As of this writing, the NAB, public broadcasters, and other broadcast organizations are actively lobbying for a law that will pass constitutional muster.

Can't-Carry Rules "Can't Carry" was a critical issue for broadcasters. The importation of signals from distant stations meant increased competition, and they would have preferred that carrying these signals by local cable system be totally forbidden. The anticompetitive consequences of such an action made it politically impossible, but broadcasters have lobbied vigorously for the FCC to control cable system program services in a manner that softens their competitive impact.

A compromise policy was adopted that generally permits the importation of distant stations but that forbids the duplication of programs carried on local stations. In such cases, the local station can require the cable system to delete the duplicate program. The rule adopted to effect this policy dealt mainly with network programs, and it is usually impractical now for a cable system to carry the programs of a distant station with the same network affiliation as a local station.

On May 18, 1988, the FCC reinstated the syndicated exclusivity, or syndex, rule. If the rule passes court tests, it will place additional limits on the right of cable systems to duplicate local programs with distant signals after it becomes effective in 1989. Program syndicators customarily give stations exclusive rights—that is, syndicated exclusivity—to show their programs in the station's primary coverage areas. This contract right is violated if a distant signal duplicates a local program having syndex. The FCC rule will permit the local station to require that the duplicate program be deleted from the cable system's signal.

Copyrights Broadcast stations pay fees for the use of copyrighted program material. These fees are based on the population within their normal coverage areas. When stations' signals are imported to distant cities, their audiences become larger, and the original formula is no longer equitable. This issue became even more difficult with the distribution of programs by satellite. Finding a solution to this problem led to bitter disputes among the parties, as well as jurisdictional disputes among Congress, the FCC, and the courts. On September 30, 1976, Congress finally resolved the issue by revising the copyright law.

The revised law mandates compulsory copyright fees for cable systems. Payment of these fees gives them the right to distribute all copyrighted programs. Authority to set the license fees and to divide the proceeds among the various classes of copyright holders, such as the movie industry and professional sports, is vested in a three-member administrative body, the Copyright Royalty Tribunal (CRT). The CRT has a complex and important mission—the assessment and distribution of tens of millions of dollars in royalties each year—and its operations have not been universally admired. One of the frequent criticisms is inadequate staff support and lack of expertise. Broadcasters also have alleged that the law gives cable systems copyright rights for a bargain price and have lobbied for its repeal. This would require cable operators, like broadcasters, to negotiate license agreements.

The Birth of Multiple System Operators

The opportunities afforded cable television systems by distant signal importation attracted a new class of entrepreneurs. Cable systems could now prosper in larger cities that already had one or more local stations. The business was becoming too big to be handled by local mom-and-pop companies, and it began to attract organizations with greater ambitions and resources. Economies of scale, greater access to capital, and more experienced management could be achieved by a single company owning a number of systems. Thus the concept of the *multiple system owner* (MSO), a company that owns systems in many cities, was developed.

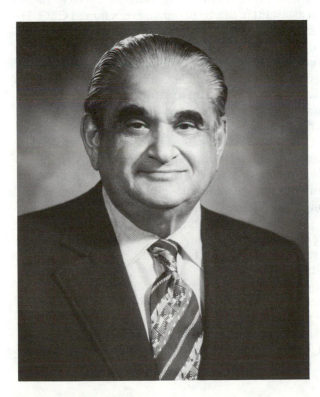

Courtesy Irving B. Kahn.

■ **Figure 7–5** Irving B. Kahn.

Irving B. Kahn   One of the first and most successful of the MSOs was the TelePrompTer Corporation, which was established by Irving B. Kahn (Figure 7–5). Kahn is a remarkable individual, and he was the most colorful of the early MSO founders. He comes from a distinguished family—his uncle was the famed songwriter Irving Berlin—and he has a variety of talents in addition to the enormous creativity and business acumen that have led to his success as an entrepreneur.

In his youth, he became so accomplished at wielding a baton that he received a drum major scholarship to the University of Alabama, and he led its band at the 1938 Rose Bowl. He is an expert table tennis player and gourmet cook. Although he did not receive a technical education, he has an intuitive understanding of technical matters, and he has been aggressive in the introduction of new technologies into the communications industry.

He first demonstrated his business ability by serving as booking agent for a number of bands while at Alabama. After graduation, he was hired as a publicist at the Twentieth Century Fox Film Corporation, where he was the first to use radio to advertise movies. Following wartime service, he re-joined Fox and became an assistant to Spyro Skouras, Fox's famed CEO. He was appointed vice president of its radio and television subsidiary and directed Fox's efforts to obtain licenses for television stations.

The 1949 freeze on new television stations was imposed before any of Fox's license applications were granted, and Skouras sought other ways to participate in the television broadcast industry under Kahn's direction. Kahn suggested a prompting device that had been developed by an actor, Fred Barton. It consisted of a roll of butcher's paper with handwritten dialogue to replace the unwieldy cue cards used by television actors. The device, which Kahn named the TelePrompTer, was enthusiastically en-dorsed by Earl Sponable, Fox's vice president for research and develop-ment, and a young member of his staff, Hubert (Hub) Schlafly.

Skouras declined to have Fox pursue the idea, but he gave Kahn a year's salary and the device's patents and told him to form his own company to "get the damn project out of your system."[2] He also guaranteed Kahn a job when he returned to Fox. With patents in hand, Kahn formed the TelePrompTer Corporation.

The TelePrompTer product, which was developed under the direction of Schlafly, has become standard equipment for television studio productions and is widely used by public speakers as well. But the TelePrompTer business, though successful, was too small to satisfy Kahn's restless nature, and he actively searched for another communications business with greater potential for growth. He found the answer in cable television systems.

---

[2] Private communication.

TelePrompTer's first purchase was a small system in Silver City, New Mexico, in 1958. From this modest beginning, TelePrompTer steadily expanded its ownership of cable systems for the next two decades, and for much of this time it was the leading MSO in terms of the number of subscribers.

In 1971, Kahn became a victim of an occupational hazard of cable companies—the power of corrupt city officials to extort payment for extending a company's franchise. TelePrompTer had purchased a small system in Johnstown, Pennsylvania, in 1961 for $500,000. At the time, it had only two thousand subscribers. TelePrompTer marketed the service aggressively, and by 1971 it had eighteen thousand subscribers. In 1966, however, the mayor and two city councilmen threatened to award the franchise to another company unless they were paid $5,000 each. Kahn yielded to this threat—the alternative being the loss of a franchise that was now worth several million dollars—and each of the three was given a check disguised as a consulting fee.

The canceled checks lay quietly in TelePrompTer's files until they were discovered by the Internal Revenue Service during a routine audit. The IRS informed the Department of Justice, which proceeded on the theory that it was bribery rather than extortion. The case was brought before a grand jury, and Kahn was indicted, tried, and convicted on October 20, 1971.

In the opinion of many, the trial was prejudiced by the judge's antagonism toward businesspeople. In any event, she turned a deaf ear to the pleas of many prominent citizens for a mitigation or suspension of Kahn's sentence. In November, she sentenced Kahn to five years in prison and a fine of $40,000. In the meantime, the city officials pleaded guilty and were given suspended sentences.

After exhausting the appeals process, Kahn began serving his sentence in March 1973. He served twenty months in minimum security prisons in Allenwood and Fort Walton.

Kahn's legal difficulties did not greatly diminish his respect among those in the industry, many of whom felt that he was the victim of a bum rap. At his first appearance before an industry group after his incarceration, his opening words were "As I was saying before I was interrupted . . ." He received a rousing ovation.

Kahn had sold his TelePrompTer stock after he was convicted, but he was soon back in business upon his release. He bought Audubon Electronics, a small cable company in a New Jersey suburb of Philadelphia, and applied for and received franchises in many neighboring suburbs. By 1980, the company had grown to 27 operating systems with 75,000 subscribers, 250,000 homes passed by cable, and 1,000 subscribers being added each week. Kahn sold the company to the *New York Times* for $119.2 million, a then unprecedented price of $1,600 per subscriber. The settlement of the final payment of $52 million was not amicable because of a dispute over contract terms, and Kahn was forced to sue the *Times* to receive the payment.

For the *Times*, however, it was a good investment. By 1988, the system had been expanded to 57 franchises and 160,000 subscribers. In October, the *Times* announced that the system was for sale, and analysts estimated that its market value was $350 million to $500 million.

Industry Growth with Distant Station Importation

The initial Belknap application for distant station importation was filed in 1951, but it was several years before this had a measurable effect on the growth of cable television. Much of the delay was caused by the regulatory problems described earlier in this chapter, and the Belknap application was not granted until 1954.

Bill Daniels, a cable television pioneer and one of the industry's leaders throughout its history, began importing signals from Denver stations into a cable system in Casper, Wyoming, in March 1953, but the regulatory situation was still fuzzy and the distant station concept continued to grow slowly until 1965. Something of a breakthrough occurred in that year (perhaps not coincidentally, this was the period of color television's breakthrough). The number of systems importing distant stations grew rapidly thereafter, and by the end of 1977 nearly four thousand systems were operating, serving thirteen million subscribers with an average of more than three thousand subscribers per system (Figure 7–6).

■ THE THIRD PHASE: SATELLITE PROGRAM DISTRIBUTION

The Concept

Geosynchronous communication satellites (see Chapter 8) have the unique ability to blanket the entire United States with a radio signal that is strong enough to be picked up by *earth stations* equipped with sensitive receivers and suitable antennas. Initially, these satellites were believed to be primarily useful for the transmission of voice and data signals, but video transmission ultimately became their most important application.

During the distant station importation phase of the cable television industry, microwave was the primary transmission medium. It was economic for point-to-point service such as the Belknap system, but it was far too

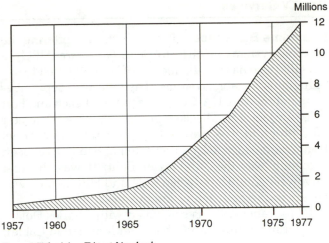

Source: *Television Digest Yearbook.*

■ **Figure 7–6** Cable television subscribers during the distant signal importation phase, 1957–1977.

expensive to provide point-to-multipoint service on a nationwide basis. The television networks used microwave and cable for program transmission to the major markets, but they had only about two hundred affiliates, and they had to resort to off-the-air pickup to reach stations in more remote areas. To transmit a signal to several thousand cable systems by microwave was clearly uneconomic.

The possibility of using satellites for nationwide distribution of programs to cable systems appears to have occurred to a number of innovative individuals simultaneously. During the late 1970s, they developed a wide variety of new program services for delivery to cable systems by satellite. Two types were offered: (1) pay-TV services for which the subscriber paid an extra premium and (2) basic services, which cable television systems included as part of their basic offering.

By the payment of an additional monthly charge, pay-TV subscribers could view a daily fare of movies and other entertainment features not available on broadcast stations or basic cable channels and without interruption by commercials. The basic service programs included *superstations* and programs devoted to news, religion, sports, public service, and general entertainment. The suppliers of basic services received their income from advertising and a modest per-subscriber fee charged the cable operators. Each service had a pioneer who was soon followed by others as its success was demonstrated.

The Pioneers in Pay-TV Services

Gerald M. Levin and Home Box Office   The distinction of offering the first pay-TV service to cable systems belongs to *Home Box Office* (HBO), a subsidiary of Time, Inc. HBO also has the distinction of being the first to use satellites to transmit programs to cable systems. The date was September 30, 1975, and the cable operators were UA-Columbia in Vero Beach and Fort Pierce, Florida, and ATC in Jackson, Mississippi. The satellite contractor was RCA (although the program was transmitted on a satellite channel subleased from Western Union, since RCA's first satellite had not yet been launched), and the program was the Ali–Frazier fight. It was the first transmission under a five-year contract between RCA and HBO, and it was the beginning of a new era in the cable television business.

The concept of pay-TV for cable systems originated with Charles Dolan, at that time the head of Sterling Manhattan Cable (later Manhattan Cable), in which Time, Inc., held an interest. It was introduced as a means of increasing the revenues of the cable system. In 1972, Dolan assigned a new member of his staff, Gerald M. (Jerry) Levin (Figure 7–8), the task of determining the feasibility of developing a network of cable affiliates for a pay-TV service interconnected by microwave.

Levin began his career as a lawyer after graduating from Haverford College and the University of Pennsylvania Law School. From 1963 to 1967, he was an attorney with the New York law firm Simpson Thacher & Bartlett. From 1967 to 1971, he was affiliated with the Development and Resources Corporation, an international investment and management company. In 1969, at the age of thirty, he became its general manager and chief operating officer, and in 1971, after its acquisition by the International Basic Economy Corporation, he served for a year in Teheran, Iran.

Levin had an abiding and long-standing interest in drama and the dramatic arts, and this led him to accept a position with Time, Inc., with the responsibility of developing the pay-TV concept. The report that he prepared for Dolan was enthusiastic about the prospects for a pay-TV network, and it led to the formation of HBO. Levin became its first programming vice president and was successively promoted to president and CEO in 1973 and chairman in 1976.

■ **Figure 7–7**   Ceremony (*top*) and plaque (*bottom*) marking the fifth anniversary of the first cable television satellite transmission. Attending the September 30, 1980, ceremony were, from left to right, Ken Gunter, executive vice president, UA Columbia; Andrew F. Inglis, president, RCA Americom; Sidney Topol, CEO, Scientific-Atlanta, Inc., the manufacturer of the earth station; and Robert Rosencrans, president, UA Columbia. The earth station antenna shown in the picture had a diameter of ten meters, which was then required by the FCC. This restriction was removed later, and smaller and more inexpensive antennas are now used.

Courtesy Scientific-Atlanta.

THE FIRST CABLE TELEVISION EARTH STATION
IN HISTORY WAS INSTALLED AT THIS SITE BY
UA-COLUMBIA CABLEVISION, INC.

HOME BOX OFFICE PROVIDED THE ORIGINAL
VIDEO PROGRAM BY RELAYING LIVE SIGNALS
FROM THE PHILIPPINES ON SEPT. 30, 1975. ON
THAT NIGHT OVER 19,000 CABLE SUBSCRIBERS
IN FORT PIERCE AND VERO BEACH WATCHED
THE ALI-FRAZIER FIGHT.

Courtesy Scientific-Atlanta.

Courtesy Time, Inc.

■ **Figure 7–8** Gerald M. Levin.

The success of HBO led to further promotions, and in August 1979, he was given responsibility for all of Time, Inc.'s video businesses, including HBO and ATC, an MSO. He became an executive vice president of the parent company with responsibility for strategic planning in 1984 and vice chairman in 1988.

HBO's first program was a National Hockey League (NHL) game in November 1972, which was fed by microwave to the Service Electric Cable system in Wilkes-Barre, Pennsylvania. There were 365 subscribers.

For the first three years, HBO's market area was limited to the northeastern states by the cost of microwave transmission, and it struggled to develop an adequate subscriber base. It received a boost when Time's Manhattan Cable became an affiliate in 1974, but at the end of 1975—the beginning of the satellite era—it had only 280,000 subscribers, far below the number required for a profitable business.

Although satellite distribution did not immediately solve HBO's problems, it became the key to its ultimate success. The number of subscribers doubled each year for the next four years, and by the end of 1979 there were four million. The growth continued through 1983, when HBO's subscriber count passed the fourteen million mark. Since then, its growth rate has leveled off somewhat, but in early 1987 it had more than sixteen million subscribers.

Ralph Baruch and Showtime   By late 1977, HBO had demonstrated the viability of the pay-TV concept for cable systems, and it soon had competitors. The first was "Showtime," a program service offered by Viacom, a company formed in 1971 by the spinoff of CBS's syndicated programming and cable television divisions after the FCC had proscribed networks from engaging in these businesses.

Viacom was headed by Ralph M. Baruch, formerly an executive vice president of CBS. He came from a distinguished European family that had fled from France to the United States in 1940 when he was only seventeen to escape Nazi persecution. He became a citizen in 1944, learned to speak English flawlessly, and rose to be a successful communications executive. His career culminated in a high managerial position at CBS and the presidency of Viacom.

Viacom was acquired in a leveraged buy-out in 1987, and Baruch has become an elder statesman of the industry, serving with distinction on a number of industry boards and committees. Showtime is a successful service, although it has been unable to overcome HBO's head start.

In early 1987, there were seven major pay-TV services estimated to have more than thirty-six million subscribers in total.[3]

Pay Cable Services	Subscribers
HBO	16,000,000
American Movie Classics	6,000,000
Showtime	5,150,000
Cinemax (an HBO service)	4,100,000
Disney	3,175,000
Movie Channel	2,900,000
Playboy	580,000

Jeffrey Reiss and Pay-per-View   On November 27, 1985, two new pay-TV services were first broadcast from satellites to cable systems over the country. They differed from earlier pay services in that subscribers were charged for viewing individual movies rather than a flat monthly rate, and they became known as pay-per-view (PPV) television.

---

[3] *Broadcasting-Cablecasting Yearbook 1988* (Washington, D.C.: Broadcasting Publications, Inc., 1988).

To offer a PPV service, a cable system must install addressable descrambling equipment. Scrambled signals are transmitted over the system, and addressable descramblers, installed at subscribers' homes, can be activated on demand. The subscriber is then billed for the program, typically $3.50 to $4.00.

A PPV service by satellite called *Request TV* was first announced by Reiss Media Enterprises, a company founded by Jeffrey Reiss after leaving the presidency of Showtime. It had an impressive group of stockholders, including Reiss's father-in-law, the well-known television producer Norman Lear. The concept was not original with Reiss because individual cable MSOs, notably Warner Amex, had offered PPV services for a number of years. Reiss, however, was the first to plan satellite distribution.

Reiss divided the time slots available on his service into ten units, and he sold nine of them to major film companies, retaining one for himself. The film companies provided the programming for the slots, and Reiss furnished the technical facilities and the marketing and billing services.

Reiss's former employer, Showtime, also announced plans to start a PPV network, *Viewer's Choice,* and the two companies soon engaged in a race to be first on the air. Both claim a starting date of November 27, 1985, which may be somewhat exaggerated. The programming of *Viewer's Choice* consisted of a single movie played repeatedly for an entire week.

The PPV concept has been modestly successful. Two years after the two original services began, each had more than one hundred cable affiliates, and each had approximately three million subscribers. Subsequently, Showtime added a second PPV service, *Viewer's Choice II* (which it sold in 1988), several general entertainment services were initiated, and three major league baseball teams offer their games on a PPV basis. In total, PPV has approximately eight million subscribers.

The Pioneers in Basic Services

Ted Turner and the Superstations   While Levin and HBO were building their pay-TV business, a widely known entrepreneur was working on another concept. The entrepreneur was Ted Turner (Figure 7–9), and the concept was the superstation.

Turner is an extraordinary person. His father died at an early age, leaving him a nearly bankrupt advertising agency. To the amazement of the business world and advertising fraternity, he turned it into a profitable enterprise. He then expanded his business interests, including the acquisition of the Atlanta Braves baseball team and an almost defunct Atlanta UHF station. These were to become the basis for the superstation concept.

Turner is a quintessential entrepreneur, and he repeatedly has made a success of businesses that more conservative and experienced business

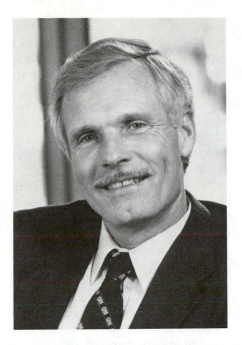

Courtesy Turner Broadcasting System.

■ **Figure 7–9** Ted Turner.

people believed were hopeless. At one time, he even attempted to acquire CBS. The attempt was unsuccessful, but it demonstrated that there was no limit to his aspirations or his courage.

He is an expert sailor and by common consent is one of the world's leading blue-water skippers. In 1977, he won the America's Cup with his yacht, the *Tenacious*. Later, he won the tragic Fastnet race in the English Channel in which a number of participants were drowned in a violent storm.

Turner has an aura of leadership and personal dynamism. With his graying hair and moustache, he is uncommonly handsome, and he was voted one of the ten sexiest men in America. He has tremendous personal drive and the ability to attract able lieutenants.

He is not universally admired, and his brash manners have earned him the titles "The Mouth of the South" and "Captain Outrageous." The latter was particularly descriptive of his conduct at the America's Cup award ceremonies at the starchy New York Yacht Club. On balance, however, he has proved himself to be an outstanding business leader, and he has made major contributions to the communications industry.

Turner's superstation concept was to transmit the signal from his Atlanta television station to cable systems all over the country by satellite, thus giving it nationwide coverage. He found that the copyright law prevented a station owner from doing this, but a loophole in the law permitted a third party to retransmit the signal. (The law has since been amended to require companies offering a superstation service to pay a license fee.)

The third party was Edward Taylor, son of John Taylor, RCA's longtime marketing vice president, who had founded Southern Satellite Systems for this purpose. Taylor leased a satellite channel from RCA, received a feed from Turner's station, and charged cable operators a modest per-subscriber fee for carrying the channel on their systems.

The arrangement was a positive-sum game in which everyone profited. Turner achieved nationwide coverage for his station at no cost to himself. Taylor developed a profitable business from the subscriber fees. Cable systems were able to add an additional service to their systems at a reasonable cost. RCA obtained a customer for its satellite services. And the public received the benefit of more program diversity.

As with any successful business, this one soon had emulators. Three companies offered similar services, one carrying WGN-TV in Chicago, another WOR-TV in Secaucus (Newark and New York), New Jersey, and the third WPIX-TV in New York. Turner's WTBS remained the most popular, and in 1987 it was estimated to have forty-one million subscribers.

Turner Again and the Cable News Network   Eager to exploit the new satellite technology and not satisfied with limiting his activity to a superstation, Turner conceived the idea of a full-time news service for cable television systems. Thus the *Cable News Network* (*CNN*) was born in 1980.

It was a daring and risky move. The annual expenses of the news departments of the major broadcast networks are hundreds of millions of dollars. This expenditure produces program time that averages little more than one hour per day. Turnver gambled that he could obtain enough revenue from subscriber fees collected from cable systems and advertising to support a news service that would equal those of the networks in quality and exceed them in quantity.

The gamble paid off handsomely. It was slow going at first, and for a time Turner faced serious competition from Westinghouse. But the Westinghouse venture was not profitable, and Turner bought out the operation. He went on to build a subscriber base of more than forty million by 1987, more than enough to sustain a profitable business.

No one has been able to duplicate Turner's success in a general news service, but specialized services such as the *Financial News Network* and the *Weather Channel* have built subscriber bases in excess of twenty million.

William Grimes and ESPN   *ESPN,* a cable sports network, was founded in 1979 by an entrepreneurial group that chose William Grimes as its first president. Like CNN, it had rough going at first, as it was forced to compete with the broadcast networks for the most popular sporting events. But starting with a fare of secondary sports, it increased its subscriber base steadily until it began to collect sufficient revenues from subscriber fees (paid by the cable systems) and advertising to compete with the networks for major sporting events. It was acquired by ABC in 1984, and in 1987 it led all the cable program suppliers with more than forty-two million subscribers.

Pat Robertson and the Christian Broadcasting Network Marion Gordon (Pat) Robertson came from a distinguished Virginia family. His father, A. Willis Robertson, was a member of the U.S. Senate for many years. He graduated from the Yale Law School in 1955 and was well on his way to a successful legal career when he experienced a religious conversion. He decided to retire from the law and devote his life to evangelism. He graduated from the New York Theological Seminary in 1959 and was ordained by the Southern Baptist Convention in 1961.

In the course of his studies and early work as a minister, he felt called to devote his very considerable talents as an evalangelist to the radio and television ministry. Starting with a small FM station in Virginia Beach, Virginia, he expanded into radio networking and television with the founding of the Christian Broadcasting Network (CBN) in 1960. The sincerity of his beliefs and his talents as a minister attracted a loyal and devout audience, which supported his work generously with freewill offerings.

Satellites gave him an opportunity to deliver his message by television to a nationwide audience. He was quick to seize the opportunity, and he was one of the first to use a satellite for this purpose. CBN now operates full-time, with more than thirty-eight million cable subscribers. Revenues are derived from subscriber fees, advertising, and freewill offerings. These have been sufficient not only to support the network but also to build a religious establishment that includes a university. As a result of CBN's cable audience, Robertson became well enough known to enter the 1988 primaries as a Republican candidate for president.

Robertson also had emulators in the use of satellites for the distribution of religious messages. Among them was the controversial Jim Bakker and his Praise the Lord or People That Love (PTL) network. Bakker was forced to resign as head of PTL as the result of sexual misconduct, and the future of PTL is uncertain at this time. Other religious cable networks that have had modest success include the ACTS satellite network operated by the Southern Baptist Convention and the Trinity Broadcasting Network.

Brian Lamb and C-SPAN A most important public service offering of the cable industry was the Cable Satellite Public Affairs Network (C-SPAN) initiated by Brian P. Lamb. The service began in 1979 and initially provided live coverage of the U.S. House of Representatives. Its programming was later expanded to include press conferences, House and Senate hearings, and other important events at the Capitol. In 1986, after prolonged debate about its propriety, the Senate permitted live coverage of its proceedings, and a second channel, *C-SPAN II*, was added for this purpose.

C-SPAN has been an important medium for giving the public a firsthand view of the operations of Congress, and it has been well received. In 1987, *C-SPAN* had more than twenty-seven million subscribers, and *C-SPAN II* had grown to more than twelve million. It is supported by corporate donations and by a charge of four cents per month per subscriber paid by the cable systems that carry it.

Kay Koplovitz and the USA Network   In addition to the satellite services with specialized programming, more than twenty entertainment services are designed to appeal to a wide variety of cultural and ethnic groups. One of the earliest and most successful was the *USA Network,* owned by the Gulf & Western Corporation and MCA and headed by Kay Koplovitz. The *USA Network* offers a general entertainment format that is not too different from that of the major television networks. It was founded in 1977, and ten years later it had thirty-nine million subscribers.

Koplovitz received a master's degree in communications from Michigan State University in 1968 and was employed as an editor for Comsat (see Chapter 8) until 1972, when she became director of community services for UA Columbia Cablevision. She became executive director of UA-Columbia Satellite Services in 1977, and the *USA Network* was born. She became its president and CEO in 1980 and has presided over its growth into a major business.

Industry Growth with Satellite Program Distribution

The proliferation of program sources made cable television service increasingly attractive to television viewers. This was reflected in the rapid increase in subscribers. From 1977 to 1988, about thirty-three million subscribers were added, nearly a fourfold increase (Figure 7–10). The significant industry statistics for the United States at the end of 1987 were as follows:

Total U.S. homes	90 million
Total TV homes	88 million
Number of systems	8,413
Homes passed by cable	74 million
Subscribers	43 million
Pay-TV subscribers	38 million

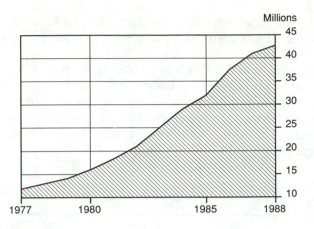

Source: *Television Digest Yearbook.*

■ **Figure 7–10**   Cable television subscribers during the satellite distribution phase, 1977–1988.

■ CABLE TELEVISION BECOMES BIG BUSINESS

Industry Revenue

The increase in cable subscribers during the satellite distribution phase was more than matched by its increase in revenues as cable operators added premium pay-TV services. Total revenues in 1976 were estimated to be about $900 million, of which only $16 million came from pay-TV. In 1988, they were estimated to be $12.8 billion, of which $3 billion came from pay-TV.[4]

Market Value of Cable Television Systems

The increase in the market value of cable television systems has been even more dramatic than the increase in revenues and subscriber count. In 1976, the going price for a system was about $500 per subscriber. By 1987, it had risen to $1,500 to $2,000. In 1988, sellers were asking $2,700 and more. Using an estimated average market value per subscriber of $600 in 1976 and $2,000 at the end of 1987, the market value of all operating systems can be calculated to have risen from $6.6 billion to approximately $90 billion. Although it is not an exact apples-to-apples comparison, this figure ex-

[4] *Television Digest*, 28 (no. 20), 16 May 1988.

ceeds the combined value of the outstanding common stock of General Motors, Ford, and Chrysler. The industry is far different from the folksy mom-and-pop systems of the 1950s.

The increase in market value was the result of a growing realization by the financial community of the potential of cable systems, particularly of their ability to generate cash. They were caught in the mergers and acquisitions wave and were subject to the leveraged buy-outs and other financial trappings of this movement. This coincided with a move by the MSOs to increase their size by acquisitions.

The Rise of the MSOs

Throughout the satellite distribution phase, there was a continuous trend toward group ownership. By the end of 1987, the five largest MSOs had nearly fourteen million subscribers, about 30 percent of the total. The ten largest had 43 percent, the fifty largest 83 percent, and the one hundred largest over 95 percent. The five largest were as follows[5]:

Company	Subscribers
Tele-Communications Inc.	5,162,793
American TV & Communications Corp.	3,700,000
Continental Cablevision, Inc.	2,180,000
Storer Cable Communications	1,471,000
Cox Cable Communications, Inc.	1,438,057

The chairman and major stockholder of Tele-Communications, Inc. (TCI) is Bob John Magness, an able entrepreneur (who has modestly limited his biography in *Who's Who in America* to four lines). Its president and CEO is John C. Malone, who directed much of its expansion and has become a powerful and influential member of the industry.

■ EXTENSIONS OF CABLE SERVICE BY RADIO

At the end of 1987, cable service was available to approximately 85 percent of the homes in the United States (that is, they were passed by cable). Most of the others will not have cable in the foreseeable future because they are in sparsely populated areas or congested urban districts where the cost of installing cable is prohibitive in relation to the population that would be

---

[5] *Television Digest*, 28 (no. 10), 2 May 1988.

served. There are two alternatives for using radio to distribute nonbroadcast program services to these areas—MDS/MMDS and direct pickup of satellite signals.

MDS/MMDS

The FCC has allocated channels in two bands of the microwave spectrum that can be used for terrestrial distribution of nonbroadcast television signals. The bands are called Multipoint Distribution Service (MDS) and Multichannel Multipoint Distribution Service (MMDS).

The MDS band extends from 2,150 to 2,162 MHz and includes two 6-MHz channels suitable for television service. (MDS channels also can be used for the transmission of data and facsimile.) The MMDS band extends from 2,596 to 2,644 MHz and includes eight 6-MHz channels. Alternate channels are assigned in two groups of four, the E channels and the F channels:

E Channels (MHz)	F Channels (MHz)
2,596–2,602	2,602–2,608
2,608–2,614	2,614–2,620
2,620–2,626	2,626–2,632
2,632–2,638	2,638–2,644

An MDS or MMDS distribution system consists of a central transmitter and antenna, which broadcasts to receiving antennas located on subscribers' premises. The transmitters are low power, typically 10 watts.

As compared with cable, MDS and MMDS systems have the advantage of much lower capital costs, and they can be economic in areas that are too sparsely populated for cable. Their disadvantages are limited channel capacity and the requirement for near line-of-sight transmission paths from the central transmitter to each subscriber's home. In spite of the problems, several hundred systems have been installed or applied for, some of them in areas where they compete directly with cable systems. Competition for licenses has been so keen that the FCC has sometimes resorted to lotteries to select the winning applicants.

Operators of MMDS systems have complained of difficulty in obtaining program services. They attribute this to pressure on program suppliers from competing cable services, and they have formed a trade association, the Wireless Cable Association (a contradiction of terms?), to lobby for legal support. Regardless of the outcome of this legal controversy, MDS/MMDS is filling and will continue to fill a useful role in bringing nonbroadcast program services to areas that are not served by cable.

Satellite-to-Home Transmission

During the early 1980s, strange-looking devices began appearing in the backyards of homes in rural areas across the country. They were antennas, or "dishes," for earth stations, or television receive only (TVRO) stations, installed for the purpose of pirating satellite signals intended for reception by cable television systems. These installations were the first use of satellites for direct-to-home broadcasting. Direct-to-home broadcasting had been long anticipated by the satellite industry, but the use of low-power C-band satellites for this purpose was totally unexpected.

This development created both an opportunity and a problem for the cable television industry. It was an opportunity because it provided a means of extending the service of cable systems beyond their franchised areas. It was a problem because it also provided a means for bypassing cable systems entirely. Whether it was an opportunity or problem depended on the extent to which cable operators could control the distribution of satellite signals to the home. The complex technical, legal, and economic issues involved in this are discussed in Chapter 8.

■ FIBER OPTICS AND CABLE TELEVISION

The Promise of Fiber Optics

Many of the technical problems of coaxial cable described earlier, particularly those resulting from its limited bandwidth, can be solved by fiber optics (see Chapter 10). The bandwidth of fiber-optic cable is enormous; for single-mode fiber it is typically 20 to 40 gigabits (20,000 to 40,000 megabits) per second. In analog terms, this is roughly equivalent to a bandwidth of 10,000 to 20,000 MHz, compared with less than 1,000 MHz for coaxial cable. Even in the digital mode, with its voracious demands for bandwidth, more than two hundred video channels could be transmitted with ease with plenty of space left for other services. The result would be noise-free, interference-free pictures with high-fidelity stereo sound.

The Problems of Fiber Optics

To achieve the full benefits of fiber optics, it would be necessary to use digital rather than analog (AM or FM) transmission. Unfortunately, fiber-optic technology has not developed sufficiently to make such systems economically feasible. A host of practical problems must be solved, such as designing an inexpensive device for converting the optical digital signal on the cable to an analog electrical signal at each receiver.

The problems are less daunting if AM laser beams are used for transmission on the fiber. Further simplification can be achieved by using hybrid systems in which fiber optics is used on long trunks but coaxial cable is used for branch trunks and subscriber drops.

Solving the cost and technical problems, particularly for all-fiber digital systems, will require a number of years at best. After they are solved, many more years and many billions of dollars will be required to replace the enormous coaxial plant now in place. Fiber-optic systems will not replace coaxial cable overnight.

Early Fiber-Optic Systems

A number of fiber-optic systems for cable television distribution are now under construction. One of them is a small experimental system being installed by Bell Atlantic and Helicon Cablevision, an MSO, in Perryopolis, Pennsylvania, near Pittsburgh. It will serve only one hundred customers and will carry both telephone and television service. TCI and ATC, two of the largest MSOs, have given the movement toward fiber a considerable impetus by announcing plans to begin incorporating hybrid AM fiber-optic trunk circuits in some of their systems.

It appears, then, that fiber optics will be introduced into cable systems on an evolutionary basis, starting with hybrid AM systems and progressing over a number of years to all-fiber digital systems.

■ THE FUTURE OF CABLE TELEVISION

The future of cable television as a basic medium for program distribution seems assured. The financial community seems to agree with this statement as evidenced by the high market value now assigned to existing cable systems. Its growth pattern in the future will probably be shaped as much by the resolution of a number of adversarial issues as by technical or marketing factors.

Adversarial Issues

Adversarial issues include the role of the local telephone companies, the relationship between cable systems and broadcasters, the role of direct satellite distribution, and the copyright problem.

Cable Television and the Local Telephone Companies The fact that fiber-optic cable can handle both telephone and television circuits suggests that economies of scale might be achieved by having local telephone companies (telcos) handle both. This possibility is well known, both to the telephone companies and to the cable industry, and the latter is viewing it with considerable alarm.

There are three possibilities:

1. The organizational status quo could be retained, with the telcos and cable companies using separate systems.
2. The telcos could lease the transmission circuits to the cable companies, which would continue to provide the programming and the marketing interface with the public.
3. The telcos would provide the complete service, offering strong competition to the cable companies and perhaps putting them out of business.

This issue involves complex questions of economics, public policy, and government regulation. The battle lines between the industries have been drawn, and the next ten years will be interesting for both.

Cable Television and the Broadcasters   The broadcasting and cable television industries developed an uneasy relationship as soon as cable systems were first installed. As cable systems grew in scope, the relationship became adversarial. Cable, with its proliferation of program offerings, has caused significant erosion of the broadcast audience, and this is a serious matter in an industry where networks compete for fractions of a rating point. Broadcasters have supported FCC and statutory actions limiting the freedom of cable systems, including must-carry and can't-carry rules. This pressure from broadcasters will continue.

Cable Television and Satellites   In spite of the fears the cable television industry has expressed over the prospect of competition from direct satellite broadcasting, it appears that a well-managed cable system should be able to compete successfully with satellite distribution by offering a greater variety of programming, including local services not available on satellites. Satellite-to-home broadcasting should not seriously harm cable.

Copyrights   The Copyright Act of 1976 presumably settled this issue, but there has been great dissatisfaction with it, with respect to both its principles and its administration. This has led to strong pressure for change. For example, a requirement that cable systems negotiate individual copyright licenses rather than obtain a blanket license in return for a fixed fee has been advocated. The outcome of this issue cannot be predicted at this time, but significant changes from the status quo could affect the growth rate of cable television.

Cable Television's Growth Opportunities

In spite of cable television's problems with its adversaries and competitors, and in spite of the enormous growth it has enjoyed over the past decades, it is by no means a mature industry. Approximately 65 percent of the homes in the United States now have cable available. There is great opportunity for growth, both by making cable available to more homes and by increasing the penetration—that is, the percentage of homes passed that subscribe—and the revenue per home. These objectives can be accomplished by delivering better programs with higher technical quality. These factors are within the industry's control, and if past experience is a portent of the future, the industry will achieve these goals.

A longer term opportunity for cable is the delivery of HDTV programs. Unlike broadcasters, cable is not constrained by the shortage of precious spectrum space for this service. HDTV's prospects are discussed in Chapter 10.

The revenues of the cable industry are now growing at a compounded annual growth rate of nearly 11 percent—from $8.4 billion in 1984 to an estimated $12.8 billion in 1988. It is a reasonable assumption that this growth rate will continue for the foreseeable future.

8

■ ■ SATELLITE PROGRAM DISTRIBUTION

■ GEOSYNCHRONOUS SATELLITES

During the 1970s, millions of moviegoers saw the motion picture *2001*, which was based on the best-selling science fiction book *2001, A Space Odyssey* by Arthur C. Clarke. Few of them knew that Clarke was also a distinguished engineer who first conceived the principle of *geosynchronous* satellites and envisioned their use for the transmission and distribution of communications signals, including radio and television programs.

The theory of geosynchronous satellites (Figure 8–1) is not based on an advanced or sophisticated technology, and its scientific principles are within the scope of a freshman general physics course. Once explained, the concept is readily understood. Perhaps its simplicity was responsible for its discovery by an imaginative engineer rather than a scientist accustomed to dealing with much greater complexities.

Arthur C. Clarke, whose conception of the geosynchronous satellite was to have such a profound effect on broadcasting and its related industries, was born in Minehead, Somerset, England, in 1917. During World War II, he was an officer in the Royal Air Force, working with Massachusetts Institute of Technology (MIT) engineers in the development of a system of ground controlled approach (GCA) for aircraft landings under conditions of restricted visibility. It was during this period that he developed and proposed the concept of communications satellites.

His proposal was described in a historic article in the British technical journal *Wireless World* in October 1945. Early in the article, Clarke addressed himself to the anticipated scoffers:

> Many may consider the solution [to the problems of long-distance transmission of telephone and television signals] proposed in this discussion to be too far-fetched to be taken very seriously. Such an attitude is unreasonable, as everything envisaged here is a logical extension of developments in the last ten years—in particular the perfection of the long-range rocket of which V-2 was the prototype. While this article was being written, it was announced that

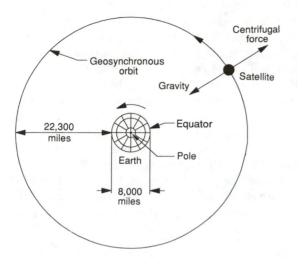

■ **Figure 8–1** A satellite in geosynchronous orbit. Geosynchronous satellites are located in circular orbits 22,300 miles above the earth and in the plane of the equator. They revolve in synchronism with the earth's rotation, one revolution per day, which gives them their name. At this height, the gravitational force pulling them toward the earth is exactly balanced by the centrifugal force pulling them outward. Since they revolve with the earth, they appear stationary from the earth's surface, and radio signals can be transmitted to or received from them with a highly directional antenna pointed in a fixed direction toward the satellite.

the Germans were considering a similar project, which they believed would be possible within fifty to a hundred years.[1]

The article included a remarkably complete and accurate description of geosynchronous satellites (although the name came later). It stated the basic requirements for the satellite's height and orbital plane. It included a calculation of the rocket velocity required to place the satellite in this orbit and observed that the then current rocket technology had already achieved half of this. It correctly forecast the geographic coverage that could be achieved from a satellite. It proposed the use of solar arrays to provide electrical energy for the satellite, and it included order-of-magnitude calculations of the amount of radio frequency power required to communicate to and from the earth. It also foresaw a number of the practical problems that would arise in operating satellites, particularly the solar eclipses, the semiannual periods near the equinoxes when the earth would shield the satellite from

---

[1] Arthur C. Clarke, "Extra-Terrestrial Relays: Can Rocket Stations Give World-Wide Radio Coverage?" *Wireless World*, October 1945.

the sun and interrupt the source of power. It even went a step beyond communication satellites and proposed manned space stations.

After his retirement from the Royal Air Force, Clarke pursued a wide-ranging career, which has resulted in his being described as a twentieth-century Renaissance man. He turned from engineering to science and received first-class honors in physics and mathematics from King's College. He became interested in oceanography and participated in a number of diving expeditions off the coast of Australia and in the Indian Ocean. He found time to write more than fifty books, some scientific and some science fiction. He was the commentator with Walter Cronkite on the CBS reports of the Apollo mission to the moon, and he was the host of his own television series, "*Arthur C. Clarke's Mysterious World.*" Truly an extraordinary individual, he retired to Sri Lanka, formerly Ceylon, where he keeps in touch with the world via the Intelsat satellite system.[2]

■ THE TECHNOLOGY OF COMMUNICATIONS SATELLITES

The 1960s were the halcyon years of the U.S. space program. As the result of *Sputnik* (see below) and competition from the Soviet Union, there was no practical limit to the funds available for the development of rockets and space vehicles. These were the years of Werner von Braun and his team of scientists brought from Germany after World War II, the astronauts with their "right stuff," and the Apollo program with the landing on the moon. Americans were hungry for heros during the trauma of Vietnam, and the space program supplied them.

They were heady years for the engineering community as well. The space program presented an almost endless list of technical challenges that taxed the resources of every engineering discipline. The excitement of these challenges was reward enough for most engineers, but as icing on the cake they received unaccustomed public recognition as indispensable participants in a glamorous industry.

The reduction of Clarke's concept to a practical form depended on research and engineering sponsored by the Department of Defense and the National Aeronautics and Space Administration (NASA). Forward-looking individuals in the communications industry recognized its potential, and the technology of communications satellites followed closely behind that of space and military programs. Communications engineers then began to share in the excitement, and a close-knit group of space communication pioneers developed. They were a band of brothers united by a common purpose, and their outlook closely paralleled that of broadcast engineers in

---

[2] See "Focus, Arthur C. Clarke," *COMSAT Magazine*, No. 15, 1984.

the early years of AM radio. Still later, the idea of space communications was embraced by the radio and television industry, and its engineers once again enjoyed the stimulus of a major new technology.

Nonsynchronous Satellites

As Arthur Clarke had noted, the rockets that were available in the years immediately following World War II did not have sufficient energy to elevate satellites to geosynchronous orbits 22,300 miles above the earth, and early communications satellites operated in lower nonsynchronous orbits.

Sputnik   *Sputnik,* the first communications satellite (this is a loose use of the term, since communication was only one-way, from satellite to earth) was launched by the Soviet Union in 1957. By modern standards, *Sputnik* was very primitive. It weighed only about one hundred pounds (as compared with two thousand pounds and up for current communications satellites), it was in a low orbit far below the 22,300 miles required for geosynchronous operation, and at night it appeared to be a moving star.

Primitive or not, *Sputnik* caused consternation in the United States. It posed a potential threat to national security, and it wounded national pride. There was incredulity that a country such as the Soviet Union, which was technically backward in so many areas, could have beaten the United States in space.

Schools were criticized for substandard education in mathematics and science, and a major campaign was launched to upgrade its quality and quantity. There was political fallout as well. President Eisenhower was just beginning his second term, and he was accused of complacency and inadequate government support of the U.S. space program. This accusation was expanded in the 1960 presidential campaign by the creation of a mythical missile gap. The most tangible effect of *Sputnik,* however, was an enormous increase in NASA's funding and an emphasis on civilian space and rocketry programs, which until then had been the province of the military.

SCORE, Echo, Courier, and Telstar I   Spurred by competition from the Soviet Union, the U.S. Air Force was able to match and even exceed its achievement a year later. Aided by the development of more powerful rockets, it launched *SCORE,* another nonsynchronous satellite, in December 1958. *SCORE* was equipped to broadcast tape-recorded messages back to earth.

The first two-way communication was achieved in August 1960 with *Echo,* a passive relay launched by the U.S. Army. It was a reflecting sphere about one hundred feet in diameter that had neither receivers nor transmitters but that reflected sufficient energy from a high-power transmitter to be detected on the ground with a sensitive receiving system.

Two months later, in October 1960, the Army accomplished two-way communication with an active receiver and transmitter carried by the satellite *Courier I*. A signal was transmitted to the satellite, where it was recorded and after a delay retransmitted back to earth.

The first real-time communication by satellite was made possible in 1962 by AT&T's *Telstar I*, which operated in a low, nonsynchronous orbit. *Telstar I* was used for transatlantic communications, and it was followed in 1963 with *Telstar II*, which provided transpacific service.

Since the *Telstar* satellites were nonsynchronous, it was necessary to use tracking antennas to follow them as they passed across the heavens. The antennas were huge—approximately one hundred feet in diameter—and the tracking mechanism had to be both powerful and precise. The high cost of these antennas, together with the limited time during which a single satellite could be used each day, made this transmission mode commercially uneconomic. Nevertheless, the Telstar satellites were an important step forward in satellite technology, and they paved the way for the introduction of geosynchronous satellites.

Geosynchronous Satellite Launch Vehicles

The technology that made geosynchronous satellites possible was the development of launch vehicles sufficiently powerful to elevate satellites into a 22,300-mile orbit. The first vehicles were expendable rockets, and the space shuttle was introduced as an alternative at a later time.

Expendable Launch Vehicles (Rockets)   Two families of launch vehicles have been used for U.S. communications satellites—the Delta series, manufactured by McDonnell-Douglas, and the Atlas/Centaur, manufactured by General Dynamics. The principal features of the *Delta 3914* (Figure 8–2) are illustrative of all expendable rockets. The launch sequence is shown in Figure 8–3.

The Space Shuttle   During the 1970s, NASA made a policy decision to replace the expendable rocket with the space shuttle as its primary launch vehicle. The shuttle is functionally equivalent to the first and second stages of an expendable rocket. After launching, it carries the satellite, together with the third and fourth rocket stages, to an altitude of about two hundred miles. The third and fourth stages and the satellite are ejected from the shuttle, and the third stage is fired to place the satellite in its elliptical transfer orbit. Later, the fourth stage, the apogee kick motor (AKM), is fired to place the satellite in the geosynchronous orbit, just as with a rocket

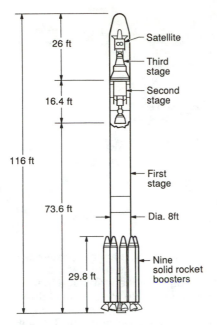

■ Figure 8–2 The Delta 3914 rocket. This rocket had three stages. The first stage consisted of a huge cylinder containing more than 300,000 pounds of liquid fuel plus nine solid fuel boosters strapped to its side. (It was the addition of these boosters that gave the rocket sufficient energy to power a geosynchronous launch.) The other two stages contained solid fuel. The fourth stage, the apogee kick motor (AKM), was located in the satellite.

launch. In the meantime, the shuttle minus its payload returns to earth, where it is refurbished and readied for the next launch.

The space shuttle has been one of NASA's most costly programs, and it has failed to achieve some of its important goals. As compared with expendable launch vehicles, the shuttle was supposed to have three major advantages:

1. Its payload capacity would be much greater—65,000 pounds versus 2,000 pounds for a Delta.
2. It could put people in space to perform experiments and other space-related functions.
3. The cost would be less—in 1980 an estimated $15 million versus $25 million for a Delta—because the vehicle could be reused.

To a large extent, these supposed advantages are contradictory and self-defeating because the reusability of the shuttle requires that it be manned. The shuttle must combine the functions of a rocket and a manned

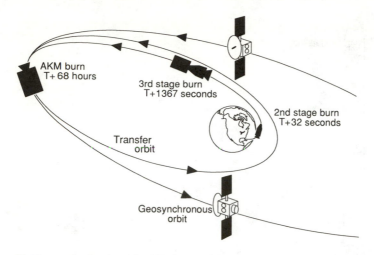

AKM burn
T+ 68 hours

3rd stage burn
T+1367 seconds

2nd stage burn
T+32 seconds

Transfer
orbit

Geosynchronous
orbit

■ **Figure 8–3** A typical Delta rocket launch sequence. In a typical sequence, the combustion of the liquid and solid fuels in the first stage required 224 seconds—that is, until T + 224 seconds. At that time, all the first-stage fuel containers had been jettisoned and the rocket and satellite were at an altitude of 66 miles.

At T + 237 seconds, the second stage was ignited and burned until T + 495 seconds. This placed the rocket in a circular orbit 104 miles above the earth. In the meantime, the fairing surrounding the satellite had been jettisoned.

At T + 1367 seconds, the third stage was ignited. This stage placed the satellite in an elliptical orbit called the transfer orbit. At its lowest point, or perigee, the satellite had an altitude of 104 miles; at its highest point, or apogee, it was at the geosynchronous altitude of 22,300 miles. All the remaining launch vehicle components had been jettisoned, leaving the satellite and AKM in orbit.

After remaining in the transfer orbit for about 68 hours, the AKM was fired to put the satellite in its circular geosynchronous orbit.

aircraft, which makes it extraordinarily complex and costly. The cost is increased further by the additional safety features needed by a manned vehicle.[3] Finally, refurbishing the shuttle is costly. Even if the program had experienced no serious technical problems, it is doubtful that the shuttle's reusability would have offset these added costs. But with a liberal use of sufficiently optimistic assumptions, NASA was able to show that it could, and the program proceeded.

---

[3] One failure in twenty-five would be an acceptable realiability for an unmanned rocket; the cost of the failed launches could be spread over the cost of the successful ones by insurance and would increase their cost by 4 percent. For flights manned by professional test pilots who have knowingly accepted the high risks of their profession, one failure in a hundred might be considered acceptable. For nonprofessional passengers, one failure in a thousand would probably not be good enough. The cost of the shuttle rises geometrically as redundant components and other safety features are added to improve its reliability.

Unfortunately, the shuttle (Figure 8–4) did encounter many serious problems, which required many years and billions of dollars to solve. The delays and costs were punctuated in January 1986 by the loss of the shuttle *Challenger* with its crew and passengers. The loss occurred shortly after lift-off as the result of a defective solid-fuel booster. NASA had included a schoolteacher in its passenger manifest to demonstrate that a shuttle flight was a safe and routine operation. Seldom has a public relations effort been such a tragic failure.

The problems of the shuttle and the *Challenger* disaster had an adverse effect on the satellite industry. After delays totaling two years, the first shuttle launch of commercial communications satellites, *SBS-3* and the Canadian *Anik C-1*, occurred in November 1982. A number of successful launches followed, underscored by the in-orbit rescue in November 1984 of two satellites, *Palapa B2* and *Westar 6*, which had been put in the wrong low-altitude orbit the previous February.

Commercial satellite launches were brought to a halt by the *Challenger* disaster. Having committed itself to the shuttle, NASA had neglected the expendable rocket program and as of this writing had been unable to launch any commercial satellites since *Challenger*. Rather, it has removed all commercial payloads from the shuttle manifest in the hope that a commercial rocket-launching industry will develop. For the time being, however, the *Ariane*, a rocket built by a European consortium led by the French and launched from French Guiana, is the only credible launch vehicle available for commercial satellites.

The First Geosynchronous Satellites

With the development of sufficiently powerful rockets, it became possible to begin launching geosynchronous satellites. The first successful geosynchronous satellite, *SYNCOM-II*, was launched by NASA in 1963. A series of experimental satellites launched during the next ten years included *SYNCOM-III* and the NASA Advanced Technology Satellites (ATS) series. The experimental programs conducted with these satellites were essential to the initiation of commercial service. They included development of techniques for stabilizing the satellite and maintaining it in position, development of methods for increasing the reliability and life of the satellites, and evaluation of the effect of meteorites and ionizing radiation.

The first geosynchronous satellite used for commercial communications was *Intelsat I*, or "Early Bird," which was launched in 1965 for international service. It had a limited capacity of only 240 voice circuits, and its life was only one and a half years. But it was a first, and it was followed by *Intelsat II* in 1967, *Intelsat III* in 1968, and *Intelsat IV* in 1971, each with greater capacity and longer life than its predecessor.

Courtesy NASA.

■ **Figure 8–4** The space shuttle shortly after lift-off. The launch of a space shuttle is an awesome spectacle accompanied by ear-splitting noise. The shuttle is dwarfed by the liquid fuel tank and the smaller cylinder containing solid fuel that is strapped to it.

By 1972, the technology of launch vehicles and satellites had progressed sufficiently to make them practical for use in domestic service within the United States. The FCC promulgated the open-skies policy (described later), and the U.S. satellite communications industry was under way.

Communications Satellites

The design of communications satellites is a marvel of technical ingenuity that challenged the best engineering minds in the United States and later in Europe and Japan. Figure 8–5 shows the satellites' principal components.

Most C-band satellites have twenty-four transponders, each with a bandwidth of 40 MHz and a power of 4 to 10 watts. Ku-band transponders have much higher power, typically 45 to 60 watts. To date this has limited the number of transponders to sixteen, but with improved power sources this may be increased to twenty-four. Satellites for direct broadcast (DBS), if and when launched in the United States, will probably have transponders with powers of 100 to 200 watts.

Satellite "Footprints"   From the user's point of view, the most important characteristic of a satellite is its "footprint" (Figure 8–6). This defines the area in which satisfactory service from the satellite can be obtained.

Signal Transmission Modes   FM is ideally suited for the satellite transmission of television signals, and it is almost universally used. It has three important advantages over AM for this purpose:

1. It does not require highly linear amplifiers.
2. It has a substantial *noise improvement factor* (see Chapter 3) so that the signal-to-noise ratio in the output video signal is higher than in the radio frequency carrier.
3. The transmission energy can be more uniformly distributed across the channel bandwidth with a technique called *energy dispersal*.

Energy dispersal is particularly important in the C band because the FCC Rules establishing the downlink power limitations of satellites are stated in terms of power per kilohertz of bandwidth. With AM and undispersed FM, there is a high concentration of energy in the vicinity of the carrier, which would put a severe limit on the permissible radiated power. This is avoided with FM by superimposing a low-frequency triangular waveform on the video signal. (More sophisticated systems use dynamic frequency dispersal in which its amount is varied in accordance with the degree of concentration of sideband energy.) This moves the carrier frequency back and forth and spreads the radiated energy more uniformly over the entire channel. The

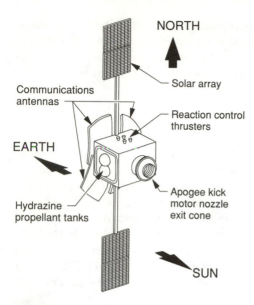

NORTH

Solar array

Communications
antennas

Reaction control
thrusters

EARTH

Apogee kick
motor nozzle
exit cone

Hydrazine
propellant tanks

SUN

■ **Figure 8–5** The principal components of an RCA communications satellite.

Communication antennas. The communication antennas are part of the payload of the satellite, and they transmit and receive signals to and from terrestrial earth stations. The pattern of the transmitted beam on the surface of the earth is called the footprint (see Figure 8–6). Satellites often have several antennas with different footprints to provide coverage to different regions of the earth.

Transponders. The transponders, located inside the satellite, are its active payload. They amplify the signals received from the ground by the antenna system, shift their frequency (hence the term *transponder*), and return them to the antenna system for retransmission to the earth.

Command and control equipment. The command and control equipment sends information back to earth for monitoring the condition of the satellite. It also receives the commands that control its operation.

Solar panels. These are arrays of solar cells that convert radiant energy from the sun into electrical energy to power the satellite. In the RCA satellite, these are mounted on a flat panel that rotates so that it is always facing the sun. The cylindrical surface of the Hughes satellite is covered with solar cells.

Hydrazine tanks. These tanks contain several hundred pounds of hydrazine gas stored under pressure. This is used to maintain the satellite in a fixed position. While in orbit, the satellite is subjected to a number of minor gravitational forces in addition to the principal vertical force, and these tend to move it from its proper position. They include the gravitational effect of irregularities on the earth's surface, such as mountain ranges, and of the sun and moon. To offset these forces, gas is periodically ejected from orifices in the satellite to push it back in position in a process called station keeping. Exhaustion of the supply of hydrazine gas is one of the factors that may limit the satellite's life. Typically this requires ten years.

Batteries. Batteries store energy to provide power for the satellite during solar eclipses, when the earth comes between the sun and the satellite.

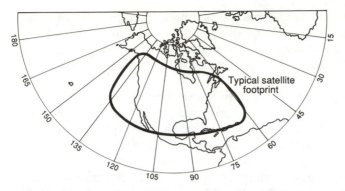

■ **Figure 8–6** A satellite signal footprint. The boundary of the footprint is a contour along which the intensity level of the signal from the satellite is equal to the minimum required for satisfactory service. This level depends on the type of service, and it is usually expressed in effective isotropic radiated power (EIRP).

bandwidth requirements for digital transmission are excessive for television, but digital technology is well suited to transmitting audio for radio broadcasting.

Earth Stations

Earth stations are the installations that transmit and/or receive signals from satellites. They vary widely in cost and complexity.

The primary earth stations of satellite communications carriers are major installations that include an array of antennas, uplink transmitters, downlink receivers, signal-processing equipment, and the system for monitoring and controlling the satellite's operation (called the TT&C for tracking, telemetry, and command). At the other extreme are small backyard receive-only stations used for the direct reception of television programs from the satellite.

Round, dish-shaped antennas that focus the energy into a narrow beam are used in all stations. The directivity of the beam, whether transmitting or

---

The satellite in the drawing is an RCA design. Most of the satellites manufactured by Hughes have been cylindrical rather than rectangular. The difference results from the manner in which their positions are stabilized. Both use the gyroscopic effect of a rotating mass, but in the RCA design the mass is an internal inertia wheel (called three-axis stabilization), while in the Hughes design the entire satellite rotates (except for the antenna, which remains pointed at a fixed point on the earth). More recently, Hughes has also used three-axis stabilization.

receiving, is determined by the ratio of the antenna diameter to the wavelength: the larger the antenna, the sharper the beam. To a first degree of approximation, the effectiveness of a receiving antenna in extracting energy from the incoming radio signal is proportional to its area, regardless of the wavelength.

From the standpoint of this narrative, the most striking development during the past ten years has been the great reduction in the size and cost of small receive-only earth stations. FCC rules originally specified a minimum antenna diameter of ten meters, and they required that the site be cleared— that is, certified to be free of interfering microwave signals. In return, the station was protected from interference from future microwave or satellite systems. With these requirements, the cost of a typical receive-only station as used at a cable system head/end was about $75,000.

The FCC deregulated receive-only stations in 1979. No showing of noninterference is required, and no minimum antenna size is specified. Most cable systems now use five-meter antennas, and the size reduction, together with price competition and economies of scale, have reduced the cost of cable earth stations to $5,000 or less. Backyard earth stations for home use have lower performance requirements, and the use of six-foot antennas is common. A typical home installation now costs less than $1,000.

■ SPECTRUM ALLOCATION AND USE

Frequency Allocations

The FCC has allocated three frequency groups for commercial communications satellites—the C- and Ku-band fixed satellite services (FSS) and the direct broadcast by satellite (DBS) service. Each group includes uplink and downlink frequency bands:

	Frequency (GHz)	
	Uplink	Downlink
Fixed satellite service (FSS)		
C band	5.925–6.425	3.700–4.200
Ku band	14.0–14.5	11.7–12.2
Broadcast satellite service (DBS)	17.3–17.8	12.2–12.7

The C-band and Ku-band channels can be used for all types of communications services—voice, data, radio, and television. The DBS channels are reserved for direct-to-home television service.

A Comparison of C-Band, Ku-Band, and DBS Satellites

The C band was the first to be used in the United States for commercial communications, and it is still the most widely used. By the time Ku-band satellites began operating, the industry had made major investments in C-band earth stations. C-band equipment was available sooner, and its electronic components were somewhat cheaper. C-band transmissions do not suffer blackouts during torrential rainstorms as do Ku-band transmissions.

The major disadvantage of C band is that its downlink frequencies are shared with terrestrial microwave systems. To prevent mutual interference with microwave transmissions, constraints are placed on the location of C-band earth stations, and the downlink power of C-band satellites is limited. In contrast, Ku-band antennas can be located without consideration of the presence of nearby microwave systems, even in congested urban areas. In addition, Ku-band earth stations can use smaller transmitting antennas because of the shorter wavelength of Ku-band radiation and smaller receiving antennas because of the higher power of Ku-band satellites. The different characteristics of C-band and Ku-band satellites have a major effect on their use in the television and radio industries.

The DBS downlink channels are adjacent to the Ku band in the spectrum, but the FCC Rules specifying their technical characteristics are quite different. The intent of the Rules is to provide television service in the home directly from the satellite with very small receiving antennas. To achieve this objective, DBS satellites are allowed much higher power, and their orbital spacing is greater.

Bandwidth Usage

The FCC has allocated a bandwidth of 500 MHz to satellites in all three bands. Flexibility is allowed in dividing this bandwidth among transmission channels. *Westar I* was the first satellite for domestic communications in the United States. Launched by Western Union in April 1974, it had twelve channels, each 36 MHz wide (compatible with microwave systems) and separated from adjacent channels by guard bands. RCA demonstrated, however, that the satellite capacity could be doubled by using both vertical and horizontal polarization, with twelve channels on each. Additional isolation between the vertically and horizontally polarized channels was provided by staggering their center frequencies (Figure 8–7). Frequency reuse is now required by the FCC, and twenty-four 40-MHz channels, each requiring a separate transponder, is the standard for C-band satellites.

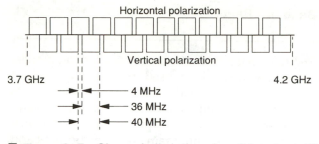

Horizontal polarization

Vertical polarization

3.7 GHz   4.2 GHz

4 MHz

36 MHz

40 MHz

■ **Figure 8–7**   Channel allocations in a C-band satellite.

■ THE ORBITAL ARC AND ORBITAL SPACING

The Prime Orbital Arc

The orbital arc for satellites is an imaginary segment of a circle in the plane of the equator at a distance of 22,300 miles above the earth. The *prime* orbital arc meets the requirement that the angle above the horizon of the path from an earth station to a satellite on the arc (the look angle) must be at least 5 degrees for C band and 20 degrees for Ku band. (A higher look angle is needed for Ku band because of the need to shorten the path through the atmosphere as much as possible to minimize atmospheric attenuation during rainstorms.) The limits of the prime orbital arc for 5 degree horizon clearance for locations in the continental United States (CONUS) and selected points in Hawaii and Alaska are given below. (The locations of the parallels of west longitude are shown in Figure 8–8.)

	Degrees West Longitude
CONUS	55–138
CONUS plus Hawaii	80–138
CONUS plus Anchorage	88–138
North Slope (Alaska)	120–190

The prime arcs for Ku band are shorter because of the requirement for greater horizon clearance.

The table shows the problem of operating geosynchronous satellite systems with earth stations at northerly latitudes. The antenna elevation angles are low throughout Alaska, and the satellites are below the horizon at the north pole.

International agreements are another constraint on the availability of the arc. The FCC assigns orbital locations between 62 and 146 degrees west longitude for C band and 62 and 136 degrees for Ku band. By agreement with Mexico and Canada, orbital locations between 104.5 and 117.5 degrees are reserved for satellites serving those countries.

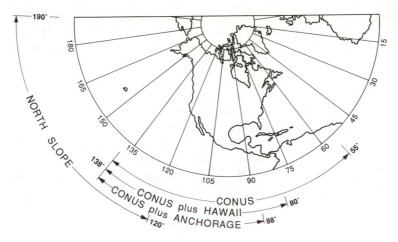

■ **Figure 8–8** The prime orbital arcs.

Orbital Spacing

The spacing between satellites has succumbed to the heavy demand for satellite capacity. It is constrained technically by the directivity of antennas and their ability to discriminate against adjacent satellites while transmitting or receiving. Rigid technical considerations, however, yielded to some extent to the capacity shortage.

C-band orbital locations were initially spaced 4 degrees apart (Canada was even more conservative with 5-degree spacing). Ku-band slots were spaced at 3-degree intervals, reflecting the ability to achieve greater antenna directivity at the shorter wavelength.

It was found, however, that the number of satellites that could be authorized with 3- and 4-degree spacing was insufficient to meet the growing demand for satellite capacity, particularly for television service. Accordingly, the FCC reduced the spacing to 2 degrees in both bands. Few problems have resulted from the reduced spacing, probably indicating that the original standards were too conservative.

■ THE REGULATION OF SATELLITE COMMUNICATIONS

The development of satellite communications technology presented government regulators with problems that had no precedent. It was inherently a long-distance communications system, which gave it an international character. Although regulators customarily thought in terms of point-to-point service—and the initial applications of satellites were for this purpose—satellites had the technical characteristics of point-to-multipoint and broadcast services. Finally, there was the issue of control of the orbital arc in which the satellites were located and which was not subject to national sovereignty. Clearly, new international agreements were necessary.

The situation was further complicated in the United States by the court-ordered breakup of AT&T during the late 1970s and the growing trend toward deregulation. Under previous regulatory policies, AT&T would probably have been the only organization permitted to use satellites for communications, and they would have been integrated into the vast AT&T system. Under the new policies, this was by no means automatic, and the FCC had to decide who could be authorized to engage in satellite communications, either privately or on a common carrier basis.

International Satellite Agreements

The International Telecommunications Union (ITU) governs radio services on an international level. It meets infrequently, and its year-to-year policy-making is handled by periodic meetings of the World Administrative Radio Conference (WARC) and the Regional Administrative Radio Conference (RARC). To have the force of law, the decisions of these bodies must be specified in treaties that are ratified by the member countries.

Satellites were made subject to this international regulatory framework, and intensive negotiations followed, involving most of the world's major powers. The objective was to develop a means of using satellites for international communications. The result was the founding of Intelsat in 1964, an international telecommunications organization that has become the chosen instrument of most countries' international communications.

Satellite Regulation in the United States

Having been given the authority by treaty to assign orbital locations to domestic communications carriers in a portion of the orbital arc and to participate in an international consortium, the U.S. government moved forward to exercise this authority.

Legislative and Regulatory Actions   The Communications Satellite Act of 1962 established the policies governing satellite communications in the United States. It was a declaration of national purpose, and its thrust was expressed in its opening paragraph:

The Congress hereby declares that it is the policy of the United States to establish, in conjunction and in cooperation with other countries, as expeditiously as practical a commercial communications satellite system, as part of an improved global communications network, which will be responsive to public needs and national objectives, which will serve the communication needs of the United States and other countries, and which will contribute to world peace and understanding.

To carry out this purpose, the Act authorized the establishment of a private corporation, later to become Comsat, to be the government's chosen instrument in international communications.

Ten years later, in June 1972, the FCC issued a historic Order, establishing what became known as the "open-skies" policy for domestic commercial satellite carriers. This was a major expression of the broad movement toward open competition that was occurring in U.S. communications policy. It declared that any qualified individual or organization could receive authorization to launch and operate a satellite. As a result, satellite communications became a competitive business (although initially a highly regulated one) rather than a regulated monopoly.

The Order was one of the many regulatory, legislative, and judicial defeats that AT&T suffered during this period. It removed AT&T's potential for monopolizing space communications, and it even went a step further. In order to protect new entrants from AT&T's power, it prohibited AT&T from using satellites for any purpose except as part of its primary switched voice and data network. The provision barred their use by AT&T for any type of private-line service, including the transmission of radio and television programs.

The Assignment of Orbital Locations   The FCC was given the responsibility for specifying orbital locations, called slots, and assigning them to specific carriers. By the early 1980s, the satellite communications industry had become sufficiently profitable to attract an increasing number of would-be participants, thus creating competition for the limited number of available slots, particularly the more desirable ones. The FCC's reduction of orbital slot spacing greatly alleviated the competition, but it did not completely solve the problem because some slots are more desirable than others.

In making its choices, the FCC endeavored to act as a handicapper by equalizing the number of slots assigned to each carrier. It followed a policy of "To those who have not shall be given," and existing carriers sometimes did not fare as well as newcomers. In addition, the FCC made it clear that, unlike broadcast channels, the assignment of an orbital slot included no presumption of a right to hold it indefinitely.

The FCC's success in equalizing the number of orbital slots among the carriers is shown by the assignments and authorizations at the end of 1988:

	C Band	Ku Band
GE (formerly RCA) Americom	5	4
Hughes	5	4
GTE	4	7
Western Union	4	2
Contel ASC	3	3
AT&T	3	0
Alascom	1	0
Comsat General	1	0
Satellite Business Systems	0	6
Federal Express	0	2
National Exchange	2	2

These assignments and authorizations essentially fill the orbital arc available to U.S. communications carriers under present policies.

Rate Regulation   The FCC had a new problem with respect to rate regulations. Having opted for open competition in satellite communications, it had to develop a policy for regulating the rates of an industry that was also subject to the competitive price constraints of a free market. This was not a problem in the early years of commercial satellite communications when all the carriers were losing money, but as carriers became profitable in the early 1980s, it became a major issue that sharply divided the FCC.

On one side were commissioners who were sympathetic to the Reagan administration's drive toward deregulation. They believed that market prices should prevail, subject only to the conditions that all customers be charged the same rates and that no extraordinary market conditions existed. On the other side were the regulatory hard-liners, led by Commissioner Joseph Fogarty, who believed that rates charged by satellite carriers should be capped at a level that would produce a rate of return only marginally higher than that allowed for monopoly carriers.

Fogarty's position seemed unfair to the industry. It is a basic business principle that the potential profit margin for an enterprise should be directly related to its risk. The application of this principle, in fact, is necessary for the successful operation of the free-enterprise system. One could hardly expect a prudent person to engage in a high-risk business unless it would produce a high rate of return if successful.

Communications common carriers had historically been regulated monopolies. In return for protection from the risk of competition, they were limited to modest rates of return on their investment.

Satellite communications carriers, however, did not have protection from competition, and it was a new and risky business. It required huge capital investments. The markets for the service were not well defined and had to be developed. The system capacity had to be committed far in advance of its availability, and it could not be quickly adjusted to demand thereafter. As RCA was to discover with the loss of *Satcom III* (discussed later in this chapter), there were real technical risks. In return for taking these risks, the satellite communications industry felt it was entitled to the profit margins of successful entrepreneurial companies.

The regulatory hard-liners were not impressed by this argument. Accustomed to the traditions and practices of regulated monopolies, they found it difficult to adjust to the concepts of competition and marketplace pricing. As a result, the rates that satellite carriers were allowed to charge were often significantly below market levels.

The issue first arose in 1979 when RCA's success in obtaining cable customers for its system (see Chapter 7) pointed the way to profitability for the industry. To avoid the rate constraints of the conventional tariffing process, a number of innovative schemes were attempted.

RCA tried first. It offered to lease seven of the twenty-four transponders on *Satcom IV*, a new satellite to be launched in 1982, at a price established by the market rather than the cost. The lease term would be the estimated life of the transponders, ending on December 31, 1989, and the market price would be established by an auction. The auction was conducted by the distinguished auction house Sotheby Parke Bernet on November 9, 1981. It was a huge success. All seven transponders were leased at prices ranging from $10.7 million to $14.4 million. The total lease price for the seven transponders was $90.1 million, more than the cost of the complete satellite.

Unfortunately for RCA, it was still bound by common carrier rules, and the FCC disallowed the leases, not on the ground that the prices were market based but on the ground that they were not uniform to all customers. To cure this defect, RCA proposed another scheme. It established a uniform lease price of $13 million per transponder on *Satcom IV* for a term extending from its impending launch until December 31, 1989, its projected life. This was judged to be the market value on the basis of the Sotheby auction. It submitted this plan as a tariff, and a sharply divided FCC approved it. The deregulatory policies of the Reagan administration were being vigorously carried out by the FCC's new chairman, Mark Fowler, and they were beginning to influence its decisions. Commissioner Fogarty dissented strongly, however, characterizing the plan as the worst tariff he had ever seen. He could not accept the principle of a tariff that was not cost based.

RCA won the regulatory battle, but it was not so successful in the marketplace. The transponders were offered for lease in mid-1982, but during the six months that had elapsed following the Sotheby auction, the market had softened perceptibly. As a result, only three transponders were leased on this program.

A more successful plan was proposed by Hughes Communications in 1982 under the leadership of its president, Thomas Clay Whitehead, formerly head of the National Telecommunications and Information Administration (NTIA) under President Nixon. It differed from RCA's in that Hughes offered the transponders for sale rather than lease. The commission approved the principle of transponder sales at market prices, and the transponders on Hughes's first satellite, *Galaxy I,* which was to be launched in 1983, were placed on the market. To increase the value of the transponders, Hughes offered a special price to HBO, the key cable programmer, correctly assuming that this would induce other programmers to follow.

It was a brilliant move. HBO accepted Hughes's offer, and other programmers did follow. In one stroke, Hughes established *Galaxy I* as the prime cable satellite, wresting this position from RCA, it eliminated the threat of losing the business to competitors at a future time, and the transponder prices were far higher than the net present value of the cash stream that Hughes could have obtained by leasing them as a regulated common carrier.[4]

Fogarty's dissent to RCA's plan was one of the last hurrahs of the hard-liners. The administration's deregulation policies were applied with increasing vigor, and when President Reagan left office in 1989, the rates and conditions of sale and lease in the satellite communications industry were determined primarily by market forces.

■ INTERNATIONAL SATELLITE COMMUNICATIONS SERVICE

International voice and message traffic was the first commercial application of communications satellites. The use of satellites for this purpose was a natural. Transoceanic radio circuits suffered from marginal quality and reliability, particularly for voice, and the capacity of the cables then available was limited and inadequate for television transmission. The cost of international voice service was high. There was an urgent need for a higher quality, higher capacity, and lower cost system that also could handle television signals. Satellites had the potential for meeting all these requirements.

---

[4] Hughes's plan would not have been as attractive to a publicly held company, since the profit and loss effect was to produce a large one-time profit followed by years of breaking even. This pattern does not enhance the market value of the stock.

Comsat

Role and Organization   The Communications Satellite Act of 1962 specified a mechanism for U.S. participation in international satellite communications by authorizing the formation of a corporation, Comsat, to be the U.S. instrument. Comsat had the unique characteristic of private ownership but with incorporators appointed by the President. In 1962, President Kennedy appointed the incorporators, a distinguished group of thirteen business and communications company executives, bankers, labor leaders, and government officials.

Their first meeting was held on October 22, 1962. One of their first tasks was to determine the amount of capitalization for the company and to establish the policy for stock ownership. It was decided that capital of $200 million was required, an amount that proved to be more than adequate. The Act required that stock ownership be broadly distributed, and it was initially divided among AT&T, the three international message carriers (RCA, Western Union International, and ITT), and the public.

Business Growth   As a business, Comsat was an act of faith during its early years. The concept was fraught with technical, political, and economic uncertainties. The first successful experimental geosynchronous satellite, *SYNCOM II*, had not even been launched when the initial stock offering was made in June 1963. (The offering, nevertheless, was oversubscribed.) If geosynchronous satellites had not been successful, it would have been necessary to use low-orbit satellites and enormously costly tracking antennas. The cooperation of foreign governments was by no means assured, and only the roughest estimates of the cost of offering the service could be made.

Backed by the confidence of its public and corporate stockholders, the organization of Comsat's staff proceeded. Its first chairman and CEO was Leo Welch, formerly chairman of Standard Oil and one of the original incorporators. Its first president was Dr. Joseph V. Charyk.

The stockholder's confidence was not misplaced. In its monopoly role as the provider of the ground segment of international circuits originating in the United States, Comsat has been successful in every respect.

The Comsat Earth Station Network   Soon after Comsat's staff was organized, it began the construction of a network of earth stations. The first, put into service in 1965, was located in Andover, Maine. This was followed in 1966 by stations in Brewster, Washington, and Paumalu, Hawaii. Comsat currently operates six stations in the United States and its possessions and six more under contract on a number of Pacific Islands.

The Comsat Laboratories Comsat has made a major and continuing contribution to the technology of satellite communications through its Comsat Laboratories. The Laboratories, located in Clarksburg, Maryland, opened in September 1969. This is one of the few research centers in the world devoted solely to satellite communications. A sampling of the diverse technologies in which it has conducted research include radio wave propagation, solid-state satellite transmitters, low-noise receiver amplifiers, and digital transmission systems. Its investigations include both satellite and ground equipment. As of 1983, its scientists and engineers had been awarded 226 patents.

Intelsat

To complete an international satellite communications system, it was necessary to have satellites as well as earth stations. To fulfill this requirement, Comsat developed the concept of an international organization, later to become Intelsat, to own and operate the satellites that would constitute the space segment of the system. The ground segment, the earth stations, would be owned by the member countries or their designated representatives, such as Comsat in the United States. Intelsat would be owned jointly by the participating countries, with their ownership roughly in proportion to their use of the system. It would be a nonprofit organization, and its rates would be adjusted to cover its costs.

Armed with this concept, Comsat delegations, headed by Leo Welch and Frederick Kappel, AT&T's chairman, began negotiations with the Europeans. It was a hard sell at first. The European governments, strongly influenced by their state-owned communications companies, which had massive investments in cable, resisted. There also may have been resentment that the United States, with its wide lead in space technology, was in a position to dictate policy. But with a combination of persistence, salesmanship, and even some threats, together with the support of Canada, Japan, and Australia, Comsat was able to persuade the Europeans.

After eight months of very difficult negotiations, an interim agreement was signed by 15 countries—the major countries in Western Europe, the United States, Canada, Australia, and Japan—in Washington in August 1964. Other countries soon expressed a desire to join, and ultimately the membership encompassed 108 countries.

Technical Progress Intelsat's technical progress can be attributed mainly to its management contract with Comsat, which provides for technical consulting services. These include working with satellite manufacturers to develop the generations of satellites Intelsat needs to handle its steadily increasing traffic. The success of this effort can be measured by the specifications of these satellites:

	Intelsat Series					
	I	II	III	IV	V	VI
First launch	1965	1967	1968	1971	1980	1981
Bandwidth (MHz)	50	150	500	500	2,150	2,640
Capacity (telephone circuits)	240	240	1,200	4,000	12,000	33,000
Design life (years)	1.5	3	5	7	7	10

Business Growth   Intelsat began service on *Intelsat I* in May 1965. The inaugural transmission was a two-way telecast, "Live via Satellite," to Europe. Regular voice service, albeit on a very small scale, followed soon after. From this modest beginning, Intelsat's traffic has grown steadily, and it provides service, directly or indirectly, to 170 countries. In 1989, nearly half the world's transoceanic communications service was handled by satellite.

The economies of scale, together with technical progress, have resulted in a steady decrease in the rates charged the member countries for the use of Intelsat's space segment. The cumulative reduction was quite dramatic. In 1965, the annual charge for a one-way voice circuit was $32,000. By 1983, it had been reduced to $4,680.

The Intelsat Competitive Issue   In spite of its technical and economic success and its steady reduction in rates, Intelsat has encountered competitive pressures. Competition from cable has always been present, and in recent years there have been several requests for authorization of competing satellite services. Competing applicants claim that they could provide more flexible service at substantially lower rates based on several characteristics of the Intelsat system.

First, the basic routing of international traffic through Intelsat is complex, and this adds to its cost. A telephone call from Buffalo to London, for example, passes through the facilities of the local telephone operating company, AT&T, Comsat, Intelsat, and the British Post Office. Indirectly, the customer pays each of these organizations for its service.

Second, the size and number of organizations involved in the Intelsat system make it difficult and costly for it to provide low-volume, specialized services. Television transmission often falls in this category.

Third, without denigrating the quality of Comsat's and Intelsat's managements, which have been quite competent, an organization can rarely achieve optimum efficiency without the spur of competition.

Finally, there is Intelsat's policy of rate averaging (similar to AT&T's). High-density routes such as the United States to England have inherently lower costs per circuit than thin routes such as the Marshall Islands to Hawaii. In order to provide services to thin-route terminals at more reasonable prices, the same rates are charged for all routes. In effect, the high-density routes are subsidizing the others. A carrier that did not engage in this practice could charge lower rates on the high-density routes, a practice described by the major carriers as "cherry picking."

Certain international television circuits have appeared to be among the most overpriced as compared to the potential of smaller integrated systems. For example, the program could be uplinked directly to the satellite from its source without using any terrestrial facilities. Providing international television service by competing satellite systems has been proposed by a number of applicants.

Both Comsat and Intelsat have fought these proposals vigorously, citing many of the arguments used by AT&T in its unsuccessful effort to block competition in the United States. Like AT&T, Intelsat has opposed any use of non-Intelsat satellites for international communications, even the smallest and most specialized. The arguments emphasize the benefits of rate averaging and the adverse effects of cherry picking. Richard Colino, the director general of Intelsat in 1984, attempted to raise the issue to a moral plane in a published interview.[5] He called the motive of the founders of Intelsat "nobility" and the motive of Intelsat's would-be competitors "greed."[6]

The pressure for independent international satellite systems was so great, however, that in November 1984 the Reagan administration declared that these systems were "required in the national interest." To protect Intelsat's economic interests, the ruling imposed strict conditions, the most important of which were a prohibition against providing a public switched telephone service and a requirement for "consultation" with Intelsat before a system could be authorized. As of this writing (1989), two independent services have been authorized, PanAmSat to Latin America and Orion Satellite Corp. to Europe.

■ U.S. DOMESTIC SATELLITES

By 1972, the technical feasibility of communications satellites had been proven by the Intelsat system. The regulatory barriers and uncertainties that had surrounded the use of satellites for communications within the United States were removed by the FCC's open-skies ruling. Communications companies were now free to apply for authorization to launch and operate communications satellites. Western Union, RCA, and AT&T were the first to respond.

---

[5] *COMSAT Magazine*, No. 13, 1984.
[6] In September 1987, Mr. Colino was sentenced to six years in prison for defrauding Intelsat of more than $5 million.

The First Round of Launches, 1974–1979

Nine C-band satellites were launched in the first round from 1974 to 1979:

Satellite	Launch Year	In Orbit in 1988
Western Union		
Westar I	1974	No
Westar II	1974	No
Westar III	1979	Yes
RCA Americom		
Satcom I	1975	No
Satcom II	1976	No
Satcom III	1979 (launch failure)	
AT&T		
Comstar I	1976	No
Comstar II	1976	No
Comstar III	1978	Yes

All these satellites used C band because equipment was more readily available for it than for Ku band and it did not have the problem of rain attenuation, which was thought to be serious in Ku band.

The use of C band solved the rain problem, but it required carriers to locate their earth stations well outside major urban areas to avoid interference from terrestrial microwave systems. The Western Union and RCA earth stations for New York, for example, were located in Vernon Valley, a mountain town about sixty miles from the city. It was connected to the New York operating center by microwave (with Ku band employed for the final hop into the earth station). All the carriers had the same problem, but it increased the cost of providing the service.

Western Union Western Union had a long history as the dominant U.S. telegraph company, but this market was declining and its management sought ways to expand and diversify. Satellites were believed to offer such an opportunity, and Western Union received authorization to launch two satellites, which it named *Westar I* and *Westar II*.

Anxious to be the first to launch a domestic satellite, Western Union rushed its program, and *Westar I* was launched in April 1974. Hurrying the launch was probably a mistake because doing so limited the number of transponders to twelve on each satellite. By waiting a little longer, RCA and AT&T were able to take advantage of a new technology that doubled the number of transponders by using horizontal and vertical polarization. As a result, Western Union's satellites had half the capacity but more than half the cost, and this put it at a serious competitive disadvantage.

Identifying the market for satellite communications circuits was a major problem for both Western Union and RCA. The unique feature of satellite circuits is the independence of their costs from their length; every circuit requires two earth stations and a space segment, and it costs as much to communicate across the street as across the country. This characteristic limits the point-to-point satellite communications market to long-haul circuits, where there is a cost advantage over terrestrial systems.

In consideration of this property of satellites, Western Union pursued the markets for long-haul, private-line voice and data circuits for business customers and intercity television circuits for the broadcast networks. (It also leased three transponders on a long-term basis to American Satellite, another specialized carrier, which in turn leased individual private-line circuits on a retail basis.)

The amount of private-line voice and data business Western Union was able to obtain was disappointing. Although the company pursued the television broadcast market aggressively and installed cable connections from the New York switching centers of the major networks to its own operating center, the networks were not initially enthusiastic about satellites, and the growth of the broadcast business was slow. Nevertheless, the prospects were sufficiently promising to encourage Western Union to launch *Westar III* in 1979. Western Union said that this satellite would be used primarily in the broadcast markets. It relocated *Westar I* and *Westar II* in the same orbital slot, thus giving the slot a twenty-four transponder capacity.

RCA Americom   Like Western Union, RCA was engaged in a mature communications business, international telex, which had little growth potential, and it also sought a means of diversifying. An opportunity presented itself in Alaska, where the U.S. government decided to privatize the intrastate intercity telephone system it had previously operated. RCA decided to bid for it.

AT&T also bid for the system, and the competition was intense. The State of Alaska, accustomed to the largesse of the federal government, was not anxious to have the system sold to any private company but probably would have preferred AT&T. RCA, however, was the successful bidder, and it formed a subsidiary, RCA Alascom, to own and operate the system.

In order to win the bid in this competitive environment, RCA was forced to make very liberal promises for increasing service, one of them being to provide telephone and television service to every bush village with a population of twenty-five or more. Given the severe climate, rugged terrain, and great distances in Alaska, this would have been impractical with microwave, and the satellite was the obvious solution.

AT&T offered to lease the space segment on its upcoming satellites to RCA, but Howard Hawkins, then in charge of RCA's communications business, proposed that RCA launch its own satellites. His proposal was contested by AT&T, which filed complaints with the FCC and the State of Alaska, but RCA was again successful in the competition. RCA formed a subsidiary, RCA Americom, to engage in the satellite communications business, and it launched its first two satellites in 1975 and 1976.

RCA had a somewhat better business situation than Western Union because it had an assured customer in RCA Alascom. With twenty-four transponders on each satellite, however, it had twice the capacity to fill. Like Western Union, it was only marginally successful in obtaining private-line business customers.[7] It also began to establish networks of wideband data circuits for NASA and the Department of Defense, but this business developed slowly. At the end of 1976, RCA Americom had a great deal of unused capacity, it was deeply in the red, and its prospects were bleak.

RCA's opportunity came from a new and unexpected source—cable television. Most of RCA Americom's executives had voice and telex communications backgrounds, and they did not rate television highly as a potential market. Further, the economic basis of distributing programs to cable systems via satellite had not been established. Hawkins, however, was far-sighted enough to negotiate a contract with HBO for the distribution of its pay-TV programs to cable systems. The initial program transmission on September 30, 1975, was three months before *Satcom I* was launched, and RCA was forced to lease capacity from Western Union to provide the service. Although it was not clearly foreseen at the time, this was the beginning of a very profitable business, not only for RCA but also for the entire satellite industry.

AT&T AT&T was the last of the three pioneering companies to launch a domestic satellite system. Its situation was unique in that it was forbidden by the FCC's open-skies policy to provide private-line service by satellite (except to the Department of Defense in its Autovon network). It was limited, therefore, to integrating its satellite circuits into its vast intercity telephone network. Because of its concerns about the technical performance of satellite circuits (although Intelsat's experience had allayed most of these fears and the questionable need for additional capacity, AT&T was not enthusiastic about the use of satellites in its long lines telephone system.

It would, however, have been contrary to AT&T's corporate policy to ignore a major new communications medium. Further, AT&T expected that the prohibition on its use of satellites for private-line circuits would eventually be lifted. (Its expectation was fulfilled; the prohibition was removed in

---

[7] RCA's business problem was exacerbated by the discovery, after its satellites were launched, that its estimate of the private-business market was based on a computer printout that had failed to eliminate short-haul circuits, such as New York to Washington, from the totals.

July 1979.) Accordingly, it decided to proceed with a three-satellite C-band system.

Its ambivalence toward satellites was demonstrated by its decision to lease its satellites from Comsat General, a subsidiary of Comsat, rather than build its own. In recognition of Comsat's ownership, they were called Comstars.

The Second Round, 1980–1983

The demand for satellite capacity was strong and growing rapidly in 1980, at the beginning of the second round, owing in large part to the strength of the cable television market. This stimulated a great increase in the number of satellites in orbit, both by existing carriers and by two newcomers, Satellite Business Systems (SBS) with its Ku-band system and Hughes Communications with it C-band Galaxies. An excess of demand versus supply reached a peak at the Sotheby auction of the RCA transponder leases in late 1981. Demand then fell off rather sharply for a year, but the ever-increasing need of the cable industry, the adoption of satellite program transmission by broadcasters, and some growth in the commercial and government voice and data markets brought supply and demand back in balance.

Fourteen domestic communications satellites were launched during this round, three of them to replace satellites from the first round that had been lost or had reached the end of their useful lives. The second-round launches, all of which were still in orbit in 1988, include the following:

Satellite	Launch Year
Western Union (C band)	
Westar IV	1982
Westar V	1982
RCA Americom (C band)	
Satcom III-R	1981 (replaced lost *Satcom III*)
Satcom IV	1982
Satcom I-R	1983 (replaced *Satcom I*)
Satcom II-R	1983 (replaced *Satcom II*)
Alascom (C band)	
Aurora	1982 (launched and operated by RCA)
AT&T (C band)	
Telstar 301	1981
SBS (Ku band)	
SBS I	1980
SBS II	1981
SBS III	1982
Hughes (C band)	
Galaxy I	1983
Galaxy II	1983

Western Union   Western Union completed its satellite system by the launch of *Westar IV* and *Westar V* in 1982. This was partly in recognition of the approaching demise of *Westar I* and *Westar II* and partly in expectation of obtaining substantial broadcast and cable business.

RCA Americom   RCA was prospering, and it launched five new satellites, one to replace the lost *Satcom III*, two to replace the aging *Satcom I* and *Satcom II*, one under contract to Pacific Telcom, and one addition to its own system.

Alascom   Pacific Telcom bought RCA Alascom, and rather than continuing to lease capacity from RCA, it chose to buy a satellite as soon as it was safely in orbit.

AT&T   AT&T started its second-generation satellite system with the launch of *Telstar 301*. The Comstar designation was not used for its new satellites.

SBS   The planning of the SBS system was a triumph of technical sophistication over business and marketing judgment. The system was a financial disaster, which was all the more amazing because SBS had such distinguished parentage, its owners being Comsat, IBM, and the Aetna insurance company.

The system was designed to meet a perceived need for a nationwide network of wideband data communications circuits by the burgeoning data processing industry. AT&T's facilities were judged to be inadequate for this purpose, particularly their lack of wideband local circuits between customers' premises and the AT&T switching centers. A highly touted feature of the SBS system was its connectivity—that is, the ability to communicate directly with the satellite and thence to other company locations through an earth station located at the site.

There appears to have been a vast overestimation of the size of the market for wideband data circuits. Most business communications traffic was and continues to be voice and narrower band data, and the SBS system was not cost-effective for these services. SBS continued deeply in the red during the 1980s in spite of determined and costly efforts to develop markets for other services, and its remnants were finally bought or taken over by MCI and IBM.

Hughes Communications   In contrast to the SBS system, the Hughes venture was a success. The company's innovative approach to the cable market was described earlier. It made another bulk sale of its transponders to MCI for integration into MCI's intercity telephone system. In total, the system was highly profitable.

The Third Round, 1984–1988

The third round brought two new companies into the satellite business—GTE and Contel/ASC. Both of these companies operated major telephone systems, and their interest was primarily in voice and data, although GTE eventually achieved a modest share of the television market.

Of greater technical significance was the trend toward the use of Ku band. RCA launched two high-power Ku-band satellites, primarily to provide program distribution for NBC and for the cable industry. GTE and Contel/ASC launched hybrid satellites that contained both Ku-band and C-band transponders. These were the third-round launches:

Satellite	Launch Year
RCA Americom (Ku band)	
Satcom K-1	1985
Satcom K-2	1986
AT&T (C band)	
Telstar 302	1984
Telstar 303	1985
SBS/IBM (Ku band)	
SBS 4	1984
Hughes (C band)	
Galaxy III	1984
GTE	
GStar I	1985 (Ku band)
GStar II	1986 (Ku band)
Spacenet I	1984 (hybrid C and Ku band)
Spacenet II	1984 (hybrid C and Ku band)
Spacenet III	1988 (hybrid C and Ku band)
Contel/ASC	
ASC I	1985 (hybrid C and Ku band)

NASA's deemphasis of its expendable rocket program and its many difficulties with the shuttle led to a hiatus on launches via U.S. vehicles that began after the launches of *GStar II* and *Satcom K-1* early in 1986. With the hope of encouraging the growth of a commercial satellite-launching industry, NASA removed all commercial communications satellites from the shuttle launch manifest, and the next launch of a U.S. communications satellite, GTE's *Spacenet III*, was on the French rocket *Ariane* in March 1988. As of this writing, the commercial launch industry has not developed, and the problems of NASA's space program have put the U.S. satellite communications industry in jeopardy.

The orbital locations of the satellites that were in active service in mid-1988 are shown in Figure 8–9.

■ SATELLITES AND CABLE TELEVISION

The meteoric growth in the use of satellites for program distribution by the cable industry was accompanied by a corresponding growth in demand for satellite capacity. The growth was tumultuous, and it was marked by wide swings in the supply–demand relationship, a plethora of legal and regulatory issues, and the loss of a newly launched satellite, RCA's *Satcom III*, at a critical time for the industry.

The Single-Satellite Era

Much of the turmoil and controversy surrounding the distribution of cable television programs in cable's early years resulted from the desire of all programmers to be on the same satellite. This, in turn, was caused by the high cost and initial regulatory restrictions of receive-only earth stations

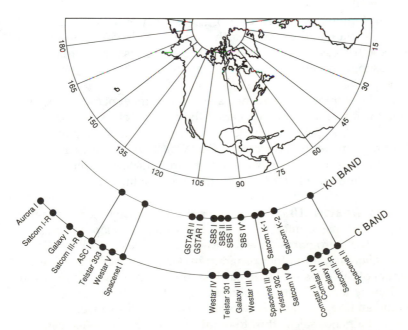

■ Figure 8–9   The orbital locations of communications satellites in 1988.

installed at the cable systems' headends. As a result, very few cable systems were willing to install more than one antenna, and this prevented them from receiving programs from more than one satellite.

From the point of view of the cable programmers, this put an enormous premium on having a transponder on the same satellite as the most popular program sources. Consequently, all cable programs were transmitted from a single satellite for the first several years of satellite program distribution.

From the point of view of the satellite carriers, this situation put an equally great premium on operating the satellite of choice for the cable industry. During this era, a carrier either had all the cable business or none. To become the satellite of choice required that its carrier obtain the business of the most popular programmer to serve as a magnet to attract other programmers.

For a number of years, HBO was the magnet. It was the first pay-TV programmer for the cable industry, and after a somewhat slow start, it became a very popular service. Most cable operators wanted to carry its programs, and they aimed their earth station antennas at the satellite that transmitted these programs. It would have been economically suicidal for another cable programmer to transmit its signal from a different satellite.

With the passage of time, this situation changed, and cable programming transmissions are now spread over several satellites. Two developments brought this about.

The first development was the increasing prosperity of cable systems and the dramatic reduction in the cost of earth stations. With earth station costs reduced to $5,000 or less and the market value of cable systems increasing to $2,000 per subscriber or more, it became economical for them to install more than one earth station.

The second development was an increase in the number of cable program services until the total number far exceeded the capacity of a single satellite. At the end of 1988, approximately seventy-five basic and forty-five premium cable program services (including East Coast and West Coast feeds and backup channels) were being transmitted from thirteen different satellites. Clearly, the industry had outgrown the one-satellite era.

RCA versus Western Union   As a result of its initial contract with HBO, RCA had the inside track to become the satellite carrier of choice for the industry. HBO's programs were fed to the East Coast and West Coast time zones on *Satcom I,* and other program services followed (see Chapter 7). The number of cable systems with earth stations grew slowly in 1976 and early 1977, but by late 1977 it had become clear that satellite distribution would become an integral part of the cable business. This caused Western Union to mount a major sales effort to obtain HBO's satellite business.

The stakes for RCA and Western Union were enormous because the entire satellite cable market hinged on HBO's decision. For Western Union, it presented an opportunity to make its satellite business profitable, and its entire top management became involved in an effort to persuade HBO to move. For RCA Americom, losing the HBO business would be disastrous, and its top management likewise made every effort to persuade HBO to stay on *Satcom I*.

In the spirited competition that followed, RCA emerged the winner, primarily because of the greater capacity of its system (Western Union's satellites had only twelve transponders). *Satcom I* was firmly established as the "cable bird" of choice.

The Loss of RCA's Satcom III   With the cable television business in hand, RCA Americom enjoyed two prosperous years, and all but four of the transponders on *Satcom I* were leased to cable programmers. Two of these four were inoperable, and the other two were reserved for high-priority voice traffic. With its business profitable and growing, Americom decided to launch its ground spare as *Satcom III*, which was intended to become the cable bird. *Satcom III* was to have all twenty-four transponders devoted to cable traffic. In addition, the added capacity of the system made it possible to designate backup transponders on other satellites that could be used in the event of a failure on *Satcom III*.

Satcom III was launched from Cape Canaveral on December 6, 1979. The first three stages performed flawlessly, and the satellite was placed in its elliptical transfer orbit on schedule. The event was celebrated at a boisterous party hosted by Americom for its customers and insurance underwriters. Unfortunately, the celebration was premature and short-lived. On December 10, the final AKM stage was fired, and the satellite disappeared.

This was the beginning of a hectic time for RCA and a period of confusion for the cable industry. RCA had to deal with its underwriters, with other carriers in an effort to lease replacement capacity, with the press, with the FCC, and, most of all, with its customers.

Collecting the insurance was the easiest part, and it was paid promptly. Eleven transponders were leased temporarily from AT&T as a partial replacement for the twenty-four that were lost on *Satcom III*. (The negotiations for these transponders were not made easier by the fact that RCA was engaged in a hard-hitting advertising campaign that attacked AT&T's private-line rates.) The loss was front-page news, and Americom's headquarters were inundated with more than two hundred calls from the press the day after. The most difficult problem was customer relations; some customers had to be disappointed, and the question was which ones. One of the most disappointed was Ted Turner, who eventually was successful in persuading the courts to order Americom to lease him his transponder of choice.

The satellite loss was a setback for the cable industry, albeit a temporary one. The demand for transponders was growing rapidly, and it exceeded in total the capacity of a single satellite. Had the *Satcom III* launch been successful, RCA would have dedicated a second satellite for cable service, designating it "Cable Net 2." With its loss, *Satcom I* remained the primary cable satellite, or "Cable Net 1," and RCA tried to use the AT&T transponders as Cable Net 2. The uncertainties of the temporary AT&T lease were so great, however, that these efforts were only marginally successful. The establishment of additional cable birds had to await the launch of additional satellites and further growth in demand.

The Multiple-Satellite Era

The multiple-satellite era for the distribution of cable programs began in the early 1980s. Although Hughes's coup in wresting leadership from RCA (described earlier in this chapter) was a serious blow to RCA, it was not ruinous. The demand for transponders grew so rapidly that there was room for everybody. By the end of 1988, 120 transponders were leased or had been sold to cable programmers by five carriers. GE (formerly RCA) and Hughes had retained their dominance, however, with nearly 90 percent of the market:

Satellite	Number of Transponders
GE Americom	
Satcom I	11
Satcom II	5
Satcom III	20
Satcom IV	20
Satcom K-1 (Ku band)	4
Total	60
Hughes	
Galaxy I	24
Galaxy II	4
Galaxy III	17
Total	45

■ DIRECT BROADCAST BY SATELLITE SERVICE

Few technical developments have seized the imagination of the public and the engineering fraternity more intensely than the broadcast of television programs directly from satellites to the home. Receiving a signal from a tiny satellite 22,300 miles in space with apparatus that can be installed in the home seems miraculous even in an age of miracles. And in a somewhat unexpected way, this miracle is now occurring.

The Power—Antenna Size Trade-Off

The design of direct broadcast satellite (DBS) systems is dominated by the trade-offs between satellite power, receiving antenna size (and hence cost), and picture quality. The FCC originally required a minimum antenna diameter of ten meters for C-band antennas, a size that was impractical and uneconomical for home use. This rule was eliminated in 1979, and homeowners can now legally install an antenna of any size as determined by this trade-off. The practical effect of this trade-off for typical systems, present and planned, is as follows:

	Band		
	C	Ku	DBS
Transponder power (w)	4–10	45–60	150–250
Footprint coverage	CONUS	CONUS	1/2 CONUS
Antenna diameter (ft.)	4–10	2–5	0.75–2.00

C-Band Homesats (Home Satellite Receivers) and Scrambling

The unexpected manner in which direct satellite broadcasting began has already been described (page 238). Engineers believed that the power limitations of C band made it unsuitable for broadcasting. But while regulators, program suppliers, and satellite engineers were studying the opportunities for direct-to-home program distribution by Ku-band satellite and DBS, the public took matters into its own hands and began the unauthorized reception of cable programs from low-power C-band satellites. Contrary to the expectations of engineers, many people were satisfied with the quality of reception that could be obtained from an antenna six feet in diameter or even smaller, and a complete earth station for home use could be purchased for under $1,000. Those who lived in rural areas were accustomed to marginal picture quality from distant broadcast stations, and the homesat picture quality was often better. In addition, of course, homesats gave the owner a choice of twenty or more programs.

A trickle of homesat installations began in 1980, and by 1982 the trickle had become a flood. By the end of 1985, it was estimated that more than 1.5 million homesats had been installed, and new installations were being made at a rate of more than 300,000 per year. By 1987, the number had increased to nearly 2 million. This was extremely alarming to program suppliers, copyright owners, and cable system operators, all of whom had property rights in the programming the public was getting for free.

In 1983, the owners of program rights decided that steps must be taken to end piracy. Legal action against hundreds of thousands of television receive only (TVRO) owners was clearly impossible, and the only practical answer was to scramble the signals so that they could not be received without a descrambler. By controlling the distribution and operation of descramblers, program suppliers could limit viewers to those who were paying their monthly fees. Scrambling also solved the cable operators' problem because it gave them a means of controlling transmissions to the home.

Scrambling A wide variety of techniques has been developed for scrambling audio and video signals so that they cannot be received by unauthorized viewers.[8] They vary widely in cost, complexity, and ease with which the code can be broken and the signal pirated. The challenge of the industry was to choose a system that presented the optimum trade-off between cost and ease of pirating.

As the leading pay-TV company with the most at stake in scrambling, HBO was in a position to establish a de facto standard for the industry. The company offering the system chosen by HBO would be in an enviable position in the large market for backyard TVRO receivers and descramblers. The choices were ultimately reduced to two companies, Scientific-Atlanta and M/A-Com. The competition was keen, and the decision was for M/A-Com's VideoCipher system, which became the standard of the industry.

In August 1986, General Instrument (GI) acquired M/A-Com's Cable Home Communications Group for $220 million. GI had previously purchased Jerrold, and the addition of the M/A-Com descrambler line made it a leading supplier of TVRO equipment.

The program suppliers did not move quickly to scramble their signals. The equipment had to be field-tested and perfected, and enough descramblers had to be made available to supply all the cable system head ends. In addition, a political storm was brewing over the right of program suppliers to scramble their signals (see the next section). This pressure made it politic to await the availability of sufficient descramblers to meet the expected demands of home TVROs. Finally, there was the problem of establishing policies and hardware and software systems for controlling the use of descramblers by home users.

All of this took time, and nearly two years elapsed between the an-

---

[8] For a summary of scrambling techniques, see A.F. Inglis, *Electronic Communications Handbook* (New York: McGraw-Hill, 1988), chapter 17.

nouncement and the beginning of scrambling by HBO and others. HBO was the first, starting on January 15, 1986. It was soon followed by Viacom and other pay-TV services.

Basic program suppliers that depended on advertising for a large part of their revenues did not have as direct an interest in scrambling as pay-TV services, but were incentives for them as well. There was the ever-fuzzy copyright problem, and many cable systems, fearing TVRO competition, refused to carry unscrambled signals. In response to these pressures, most of the larger basic program services began scrambling. By 1988, all the pay-TV services and about half the basic services were transmitting scrambled signals.

Descramblers also began to be sold to the public. In August 1986, GI reported that it had manufactured ninety thousand descramblers but that only twenty-five thousand had been sold. TVRO owners seemed to be skeptical that scrambling would be permanent, and a number of major program services were still unscrambled. As more signals were scrambled, however, the rate of sales rose rapidly, GI licensed a number of manufacturers to produce VideoCiphers, and supply and demand were soon in balance.

January 1986, the month in which HBO began to scramble its signals, was disastrous for TVRO manufacturers, as the realization that free programs might no longer be available began to reach the marketplace. As a result of public uncertainty, sales dropped to a small fraction of their prescrambling level. They did not begin to recover until a year later, when descramblers became available in quantity and the public's uncertainty was relieved.

GI built a central station for the technical control of descramblers that could provide or deny access to individual services for individual subscribers. The regulatory problem of establishing an authorized middleman between the program service companies and the public—the role played by cable operators for cable systems—was more difficult to solve and involved not only the FCC but also the U.S. Congress.

Attempts to Regulate Scrambling   An impartial observer would have to view the reaction of the TVRO manufacturing industry to HBO's scrambling announcement with a high degree of incredulity. Notwithstanding the fact that their products were being used in a manner that was possibly illegal and certainly a violation of property rights, manufacturers were righteously indignant. TVRO owners were equally outraged, but their reaction was more understandable. They were not versed in the fine points of the law, and many of them had been advised by their dealers that the government would never allow scrambling. The feeling of hostility toward scrambling was so great that a major TVRO trade publication refused to carry an HBO ad describing the tests it was conducting prior to putting scramblers in service.

Cooler heads realized, however, that providing services to households that did not have cable could be profitable for everyone. It was unfortunate that the parties could not reconcile their conflicting interests without the threat of government intervention.

But government intervention there was. Members of Congress became aware that their constituents were angry about being deprived of their perceived "right" to receive satellite programs without charge. The FCC also was sensitive to these complaints. The result was a series of hearings in both houses of Congress and in the FCC and the introduction of a number of proposed bills for the regulation of scrambling.

The desire of the TVRO industry—to forbid scrambling altogether—was clearly unreasonable, and it never received serious consideration. Instead, the proposed legislation attempted to regulate scrambling by (1) requiring that homeowners in all parts of the country have access to scrambled signals at a reasonable cost, and (2) that program distribution be handled by a third party—that is, neither a program supplier nor a cable operator.

The cable television industry may have had itself to blame for the anticable atmosphere that surrounded the hearings. Many cable operators took the position that they should have the responsibility for controlling distribution not only to the TVROs within their franchised areas but also to those outside these areas. This created a perception of monopolistic power and an insistence that distribution be controlled by third parties.

Hearings continued, and bills were introduced and debated throughout 1986 and 1987. Faced with this political pressure, the industry proceeded to achieve the objectives of the legislation before any of the bills were passed. The ready availability of descramblers and a system for controlling them achieved the first objective, and progress was made in achieving the second by the establishment of third parties such as the National Rural Telecommunications Co-op, which undertook to provide distribution to TVRO stations. In July 1987, the FCC and the Department of Justice advised Congress that there was no need for government intervention. This did not completely convince Congress, but its recent activity has been directed more toward ironing out the ambiguities that remain in the copyright laws.

The Promise of Ku Band and DBS

The unexpected success of C-band direct-to-home broadcasting encouraged satellite companies to believe that it would be even more successful with Ku band and DBS. It was expected that the use of much smaller antennas, which was possible in these bands, would significantly broaden the market for this service. As of 1989, however, Ku-band and DBS systems have not been successful. In large part this is attributable to the characteristics of the direct-to-home market.

The Direct-to-Home Broadcasting Market

The potential market for direct-to-home broadcasting is primarily limited to homes in areas where cable is not available. Cable can offer a greater diversity of programs than satellites alone at a cost that is comparable to a homesat. There is little incentive for a homeowner to install a homesat if service from a well-managed cable system is available.

It has been estimated that about fifteen million homes in the United States do not have the prospect of cable service in the foreseeable future, and this is the present market for direct-to-home broadcasting. In Europe and Japan, the use of DBS for the transmission of HDTV (see Chapter 10) is being planned. Satellites are not subject to the same spectrum limitations as the standard television broadcast bands and thus have an important advantage for wideband HDTV. If and when HDTV becomes popular, the potential market for direct broadcast service from satellites will increase. At the present time, however, it is a limited market.

DBS in the United States

The limited market creates serious questions about the economics of DBS. Satellites licensed for DBS can be used only for television, and their cost is so great that they cannot compete on the basis of price to provide service to cable systems. Therefore, the cost must be borne by a limited number of homesat owners. Further, because of the power required by each transponder, the number of transponders in each satellite is limited, typically to six, and the cost of the satellite must be borne by a small number of program services. (Recent developments in homesat technology have, however, made it possible to consider a larger number of lower power transponders. Hughes has proposed two 16-transponder satellites in a single orbital slot for a total of thirty-two.)

The WARC and RARC DBS Meetings   In spite of the economic problems of DBS satellites (which are less significant in countries with state-owned television systems), the establishment of international allocations and standards raised a number of highly contentious issues. Perhaps the most difficult was the demand of tiny countries such as Luxembourg to be assigned nearly as many orbital slots as their larger neighbors. These issues were first raised formally in a meeting of WARC that convened in Geneva in September 1979. The resolution for Western countries came in an RARC meeting in June 1983. The United States was ably represented in this meeting by Abbott Washburn, who had recently retired from the FCC. After five weeks of discussion, he was able to obtain agreement on most, though not all, of the U.S. objectives.

Initial U.S. Attempts   In the interval between the WARC and RARC conferences, the FCC decided to accept applications for DBS satllites on an experimental basis. Eight companies, including RCA, STC (a subsidiary of Comsat), and Western Union applied for launch authority in 1981 and 1982. Soon afterward, however, economic reality took precedence over legal and regulatory issues, and STC was the only one of the eight actually to build any satellites. It placed an order with RCA in 1982 to build and launch two satellites plus a ground spare.

The cost of building and operating a DBS system was beyond the resources of Comsat, and for the next two years, it sought to obtain partners to share the burden of financing. Comsat was ultimately unsuccessful, and the satellites were completed but never launched.

The economics of DBS are still in doubt, but it has important supporters. In an interview with *Broadcasting* magazine, Steven Petrucci, president of Hughes Communications, expressed confidence that the two DBS satellites for which it has authorization will be the basis for a profitable business.[9]

Direct Satellite Broadcasting at Ku Band

The Ku band provides a different kind of opportunity for direct-to-home broadcasting. Ku-band satellites are more expensive than C-band satellites, but the lease price is not out of reach for cable programmers. Thus programs could be transmitted on Ku-band satellites for simultaneous reception by cable systems and homesats in noncabled areas. Homesat owners would require three-foot antennas—larger than the one-foot antennas required for DBS but substantially smaller than the C-band antennas the public has found acceptable where cable is not available. The transmissions would be scrambled, and homesat owners would be furnished the descrambling code in return for a monthly charge. Cable companies could provide homesat sales and marketing services for the programmers in adjacent noncabled areas.

The potential market for cable programs transmitted by Ku band would include virtually every houshold in the country, both cabled and noncabled. The programming costs would be shared by cable and homesat viewers, and the satellite costs would be borne by a larger number of program services. On the basis of paper studies, the economics appear favorable.

As of this writing, however, Ku-band broadcasting has still to become a commercial success. RCA/GE, in partnership with HBO, launched *Satcom K-1* in 1986 with the expectation that HBO would again be the magnet that would attract other cable programmers to lease the remaining tran-

---

[9] "Special Report: Satellites," *Broadcasting*, 18 July 1988.

sponders. But at the end of December 1988, HBO was the only cable programmer on the satellite.

In a contemporaneous interview with *Broadcasting*, Kevin Sharer, president of GE Americom, analyzed the situation as follows:

> The elements are all on the table, but nobody's put them together yet. There is a clear consensus that the market is there. [But] to find the first guy who was waiting to jump out is going to be a tough thing. It will be more than a risk. A lot of people will have to simultaneously cooperate, because no one company —not GE, not GM, nobody—has got the wherewithal to establish this business system overnight. The distribution start-up cost is going to be at least one billion dollars.[10]

It remains for the future to decide whether DBS or Ku-band satellites will be the basis of a profitable program distribution industry.

■ SATELLITES AND BROADCASTING

The television broadcast industry was slow to adopt satellites as a medium for the transmission and distribution of its programs. Nearly a decade passed between the launch of the first domestic communications satellites and their extensive use by broadcasters. In part this resulted from a lack of effective communication and a mutual understanding between the cultures of the communications and broadcast industries, particularly of the major networks, which were the most likely customers for satellite transmission.

The satellite companies, with their background in voice and message traffic, were naive in their understanding of broadcasting, and this was reflected in their early proposals. For their part, the network engineering and operating staffs had the awesome problem of creating and delivering an ephemeral and highly perishable product that had an annual worth of billions of dollars. Reliability was essential because even a short interruption of service would cause a serious loss of revenue that could never be regained.

In addition, competition among the networks led to the need for increasingly intricate switching and transmission systems for the generation of programs. The Sunday afternoon broadcasts of professional football games are a good example. The broadcasts originate from more than a dozen different stadiums, they are integrated with national and regional spot announcements, and they are transmitted to individual affiliates on the basis of their viewers' interests. Each affiliate receives a customized combination of game and commercials through a complex switching and trans-

---

[10] Ibid.

mission network. Given the magnitude of these responsibilities, it is not surprising that the network technical staffs were unwilling to adopt satellites as a primary transmission medium until their reliability had been established and their capabilities and limitations were understood.

Gradually, however, the successful use of satellites by the cable industry and the Public Broadcasting Service (PBS) confirmed the results of the broadcast industry's own studies. Broadcasters increased their use of satellites, slowly at first but more rapidly as their value was demonstrated.

Satellites were initially used principally for four applications in television broadcasting:

1. Program syndication
2. Specialized and ad hoc networks
3. Electronic newsgathering (ENG)
4. PBS and the major commercial networks

In addition, the radio networks are now making wide use of satellite program distribution.

Program Syndication

Program syndicators are entrepreneurs who obtain the rights to programs and sell individual stations the right to broadcast them (see Chapter 7 for the problems of syndicated exclusivity created by cable television and the superstations). Their best customers are independent (unaffiliated) stations, and the market for syndicated programs grew as the number and prosperity of these stations increased.

The conventional method for distributing syndicated programs was to record them on tape and "bicycle" the tape from station to station. Each station made a copy for its own use before forwarding it to the next. Although it was a satisfactory method, it was time-consuming, and the repeated handling of the bicycled tape was not conducive to the highest technical quality. A number of leading syndicators, led by Ralph Baruch of Viacom, believed that satellites would provide a better, faster, and possibly more cost-effective way of distributing programs. The station would receive the program by satellite and make its copy off the air instead of from a bicycled tape.

Program distribution by satellite requires that stations be equipped with receive-only earth stations. To break this bottleneck, in 1980 RCA initiated a program called SMARTS and offered to give earth stations to television stations in return for an agreement to use them only for RCA satellites. The program was modestly successful, but as the acceptance of satellites grew and the cost of earth stations declined, it became unnecessary. The number of television stations equipped with earth stations increased from 50 at the end of 1979 to 250 in 1981 to 600 at the end of 1983. The distribution of syndicated programs by satellites is now an accepted industry practice.

Specialized and Ad Hoc Networks

Specialized and ad hoc networks are organizations that distribute specialized or occasional programs to groups of stations. They are technically similar to those used for syndication, but syndicated programs are recorded, whereas specialized network programs are usually live, primarily news and sports events. Backhauling television signals—that is, transmitting them from their source, such as a sports arena, to a group of television stations on a scheduled basis—is an example of specialized networking. The Hughes Television Network was one of the first and most successful companies to offer this service.

Ad hoc networks are formed for a specific purpose or event. They lease facilities on a wholesale basis from satellite carriers, integrate satellite and terrestrial circuits as required for a specific application, and offer end-to-end service to their customers on a retail basis. Wold Communications was the pioneer in providing this service, having offered it as early as 1975. TVSC, a subsidiary of Westinghouse's Group W, is another major supplier of this service.

Since much of the syndication and specialized network traffic is short-term, satellite service is leased under occasional use tariffs. There is no shortage of capacity. At the end of 1988, more than one hundred C-band and twenty-five Ku-band transponders were listed as being devoted to occasional service.

Electronic Newsgathering

The use of satellites for ENG has grown exponentially in recent years. It is not uncommon to see a forest of portable satellite antennas at the scene of a major news event. These antennas transmit television signals generated at the site to the stations via satellite.

The success of ENG depended on two technical developments—the development of portable cameras (see Chapter 5) and the availability of Ku-band service. Ku band is almost a necessity for ENG because uplink facilities are required on short notice and at random locations. The uplink antennas are smaller, and there are no siting problems resulting from interference with microwave systems.

A number of companies have been formed to offer uplink services for ENG on a rental basis. Notable among these is CONUS, owned by the Hubbard Broadcasting Company. Many larger stations, however, have enough participation in ENG to afford their own portable uplinks, and in 1986 more than sixty stations had these facilities. In the same year, the major networks and CNN gave impetus to the concept by offering to share the cost of satellite ENG facilities with their affiliates. The result was even more rapid growth in the number of stations so equipped.

PBS and the Major Commercial Networks

PBS was the first network to use satellites. Free of the profit and loss concerns of private enterprise and aided by a generous infusion of funds from the federal government, PBS was better able to assume the risks of a new medium. It awarded a contract for satellite services to Western Union and began the transition from terrestrial circuits in 1978. The satellite service was successful, and it set a useful precedent for the later adoption of satellites by the commercial networks.

The commercial networks were under no particular pressure to use satellites because of their overall satisfaction with the performance of the terrestrial facilities furnished primarily by AT&T. From long experience in working together, AT&T and the networks had come to understand each other's business, and their relationship was generally satisfactory.

Nevertheless, there were problems. AT&T's video circuits did not reach all the affiliates, particularly in the more remote parts of the country, and they had to rely on private microwave systems or even off-the-air pickup from other affiliates. AT&T's terrestrial facilities also lacked flexibility, and they could not provide service from locations far from its switching centers. Finally, the networks were concerned that AT&T's monopoly gave it the power to raise its rates to unreasonable levels. In view of these problems and concerns, the networks began to consider the use of their own satellite facilities.

Several more years of technical progress and advanced systems planning came to fruition in the early 1980s, when the networks began the transition from terrestrial to satellite transmission. NBC began first, and it was soon followed by ABC and CBS.

NBC After long study, NBC opted to use Ku band, concluding that its higher power and freedom from terrestrial interference outweighed the occasional problem of rain outages—a problem that could be mitigated by supplying a C-band backup. In 1983, it entered into a ten-year contract with Comsat General, a subsidiary of Comsat, to provide signal distribution services to its affiliates. Comsat provided end-to-end service, including the earth stations on the affiliates' premises. The traffic was first placed on an SBS satellite but was switched to RCA's *Satcom K-2* after it became operational. NBC's transition to satellites began in January 1984 and was completed early in 1985.

CBS CBS chose to continue the use of C band and to lease its facilities from AT&T. It began the transition in March 1984, and the changeover was essentially complete by 1987. CBS currently leases transponders for East and West coast feeds on both *Telstar 301* and *Telstar 302*, thus providing total redundancy.

ABC Like CBS, ABC chose C-band distribution with AT&T as its carrier. Also like CBS, its traffic is divided between *Telstar 301* and *Telstar 302*. ABC started a little later than CBS, beginning the transition to satellites for the Mountain and Central time zones late in 1984. Its transition for the Eastern and Pacific time zones was completed in 1987.

The Radio Networks

The impetus for the use of satellites by radio networks came from the increased demand for high-fidelity stereo transmissions (see Chapter 3). Standard telephone voice channels have a bandwidth of only 3.4 kHz. AT&T offers specially conditioned circuits with wider bandwidths for the transmission of audio signals for radio broadcasting, but their performance is still far short of the requirements of high-fidelity audio systems.

With increasing competition from high-fidelity stereo broadcasting by independent stations, it was necessary for the networks to improve their signal quality as part of their renaissance (see Chapters 2 and 3). Satellites provided the solution.

The first radio networks to use satellites were the Mutual Broadcasting System, National Public Radio, AP Radio, and the PKO Radio Network. They used analog transmission on a Western Union satellite and made the transition in 1978 and 1979.

ABC, CBS, and NBC switched to satellites five years later, beginning the transition to an RCA satellite in 1983 and 1984. Initially called the Audio Digital Distribution Service (ADDS), the service is now known as the Digital Audio Transmission Service (DATS). The use of digital transmission ensured that the satellite system was essentially transparent—that is, it added no noise or distortion to the signal. It has been a highly successful service and has been adopted by a number of regional networks as well.

■ SUMMARY

Communications satellites have had an enormous impact on the radio and television industries. They have made cable television a major communications medium. They have provided a more economical and flexible facility for network program distribution and have spawned a host of ad hoc and specialized networks. They have aided broadcasting's quest for better news coverage by providing the communications link for ENG. They have made high-fidelity radio networks possible. All this happened in a little more than a decade.

The future of satellites looks equally bright. New competing technologies, particularly fiber optics (see Chapter 10), may supersede satellites for some applications, but satellites will continue to be dominant for point-to-multipoint service and in situations that require a high degree of flexibility. To the extent that the new technologies stimulate an even more rapid growth of radio and television, satellites will benefit.

9

■■ HOME VIDEO RECORDERS AND PLAYERS

On the evening of April 4, 1984, three of RCA's top executives—Thorton Bradshaw, its chairman and CEO, and vice presidents Roy Pollack and Jack Sauter—hastily convened a meeting of its distributors in the Waldorf-Astoria Hotel in New York. The purpose of the meeting was to announce RCA's decision to withdraw from the manufacture and sale of videodisk players, devices for the playback of prerecorded television programs in the home.

It was a painful and difficult decision for the company. Its investment and cumulative operating losses in the videodisk program had totaled $580 million over a twenty-year period, but its profitability appeared to be years in the future at best, and there was no assurance of its ultimate success.

RCA's action was both illustrative and symbolic. It was illustrative of the difficulties encountered in developing, manufacturing, and marketing a competitive home video player. And since the Japanese were ultimately successful, it was symbolic of the passage of leadership in consumer electronics from the United States to Japan.

The development of the technologies for home video recorders and players was one of the great technical achievements of the electronics industry. In 1955, it was not possible to manufacture a video recorder at any price. In 1956, a monochrome-only recorder with marginal performance (although it seemed miraculous at the time) and requiring two racks of equipment plus a huge console could be purchased for $50,000 (see Chapter 6). In 1986, thirty years later, a color recorder with superior performance and packaged in a small case a few inches high could be purchased for $300 or less. In 1988, 11.7 million were sold in the United States alone (plus 1.6 million recording decks in camcorders), and cumulative worldwide sales through that year exceeded 300 million.

But RCA's unsuccessful attempt to enter the videodisk business demonstrated that progress in home video recording technology did not come easily or cheaply. In addition to the extraordinarily difficult technical and manufacturing problems that had to be solved, there was the equally difficult problem of choosing the recording medium and format that would be the most successful in an intensely competitive marketplace. As RCA discovered, the wrong choice could be financially disastrous. Its recorder was technically satisfactory, but it failed in the marketplace.

■ HOME RECORDING MEDIUMS AND FORMATS

Electronics manufacturers in the United States, Europe, and Japan explored four different mediums for video recording in the home—photographic film, embossed plastic tape, disks, and magnetic tape. Eventually, at least sixteen combinations of mediums and formats were offered:

Medium	Trade Designation	First Offered By	Format	Announced	Life in the Consumer Marketplace
Photographic Film	EVR (see Chapter 5)	CBS	8mm	1967	1968–1971
	Phototape	RCA	8mm	1967	None
Embossed Plastic Tape	Selectavision	RCA	Holography	1969	None
Disks	Teldec	Telefunken	Grooved	1970	1975–1976
	VHD	JVC*	Grooveless	1980	1983–1983**
	Selectavision	RCA	Grooved	1979	1981–1984
	Laservision	Philips/MCA	Laser	1978	1978–1983** 1987–present
	CDV	Philips/Sony	Laser	1986	1986–present
Magnetic Tape (VCRs)***	U-Matic	Sony	Helical 3/4"	1972	1972–present
	LVR	Toshiba	Linear 1/2"	1973	None
	LVR	BASF	Linear 8mm	1973	None
	EAILI	EAILI	Helical 1/2"	1974	None
	Beta	Sony	Helical 1/2"	1974	1974–1988
	VHS	Matsushita	Helical 1/2"	1977	1977–present
	Cartrivision	Cartrivision	Helical 1/2"	1978	None
	Magnavision	RCA	Helical 1/2"	1979	None
	8mm	Sony	Helical 8mm	1984	1984–present

* Japanese Victor Company; a subsidiary of Matsushita.
** The Laservision and VHD formats were offered in the industrial and educational markets from 1983 to 1987. The Laservision was reintroduced to the consumer market in 1987 (see text).
*** The home recorders in this list were preceded by a plethora of recorders offered by Ampex, Philips, Sony, International Video, and others. These were priced between professional broadcast and home recorders, typically $1,000 to $10,000. They were offered in the industrial and educational audiovisual markets. The most successful was the Sony U-Matic.

■ THE INITIAL FORMAT ELIMINATIONS

Nine of the sixteen medium–format combinations listed above were eliminated from competition rather quickly and had been removed from the marketplace by the end of 1981 as a result of technical deficiencies, high cost, and/or other unattractive features. The losers were CBS's EVR and RCA's Phototape, RCA's embossed plastic tape, Telefunken's Teldec mechanical disk, and five of the magnetic tape formats—the helical scan formats announced by RCA, Cartrivision, and EAILI and the linear recording formats announced by Toshiba and BASF.

EVR and Phototape

The unhappy life of CBS's EVR, which ended with William Paley's final disillusionment with its inventor, Peter Goldmark, is described in Chapter 6. RCA never had a strong belief in the merits of photographic film as a medium for home recorders and pursued it more as a defensive measure in the event EVR was successful than as the result of a serious conviction.

Furthermore, CBS's espousal of EVR was a disadvantage of the film medium in the eyes of RCA's management. Robert Sarnoff had recently succeeded his father as RCA's CEO, and he was anxious to continue and even exceed his father's accomplishments as an innovator. Accordingly, he was not eager to offer a product that CBS had pioneered.

By the end of 1969, RCA's engineers and scientists had made sufficient progress on two new and highly innovative medium–format combinations —holography on plastic tape and the videodisk—to encourage management that it was no longer necessary to pursue Phototape. Phototape was then quietly dropped.

Embossed Plastic Tape

The embossed plastic tape medium used holography, a sophisticated optical process based on the unique *coherent* light emitted by lasers. If a suitable pattern is embossed on a plastic tape and the tape is illuminated from the rear by a split laser beam, the *interference* effects between the two beams can be made to produce an optical image.

RCA Laboratories began research on holography as a possible medium for video recording as early as 1965. Initial demonstrations were made to RCA's management in 1967, and in 1969 it was demonstrated to a corporate staff group headed by Sarnoff's chief of staff, Chase Morsey.

The picture quality was not very good, and a number of practical questions, such as the recording of sound, had barely been addressed, but the quality was good enough to persuade Morsey. It was an imaginative, innovative concept, qualities he knew would be attractive to his boss, Robert Sarnoff. Strongly influenced by wishful thinking and probably encouraged by some of the laboratories' scientists who were equally unaware of the realities of product design and manufacturing cycles, Morsey proclaimed that holography would be the basic technology of RCA's home video player. The product was to be named *Selectavision*, and he promised that deliveries would begin in 1971, just two years in the future.

The two-year lead time was absurdly unrealistic. It would have been a challenging goal even if all the basic technical problems had been solved, which was far from the case. RCA's operating managers knew this, and Morsey's promise added to his reputation as a loose cannon careening out of control, all the more dangerous because he had the confidence of Robert Sarnoff.

With Sarnoff's endorsement, however, the holography, or *holotape*, program proceeded at full speed. An eager group of young scientists was assigned to the project, and on September 30, 1969, Sarnoff himself presented the plan to the press. He was highly enthusiastic, and he described the program as an example of the "new RCA" in which marketing judgments would take precedence over technology.

The picture quality demonstrated at the press conference was just short of awful, but the attendees, some of whom had seen the equally poor pictures shown by RCA in the first demonstrations of its color system twenty years earlier, were not surprised, nor did they take it as a sign of inherent problems in the system.

Whether, given sufficient time and money, holography could have been the basis for a practical system of home recording is not known. Long before this could be determined, however, the promise of the videodisk and magnetic tape had increased so greatly that holography faded away as a credible competitive candidate. By 1971, the year Morsey had promised to have the product on the shelf, many of the basic technical problems were still unsolved, and RCA dropped the program.

There was considerable embarrassment concerning the trade name Selectavision, which had been closely identified with holographic recording in RCA's publicity. Morsey solved this problem by averring that it was a generic term that could be applied to any home playback system. It was later applied to RCA's videodisk.

Teldec

Teldec was developed by Telefunken in a joint venture with Decca, hence the trade name. It used a grooved plastic disk, and the playing time for one side was initially only five minutes, later increased to fifteen. With this short playing time, it had limited customer appeal. It was put on the market in Germany in 1975, but sales were so poor that it was withdrawn the following year.

The reasons for the failure of Teldec were of particular interest to RCA because it was internally committed to its own version of videodisk. Were there fundamental flaws in the videodisk concept, or were the problems the result of Telefunken's approach? RCA decided that it was the latter—a combination of its short playing time and the limited number of prerecorded programs that were made available. Neither of these problems was fundamental, and RCA had plans for solving both of them.

Unsuccessful Magnetic Tape Formats

Nine different formats were announced, but only three of them survived the initial eliminations: Sony's Beta and 8mm and JVC/Matsushita's VHS. The others never reached the marketplace or were soon withdrawn from it for a variety of reasons.

RCA's Magnatape was a helical scan recorder that differed from U-Matic, Beta, and VHS in that the drum had four heads and the tape wrap around the drum was only 90 degrees as opposd to 180 degrees for the other helical scan formats. This made it unnecessary to have a mechanism for extracting the tape from the cassette when it was loaded in the recorder. As a product, it never had a chance. RCA did not have the resources to pursue both the tape and disk media, and, for reasons to be described later, its management chose the disk.

The Cartrivision product was ahead of its time, and its performance was not satisfactory. The EAILI unit was unable to compete with the Japanese products from Sony and JVC.

The Toshiba and BASF LVR (linear video recording) units were based on multitrack linear recording (see Chapter 6). The tape was mounted in a cassette in a continuous loop (as with audiocassettes) and operated at very high speeds—217 ips for Toshiba and 158 ips for BASF. The tape was automatically reversed, and the head moved to the next track at the end of each pass. With the use of multiple tracks, the Toshiba machine was capable of two-hour playing time, and the BASF was capable of three hours.

Although the Toshiba and BASF recorders embodied some innovative design features, the linear recording format was unable to compete with helical scan because of its prodigal use of tape, and these products were removed from the market.

Summary

There were five survivors of the initial medium–format eliminations ending in 1981. Three of them, Selectavision, VHD, and Laservision, were disk formats, and the other two, Beta and VHS, were VCRs (videocassette recorders) (The terms "videocassette" and later "camcorder" were first coined by the editors of *Television Digest*.) using magnetic tape loaded in cassettes. Two additional formats were added at a later time, the CDV laser disk and the 8mm VCR.

While there was vigorous competition among formats, particularly between the Beta and VHS tape standards, the original and most basic competition was between the two mediums, disks and VCRs.

■ DISKS VERSUS VCRS

The Pros and Cons of Disks and VCRs

The choice facing would-be manufacturers of home video recorders and players, disk or VCR, was an extremely difficult one. Reduced to its simplest terms, the decision had to be based on a subjective judgment of the public's desire to make recordings in the home. Except for this capability of the VCR, it appeared that the disk recorder had a sufficient cost advantage that would make it the preferred medium. Disks were cheaper than cassettes, and the cost difference increased when the cost of making copies was included. Disks could be reproduced, or replicated, by a stamping process, but replicating cassette tapes required a costly rerecording process.

Forecasting the public taste in such matters is notoriously difficult, but the disk proponents thought they had a model in audio recording. The disk was the original audio recording medium, and it survived the competition from tape, first from the reel-to-reel format and then from the more convenient cassettes. Home recording of audio programs was important to some members of the public, but most people were satisfied with a playback-only medium (disk). The two mediums coexisted, each filling a role in the total market, and the position of disks became even stronger with the advent of CDs (see Chapter 3).

The Marketplace Chooses

Initial Enthusiasm for the Disk The market potential of videodisks was initially judged favorably by most of the industry. Each of the three systems eventually developed an impressive coterie of manufacturers, program suppliers, and major retail establishments:

Selectavision (RCA)	*VHD (JVC)*	*Laservision (Philips)*
RCA	Matsushita	Philips
Zenith	JVC	North American Philips
Hitachi	Thorn-EMI	Magnavox
Sanyo	GE	Pioneer
Sharp	Quasar	Philco
CBS	Panasonic	Sylvania
Sears		MCA
Radio Shack		
Montgomery Ward		
J.C. Penney		

Enthusiasm for the concept was not limited to the leading manufacturers. Even after Sony had gone to market with its Beta VCR, its chairman,

Akio Morita, negotiated for a time with MCA to manufacture its LV disk recorder. In 1976, he stated that "in the future videodisc and the video recorder will coexist just as tape recorders and records coexist in the audio field." He stated further that "videocassette is the door opener of video-disk."[1]

Disks Fail the Acid Test Unfortunately for the manufacturers who chose disks, video recording did not follow its audio analogue. The enthusiasm that greeted the introduction of videodisks was not followed by an equally enthusiastic sales response. When videodisk players were first placed on the market, Philips's Laservision in 1978 and RCA's Selectavision in 1981, VCRs were already well established and were beginning a period of rapid growth (Figure 9–1). Competition from disks had no noticeable effect on the growth rate of VCR sales—in fact, they continued to accelerate—and the disk share of the home playback market never exceeded 10 percent.

Explanations Even with the benefit of hindsight, the reasons for the vast difference between the popularity of videodisks and audiodisks are not entirely clear. The inability to make home recordings was never a major deterrent to the sales of audiodisks, but for video it was decisive. Two commonly given reasons are timing of the introduction of the videodisk vis-à-vis the VCR and the higher cost of video programming.

Audiodisks had been on the market several decades before tape recorders were available, but by the time videodisk recorders came on the market, VCRs were well established in the marketplace. Some RCA executives attributed the failure of its disk system solely to timing; had it been available sooner, they said, it would not have been necessary for it to face entrenched competition from VCRs.

While timing may have been part of the problem, it is oversimplistic to attribute the failure of disks to this single cause. If the disk had truly been a better mousetrap, or if it met a different but equally large market demand, it should have succeeded in spite of its late arrival. There must, then, have been other reasons for its failure.

A plausible explanation is that home recording of video was more popular with the public because of the higher cost of video programs. Because of the higher cost of video programming, the replication cost is a smaller fraction of the total (programming plus replication), and it is not as significant in determining the selling price. The high cost of prerecorded programs led to the development of rental libraries. These also reduced the importance of the higher costs of prerecorded tapes. Finally, the high cost of prerecorded programs increased the attractiveness of home recording. With VCRs, programs can be recorded off the air at no cost except for that of the

[1] James Lardner, *Fast Forward* (New York: W.W. Norton, 1987).

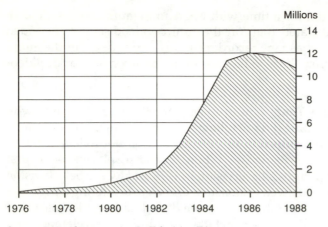

Source: Annual summaries in *Television Digest*.

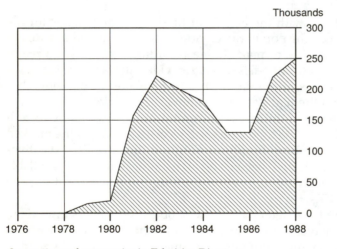

Source: Annual summaries in *Television Digest*.

■ **Figure 9–1**   VCR unit sales (*top*) and videodisk unit sales (*bottom*). The disparity in the sales volumes of VCRs and videodisks is so great that they cannot be plotted to the same scale. These graphs show the total sales of all formats, but the VCR totals do not include the recorders built into camcorders. After 1984, with the withdrawal of RCA from the videodisk market, most of the disk sales were made to industrial and educational institutions rather than to the consumer market.

blank tape, then played back at a later time. Home video recording has not been limited to hobbyists as it was with audio, and it has become popular with a substantial portion of the general public. (It also created difficult copyright problems, described later in this chapter.)

In spite of its initial failure in the home market, the videodisk has been a remarkably tenacious medium. Disks continued to enjoy a small niche in the audiovisual market, and as of this writing, laser disks are beginning to have a revival in the consumer market. The industry may be approaching the phase forecast by Morita in 1976 in which the two mediums will coexist profitably.

■ VIDEODISKS

The Competing Formats

Two of the disk formats that survived the initial eliminations, RCA and JVC, were electromechanical. They used a capacitance readout in which a stylus made physical contact with the record as in audiodisk recording. The RCA format had a spiral groove on the disk that the stylus followed. The JVC format used a grooveless disk, and automatic electronic circuitry caused the stylus to follow the spiral recording track.

The third format, Laservision, pioneered by Philips and DVA (a subsidiary of MCA), uses a laser readout of a dot pattern recorded on the disk. Unlike the CD developed for audio (see Chapter 3), which records the signal in digital form, Laservision uses analog recording with an FM carrier.

The CDV (compact disk, video) format was developed at a later time. It uses the standard 4.75-inch CD and has the capacity for recording five minutes of video and twenty minutes of audio.

RCA's Videodisk: Selectavision

RCA Chooses a Mechanical System It is ironic that RCA, having chosen the disk over magnetic tape, should have opted for a mechanical system that used a stylus rather than a laser as the pickup element. In some respects, the relationship of stylus and laser pickups is analogous to that of the CBS and RCA color systems. Like the CBS system, stylus recording used less exotic technologies (although its technical challenges were enormous), and it was capable of producing pictures of acceptable quality. But also like the CBS system, there were inherent limitations in its performance that could not be overcome by additional engineering development. Laser recording was in its infancy, and it had far greater potential. RCA had contemptuously rejected the CBS color system, in large part because it was "mechanical." Now it placed itself on the other side of the mechanical–electronic contest by chosing stylus recording.

One of the reasons for RCA's choice was timing. The basic technology of a stylus system had first been conceived as early as 1960, and organized research by RCA's scientists began in 1964. It seemed unlikely at that time that a low-cost consumer product could be based on lasers. RCA's engineers became aware of the competitive threat of lasers for videodisks when they observed a demonstration of an early Philips prototype in 1972, but by then the company already had a substantial investment in a stylus system.

The Philips demonstration gave RCA a scare, especially because the Philips recorder could produce higher quality pictures, at least under laboratory conditions. RCA, however, continued to believe that stylus recording was more reliable and economical and that these qualities would enable it to win in the marketplace.

The Technology of the RCA System Once committed to the mechanical system, for more than a decade RCA devoted a major portion of its Laboratories' efforts to developing this technology. Its basic principle was rather simple. The disk had a spiral groove on its surface that guided the playback stylus as it rotated. Undulations were pressed on the bottom of the groove, which produced an FM output signal (see illustration on next page).

Although the basic principle of the RCA recorder was not inherently complex, its design required the solution of enormously difficult mechanical problems. They are illustrated by a comparison with an LP audio recorder:

Parameter	RCA Videodisk	LP Recorder
Disk diameter	12 inches	12 inches
Groove spacing	2.5 micrometers	100 micrometers
Rotation speed	450 rpm	33 rpm
Stylus velocity	200 ips	14.7 ips
Stylus tip	2 micrometers	35 micrometers
Recording time per side	1 hour	25 minutes
Groove length per side	29 miles	0.7 miles
Stylus pressure	65 milligrams	1,000 milligrams
Max. recorded frequency	9.2 MHz*	15 kHz

* Highest carrier sideband frequency. The highest recorded video frequency was less than 3 MHz.

The difficulty of designing and manufacturing products to these dimensions is illustrated by the fact that the tip of the stylus is too small to be seen with an optical microscope, and a scanning electron microscope had to be used. But one by one, over a fifteen-year period, the basic technical problems were solved, and by 1977 there had been sufficient progress to give reasonable assurance that a salable product could be designed and built.

The Decision to Go to Market RCA was now faced with a major business decision: Should it make the enormous commitment of capital, engineering, and start-up costs required to manufacture and bring the

THE RCA VIDEODISK RECORDING SYSTEM

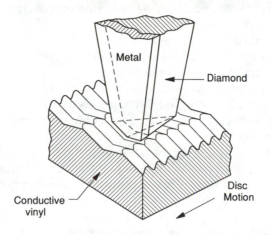

The basic component of the RCA recording system was an incredibly tiny diamond stylus plated with metal on its trailing edge. It rode over undulations pressed on the bottom of a narrow spiral groove. The undulations changed the spacing and therefore the capacitance between the metal plating on the stylus and the conducting disk, and the capacitance variations were then made to vary the output signal. Hence this was called a capacitance system.

The undulations had constant amplitude but variable spacing, so the output signal was frequency modulated. (RCA's engineers had dropped their prejudice against the use of FM that had been so disastrous in their early work on magnetic video recorders. See Chapter 6.) The frequency deviation extended from 4.3 MHz (sync tip) to 6.3 MHz (white level), producing sidebands from 2.0 to 9.3 MHz. The system was unable to record the color subcarrier at 3.56 MHz, and the color information was transmitted by means of a buried subcarrier at a frequency of 1.56 MHz.

product to market? The commitment would have to include not only the hardware program but also the software—that is, an extensive and attractive library of prerecorded disks.

It was perhaps the most difficult decision that Edgar Griffiths, RCA's chairman and CEO, faced during his entire administration, and it was not made easier by the wide differences of opinion among his principal advisers. One of the least enthusiastic was Roy Pollack, the head of RCA's consumer products businesses and a member of the board of directors. With good reason, he was particularly concerned about the competitive formats coming on the market. Sony had announced the Beta VCR in 1976, and Matsushita had followed with the VHS. The JVC grooveless capacitance disk and the Philips laser disk were about to be announced formally. This formidable competition, combined with the uncertainties of the market-place, the major engineering and production problems that still had to be solved, and the adverse effect that the diversion of resources would have on the television receiver business at a time when it was already in trouble from Japanese competition, led Pollack to believe that it was a high-risk program that might be better deferred—even though it might increase the risk by allowing the competition to become entrenched.

RCA's technical community was solidly behind the disk program. It was supported by RCA's licensing and public affairs departments, both of which believed it was necessary to maintain the company's image as a leader in advanced technology.

Griffiths had a well-earned reputation as a superb administrator, but he had been criticized for his lack of vision and his unwillingness to support programs that did not have a rapid return. This perception by his critics would have been strengthened had he decided to terminate the videodisk program, and a desire to enhance his image may have been a factor in his decision late in 1978 to proceed. The stated reason was that RCA's disk players and the disks themselves would be significantly cheaper than competitive formats. Retail price goals of $400 for the player and $20 for a typical disk were established. These were less than half the VCR prices, but paper studies indicated that they were realistic. Griffiths went to the board of directors with his recommendation, the board approved it, and a full-fledged product program was launched.

Design, Manufacture, and Marketing   Once the program was approved, a massive engineering and production effort was undertaken. Although the basic technology had been developed and proven, many difficult problems had to be solved before the player and disk could be mass-produced at a reasonable price. The engineering and preproduction programs were completed in a little over two years, a remarkable achievement considering their difficulty. In addition, RCA was determined not to repeat Magnavox's mistake with Laservision and go to market before the product was ready.

RCA's marketing organization was busy as well. The decision to proceed was announced early in 1979, and during the next two years a steady stream of press releases and speeches by RCA executives reported glowingly on the progress of the program. These reports concerned not only the player but also the catalog of titles that was being assembled.

The introduction of the product to RCA's distributors and dealers on March 2, 1981, was described in the trade press as an extravaganza. It was a professionally produced program from NBC's New York studios, and it was distributed nationwide by satellite.

RCA's top executives, overwhelmed by enthusiasm, outdid themselves in praising the product. In his announcement to the distribution, Griffiths, who had just announced his intention to retire, stated that this introduction would "stand out like a searchlight" in his memories of thirty-three years with RCA. Pollack, now a believer (at least publicly), forecast that other manufacturers would abandon competing disk formats and would join the RCA camp. And Herbert Schlosser, formerly the president of NBC and now in charge of the videodisk software program, announced that there were already 100 titles in the RCA disk library, with 135 more to follow shortly.

The Response of the Marketplace The acid test of the product, the response of the marketplace, was initially and exasperatingly inconclusive for the first two years. It was too favorable to brand the product as an obvious failure, but it was too poor to characterize it as a success. Selectavision sales were far ahead of sales of other disk formats, but they were increasingly overwhelmed by those of VCRs. The estimated factory sales were as follows:

	RCA Selectavision	VCRs
1980	0	800,000
1981	100,000	1,375,000
1982	172,000	2,034,000
1983	130,000	4,091,000

Perhaps engaging in wishful thinking, RCA's management tried to put the best face on these disappointing results. It was noted that disk player unit sales exceeded those of color receivers for the first three years after product introduction. An even more encouraging result was disk sales. The number of disks sold per player customer exceeded forecasts by a wide margin, and RCA decided it would subsidize player sales to create a market for disks. In accordance with this strategy, the retail price of players was reduced to $299 in 1982 and $200 in 1983.

The Decision to Withdraw But even the below-cost cut-rate prices failed to close the huge gap between disk player sales and those of VCRs. In fact, it continued to widen. This led Bradshaw, who had succeeded Griffiths as RCA's CEO in 1981, to decide that RCA's best interests would be served by withdrawing from the business. His decision was announced publicly on April 4, 1984, as described at the beginning of this chapter.

It was a particularly bitter blow to the hundreds of engineers and scientists who had enthusiastically devoted a large part of their careers to the technology of the disk. They had done their job, and the product met all the technical goals that had been established at the beginning of the program. Its failure had resulted from competitive factors and management decisions that were beyond these people's control.

JVC's VHD

Although JVC's grooveless disk, the VHD, was touted as using a more advanced technology than RCA's Selectavision, it had an even shorter life as a consumer product. It was first demonstrated publicly in October 1978 (shortly before RCA made a corporate decision to proceed with the Selectavision program). As described previously, a respectable coterie of manufacturers, including JVC's parent, Matsushita, agreed to adopt the VHD format. In 1980, GE, Matsushita, JVC, and Thorn-EMI formed a consortium to manufacture and market VHD products.

From 1978 to 1983, JVC continued the VHD engineering program, but its enthusiasm for the product appeared to wane, perhaps as a result of the indifferent market response to RCA's product. After many delays, shipments began in April 1983. Sales were disappointing, and by the end of 1983, the VHD format had faded from the consumer market. JVC and other manufacturers, however, continued to offer VHD players to industrial and educational users, where a modest market continued for several more years.

Laservision (Discovision)

Laservision was a potentially more formidable competitor of RCA than JVC's VHD. It used the more advanced laser technology and was capable of significantly better performance than the RCA system. It had a serious cost problem, however, and its sophisticated technology led to engineering problems that were exacerbated by a premature announcement. The cost problem was a result of the location of laser technology on the industry's learning curve. While there was great potential for cost reduction as the technology matured, this was years in the future.

The Technology of the Laser Disk   The operation of the laser videodisk is similar in principle to the audio CD described in Chapter 3 except that the signal is recorded in analog rather than digital form. As with the RCA videodisk, the dimensions are microscopic. The laser readout beam must be focused on a spot that is even smaller than the tip of the stylus in the RCA player, and the ability of the coherent laser beam to be focused on such a tiny spot is fundamental to the operation of the system.

The Philips/MCA Format   Philips began laser recording research during the late 1960s, when the basic laser technology was first being developed. By the early 1970s, laboratory prototypes performed well enough to persuade Philips's management that a practical consumer product could be designed. In an effort to obtain industry support and agreement on format standards, the prototype was demonstrated to competing manufacturers, including RCA, in 1972.

Philips soon discovered that a subsidiary of the diversified entertainment company, MCA, was working on a similar approach to laser recording. MCA's largest subsidiary was Universal Pictures, a leading film company, and its management hoped to participate in all phases of the entertainment industry, including hardware. Rather than engage in a protracted patent dispute, the companies negotiated a settlement in 1974 that included cross-licensing and agreement on a standard format that was often described as the Philips/MCA system.

The spacing between tracks in this format was even less than on the RCA disk, 1.6 versus 2.5 micrometers, and the frequency of the output pulses deviated from 7.5 MHz to 9.2 MHz versus 4.3 MHz to 6.3 MHz for RCA. The higher pulse rate permitted the full 4.2-MHz video bandwidth to be recorded as compared with only 3 MHz for RCA.

The disks were twelve inches in diameter and came in two versions. One used a constant rotational speed of 800 rpm and had a capacity of one-half hour of recording time per side. The other achieved a capacity of one hour per side by varying the rotational speed so that the linear speed of the readout beam along the track remained constant as its radius changed.

Design, Manufacture, and Marketing The Philips/MCA format attracted intense interest among consumer electronics manufacturers, and a number of them completed product design and began manufacturing. In the United States, the most aggressive were Magnavox, a subsidiary of Philips, and Universal Pioneer, a joint venture of Pioneer (Japan) and Discovision Association (DVA), which in turn was a joint venture of MCA and IBM. (DVA had been formed primarily to produce prerecorded disks.)

Magnavox was the first manufacturer to offer Laservision players in the United States. On October 19, 1978, it demonstrated a model at an ITA video programming conference, and it hinted strongly that it would be placed on sale in Atlanta in December. In November, Magnavox confirmed the December introduction and announced that the price of the constant-speed unit, one-half hour on each side, would be about $700. MCA announced that the typical prices of its disks would be $16 for major movies, $10 for documentaries, and $6 for "how-to" shows.

The Laservision playback unit, which Magnavox named Magnavision, was put on sale in three Atlanta stores on December 15, 1978. The introduction was a smash. It was described by *Television Digest* as follows:

> Never have so many done as much for so few videodisc players. They drove from distant states, phoned from Europe, waited all night in front of stores— to buy one of the first Magnavision optical disk players, which went on sale last Fri. (15th) at 3 Atlanta locations. Although stores could have sold thou-

sands, actually only about 25 changed hands—that's all there were. Frustrated would-be customers then converged on videodisc displays and bought virtually all available software (presumably an indication they intended to come back for players).[2]

The success of this introduction, which preceded RCA's Selectavision by more than two years, was good news and bad news for RCA. It seemed to indicate that there was a strong public demand for prerecorded disk players, but it also indicated that the laser disks would offer serious competition. The major surge in VCR sales had not yet begun, and at that time most RCA executives, as well as its technical community, believed that the laser disk rather than the VCR was the format to beat.

The 1978 introduction of Discovision provided a momentary thrill for Magnavox, but it proved to be a tactical error of major proportions. Magnavox simply was not ready to deliver. It was able to produce only about five thousand units in 1979, many of which had quality problems, and its failure to meet delivery promises created a severe credibility gap. In the same year, however, Universal Pioneer produced ten thousand units in the audiovisual price range for General Motors.

The market enthusiasm for laser disk players continued for a time in 1979, and as a result of the product shortage, a black market developed in which the price was as high as $2,500. But the enthusiasm was short-lived. Leading department and video stores in Atlanta indicated that player sales had virtually collapsed. The reasons given were that the Magnavox players had quality problems, the price was too high, a half-hour playing time was inadequate, most of the software was old movies that customers had already seen, and the competition from VCRs was growing.

In spite of these adverse reports, members of the laser disk industry carried out ambitious plans for growth. In March 1980, Universal Pioneer announced plans to build a plant with a capacity of nearly half a million players a year. Pioneer also announced a $10 million advertising campaign.

The ambition of these plans was not matched by the response of the market. Sales continued to be miniscule compared with those of VCRs, and they were even less than RCA's after its Selectavision came on the market. Estimated factory sales were as follows:

	Laservision Players	VCRs
1979	15,000	300,000
1980	20,000	800,000
1981	57,000	1,375,000
1982	50,000	2,034,000

[2] *Television Digest*, 18 December 1978.

The first public admission that the Laservision program was in trouble came from John Messerschmitt, president of North American Philips, in March 1981. He stated that laser recorders were too high priced for the consumer market and that their niche was the industrial market.

Messerschmitt's observation was correct. Laservision units were withdrawn from the consumer market, though without the drama of RCA's withdrawal, and for the next ten years the Laservision was offered only in higher priced versions to industry, particularly by Pioneer.

The Renaissance of Consumer Laser Recording   In 1987, there was an upsurge in the sales of laser recorders (although sales were still very small compared with those of VCRs) as a result of revived interest in their use in the home. It resulted from the introduction of the CDV, a 4.75-inch disk (the same as the CD) that could record fifteen minutes of digital audio and five minutes of analog video, and the appearance of laser combos, playback units that could handle 10- and 12-inch laser disks and 4.75-inch CDs and CDVs. In a sense, laser video was profiting from the popularity of audio CDs and riding along on its coattails. In 1988, total annual sales of laser video playback units, both audiovisual and consumer, were only slightly more than 200,000 units a year, but the industry had hopes for a rapid increase.

■ VIDEOCASSETTE RECORDERS

Made in Japan

The Japanese and VCRs   The steady and relentless invasion of the markets for radio and television products by Japanese manufacturers has been described in previous chapters. In each of these markets—radios, monochrome and color television receivers, broadcast cameras, and video recorders—there was a similar sequence of events. U.S. and European manufacturers developed the technology and the initial products and markets. Building on this base, the Japanese applied their extraordinary engineering and manufacturing skills to the production of superb products at highly competitive prices.

The sequence of events for VCRs was different. Although European and U.S. manufacturers had contributed the basic technologies, the initial product designs were Japanese. They applied their special skills, particularly miniaturization, to this task, and thus VCRs are a uniquely Japanese product. In 1988, Japanese manufacturers had about 75 percent of the world market.

Japanese Product Engineering The many reasons for Japan's postwar domination of the consumer electronics industry have been extensively analyzed elsewhere. In this book, it is appropriate to focus on the superiority of Japanese product engineering, a phenomenon that has several explanations.

Japanese engineers are extraordinarily competent. Modern Japanese society insists on a strong educational background, an emphasis that is reflected in the excellent mathematical and scientific abilities of Japanese students. The result is a deep and gifted pool of engineering professionals that greatly contributes to the country's economic success.

The Japanese devote a far higher percentage of their total scientific and engineering effort to the design of products than is the case in the United States. They have done very little basic research because the results of research in Western countries is available through technical publications and patent licenses. Little if any of their engineering talent is devoted to the development and design of military products, a function that occupies an extremely high percentage of the electronics engineering effort in the United States. The converse of this is that they devote a much higher percentage of their total technical effort to the design of commercial products—the final and payoff step. Commercial product design is often not as interesting or glamorous as pushing back the frontiers of scientific knowledge or designing exotic military hardware, but Japanese engineers have the talent and motivation to do it superbly.

Japanese management has been able to fund a higher level of engineering effort because the salaries of Japanese engineers have been far lower than those of their American counterparts, at least until the recent devaluation of the dollar. In part this is because the U.S. military establishment creates a huge demand for engineering manpower, and the law of supply and demand operates here as well as elsewhere.

Finally, Japanese management has more patience and is more willing to fund engineering programs that have no prospect of immediate return but are part of a long-range business strategy. This quality is not totally lacking in U.S. business—Sarnoff's costly pursuit of color television for more than fifteen years before it became profitable is an example—but it is far more rarely demonstrated, particularly in an age of corporate takeovers.

Akio Morita More than any other single individual, Akio Morita (Figure 9–2), the chairman of the Sony Corporation, personifies the success of the postwar Japanese electronics industry. His autobiography, *Made in Japan*,[3] gives fascinating insights into both the man and his times.

He was born in 1921, the oldest son and heir apparent of one of Japan's

---

[3] Akio Morita, *Made in Japan* (New York: E.P. Dutton, 1986).

Courtesy Sony Corporation.

■ **Figure 9–2** Akio Morita.

most prosperous sake-brewing companies, which had been in the family for fifteen generations. His father had rescued the firm from near bankruptcy, had become wealthy, and was able to raise his family in an affluent home.

In general, Morita was a poor student, but he excelled in mathematics and physics. Somewhat to the distress of his father, who was training him to take over the family business, he chose to enter Osaka University as a science rather than an economics student.

After the outbreak of World War II, he accepted a permanent commission in the Japanese Navy as a scientist. For a time, he was allowed to continue his studies, but in early 1945 he was assigned to the Office of Aviation Technology at Yokosuka, where he was first put to work in a machine shop. The navy soon decided that this was a waste of his talent, and he spent the rest of the war in various laboratories doing research in optics and thermal guidance weapons. It was during this period that he met Masaru Ibuka, a man who was to have a profound influence on his life.

Ibuka was the head of the Japan Measuring Instrument Company, an electronics firm with fifteen hundred employees, which was manufacturing magnetic detectors for antisubmarine warfare in the Nagano Prefecture. At the end of the war, he moved the company to a bombed-out building in Tokyo, changed its name to the Tokyo Telecommunications Research Laboratories, and started a new business with only seven employees.

In the meantime, Morita had been demobilized and had accepted a position on the faculty of the Tokyo Institute of Technology. He immediately looked up Ibuka and decided to work for his new company part-time while teaching part-time. Events moved rapidly thereafter. Ibuka and Morita, together with Ibuka's father-in-law and wartime minister of education, Tamon Maeda, decided to form a new company and go into business together.

The conflict of this decision with Morita's teaching career was soon ended involuntarily because, as a former military officer, he was proscribed from teaching by General MacArthur's occupation policy. But he still had a problem with his father because by Japanese tradition he was expected to take over the family sake business. Ibuka and Maeda paid a visit to Morita's father to ask that his son be released from this obligation. To their immense relief, the father not only agreed but also promised a degree of financial support to the new company. Thus the Tokyo Telecommunications Engineering Corp., the forerunner of Sony, was founded.

The triumvirate of Ikuba, Maeda, and Morita was a particularly felicitous combination. Each brought a unique set of skills to the company. Ikuba was the innovative engineer. Maeda had many contacts among Japan's financial and government elite and was able to enlist their backing. Although Morita was a trained and able engineer, his most important contribution was in marketing and particularly in opening up the huge U.S. market.

The new company struggled at first. Starting a new business in the wreckage of postwar Japan was a challenge to any management, and Tokyo Tsushin Kogyo, as their company was known in Japanese, did not even have a clear idea as to what products to manufacture. Ikuba did not want to manufacture radios because doing so was not sufficiently innovative, but it was necessary to produce something while they were looking for a truly original product. To provide some income until they found one, the company began manufacturing motors and pickups for phonographs. It also entered the broadcast equipment business when it received a major contract from NHK, the Japanese broadcasting company, for an audio mixing unit for its studios.

The company's breakthrough came with its decision to manufacture audiotape recorders. Morita learned of the Germans' successful use of tape in place of wire as a recording medium during the war, and he and Ikuba decided that this was the innovative product they were seeking.

Designing a totally new product in postwar Japan was especially difficult because of the shortage of critical materials—among other problems, it was necessary to use paper tape because cellophane was unavailable—but they persevered, and a highly successful product resulted. It was the first of a stream of new products that came from the company.

By 1953, Morita decided that the Japanese market was too small for the company's ambitions, and he made his first trip to the United States. It was during this trip that he discovered the transistor and negotiated a patent license agreement with Western Electric. He also decided that the company needed a new name to do business in the United States, and Sony was the result.[4] Most importantly, it was the beginning of more than three decades during which Sony became one of the leading electronics suppliers in the United States.

The transistor patent agreement with Western Electric was one of the great bargains of the electronics industry. The initial payment was only $25,000. Morita recognized that the Japanese had unique skills in the design and manufacture of miniaturized products such as transistors and transistor products, and this agreement was one basis for Japan's extraordinary success in electronics. Ironically, MITI, the Japanese government's commercial bureau, for many months refused to free the foreign exchange required to obtain the license.

Morita's ability to bridge the language and cultural gap between the United States and Japan was truly extraordinary. When he came to the United States in 1953, he knew very little English and was totally unfamiliar with the American way of life. But he was a fast learner, and within a few years he had a better grasp of the U.S. market than most U.S. businesspeople.

In 1960, he founded the Sony Corporation of America. It was listed on the New York Stock Exchange, and he was exposed to the tremendous complexity of the laws regulating the U.S. securities industry.

In 1963, although he was the executive vice president of Sony with great responsibilities at the home office, he decided that he would have to live in the United States for a period to understand it completely. He sublet an apartment on Fifth Avenue from his friend, violinist Nathan Milstein, and brought his wife and family to New York. It was total immersion in a new language for them, but they learned, too, and Morita pays a particularly gracious tribute to his wife's contribution as hostess and family support system during their residence in New York. Morita was promoted to president of Sony in 1971 and to chairman and CEO in 1976, a post he still holds today.

In spite of the wide cultural gap between Japan and the United States, Morita developed an intuitive understanding of the U.S. market. He was skeptical of formal market research, but his marketing instincts were almost

---

[4] Morita's autobiography, ibid., has an interesting description of his coining the name Sony (which has no particular meaning in itself). In so doing, he displayed a remarkable sense of a language that was foreign to him.

always right. There was one major exception: his failure to recognize the importance of long playing time in VCRs, which resulted in the ultimate defeat of Sony's Beta format by Matsushita's VHS in the marketplace. Even the most able businesspeople make mistakes, however, and the growth of Sony from a tiny company housed in a bombed-out building in a defeated country to one of the world's industrial giants is proof enough of Morita's stature as one of the outstanding business leaders of the postwar world.

The VCR's Ancestry

The quadruplex videotape recorder, introduced by Ampex in 1956, was the distant ancestor of the VCR, but its parent was the professional helical scan recorder first offered for use in broadcasting in the early 1970s (see Chapter 6). Like quadruplex machines, VCRs record the video signal in analog form using FM, but none of the other major technical concepts of quadruplex technology survived. Among its other drawbacks, the quadruplex format was far too complex to be used in a consumer-priced recorder.

In contrast, the helical scan format not only had the important advantages that caused it to supersede quadruplex in professional recorders, but it also was remarkably versatile and provided the basis for recorder designs with an extremely wide price range. Its development followed a progressive sequence from professional units selling from $25,000 to $100,000 to recorders designed for the audiovisual markets selling for $1,000 to $10,000 and finally to VCRs designed for home use and selling for a few hundred dollars.

VCR Technology

VCRs for home use must meet extraordinarily difficult specifications. They must be small, easy to load, easy to operate, reliable, virtually maintenance free, capable of producing picture quality equivalent to broadcast programs seen on a typical home receiver, and most of all low priced. David Sarnoff envisioned such a product in 1951 in his famous "birthday present" speech (see Chapter 6), but it was left to the Japanese to bring the concept to full fruition.

The Japanese talent for miniaturization, their meticulous care in the design of every product detail, and their cost reduction with little compromise of performance and no compromise of reliability was embodied in their VCRs. When the first successful cassette recorder, the Sony U-Matic, was examined by American engineers, they were amazed at the exceptional quality of its design and its low selling price.

It does not denigrate the accomplishment of the Japanese to note that they made liberal use of technologies originating in other countries. Although they made a greater contribution to the technical content of VCRs than of other radio and television products, their design would not have been possible without the use of key technologies developed elsewhere. The use of high-density recording techniques pioneered by the computer industry is one example.

A prototype of the U-Matic recorder was first shown to other Japanese manufacturers in 1970 in a successful effort to persuade them to agree on a standard format. (A license was also offered to Ampex, but it was refused.) It was introduced in the Japanese market in 1971 and in the United States in 1972. It was too large and expensive for the home market, but it has been sold in large quantitites in the industrial and educational markets and even to broadcasters for portable applications.

The U-Matic was the precursor of home VCRs, all of which use the same basic configuration. The tape is contained in a cassette that holds both the supply and take-up reels. The drum has two recording heads, and the tape is brought in contact with the drum in a half-wrap in the shape of a U (Figure 9–3), hence its trade name. This arrangement met the ease-of-threading requirement and also provided a convenient holder for handling and storing the tape.

The problem facing VCR designers was to make them smaller and cheaper than the U-Matic, an enormous challenge since smaller does not automatically mean cheaper. In some cases, such as watches, the opposite is true. But Japanese engineers at Sony, Matsushita, JVC, Hitachi, and

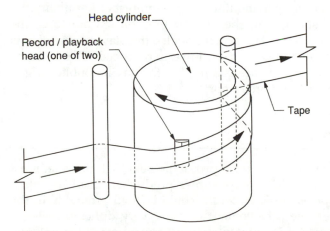

Head cylinder

Record / playback head (one of two)

Tape

■ **Figure 9–3** A half-wrap tape configuration. This configuration is used in most VCRs. When the cassette is inserted in the recorder, a mechanism extracts the tape and pulls it around the drum. The recording drum has two heads and rotates at 1,800 rpm. A complete television field is recorded by each head as it passes across the tape in a half revolution.

Mitsubishi rose to the challenge and developed three VCR formats for home recorders:

	U-Matic	Beta	VHS/VHS-C	8mm
Tape width	3/4"	1/2"	1/2"	8mm
Cassette size (inches)				
Width	4.8	3.8	4.1/2.3	2.5
Length	7.3	6.1	7.4/3.6	3.7
Thickness	1.2	1.0	1.0/0.9	0.6
Playing time (hours)	1	1/2/3	VHS, 2/4/6 VHS-C, 1	2/4

Recorders with the U-Matic format are used professionally in broadcast and audiovisual applications but are too large and costly for the consumer market. U-Matic specifications are given in this table for comparison purposes.

Each of the home VCR formats offers a choice of recording times. The playing time for the 8mm format requires the use of a thin metal particle tape. If this tape is used with the VHS format, the maximum playing time is increased to eight hours.

The VHS-C format was developed for use in camcorders. It has the same recording standards as VHS, but the cassette is smaller and holds less tape (and hence there is a shorter playing time) in the interest of portability.

From a commercial standpoint, the playing time turned out to be the most critical specification. The VHS format doubled the playing time of Beta by using a larger cassette and 20 percent lower tape speed. This increase was achieved at some loss of picture quality, but the buying public judged the longer playing time to be more important.

Sony Introduces Beta

In 1974, Sony engineers completed a prototype Beta VCR that met most of the objectives established for a consumer product. It used less than one-third as much tape as U-Matic, and it could be sold profitably in the consumer price range. The achievement of these very difficult goals resulted from extensive use of integrated circuitry and a host of incremental improvements.

With a working prototype to demonstrate, Sony decided to show it to its competitors as it had with U-Matic in an effort to persuade them to agree on a standard industry format. This strategy had been successful with U-Matic,

VCR TECHNICAL PARAMETERS

	U-Matic	Beta	VHS	8mm
SMPTE type	E	G	H	—
Tape speed, (in./sec.)	3.75	1.57/0.79/0.52	1.3/0.66/0.43	0.56/0.28
Head-to-tape speed (in./sec.)	410	276	229	148
Track width (mils)	3.35	2.3/1.15/0.77	2.3/1.15/0.75	0.8/0.4
Tape area, (sq. ft./hr.)	71	19.6/9.8/6.5	16.3/8.1/5.4	4.3/2.1
FM carrier range (MHz)	3.8–5.4	3.5–4.8	3.4–4.4	4.2–5.4

The head-to-tape speed is nominally the same within each format, regardless of the recording time. Slowing the tape speed to increase the recording time narrows the recording tracks with some loss of performance.

Super or high-band versions of these formats have a higher range of FM carrier frequencies (for example, the Super-Beta carrier range is 4.4–5.6 MHz). With high band, a horizontal resolution of 430 lines (versus 340 for standard broadcast and 240 for VHS) can be recorded. The short wavelengths of high-band recordings require the use of metal particle tape.

but it did not work with Beta. Sony's most active competitor, Matsushita, was unwilling to agree on the Beta format on the ground that the one-hour playing time was inadequate. Subsequent events showed that Matsushita's position was correct.

Impatient with his inability to persuade Matsushita, Morita decided to introduce Beta to the marketplace with or without an agreement. He held a press conference on April 16, 1975, at which he announced that it would be available in Japan at a price of 229,800 yen (about $870). Ten months later, in February 1976, it was introduced in the United States at a suggested retail price of $1,295. Its recording time was one hour, and blank tape cassettes cost $15 each.

The response in the marketplace was fabulous, both in Japan and the United States. Recorders were sold as fast as they could be produced, and a shortage of cassettes soon developed. Sony kept the market growing in the United States with an aggressive advertising campaign that emphasized time shifting, the ability to record a program at one time (perhaps while watching another one) and to play it back at a later time. All indications were that the product was a spectacular success and that it would make enormous contributions to Sony's sales and profits for many years to come.

The reality was different. Beta was indeed a profitable product, but its success was threatened by a development in the courts and dimmed by competition in the marketplace. The legal problem resulted from a lawsuit filed against Sony America by two film companies, Universal and Disney, which charged that Beta recorders gave their owners the ability to make unauthorized copies of copyrighted program material and that Sony knowingly sold them for this purpose. The lawsuit did not seriously affect Sony's sales, and Sony ultimatey won, but mounting a defense was a time-consuming and costly drain of corporate resources.

The marketplace problem was the introduction of another VCR format, the VHS, by Sony's competitor, Matsushita. It was a much more serious problem, and it has had a major adverse affect on Sony's sales and profit, even to the present time.

The Universal Pictures–Sony Copyright Battle

Universal Pictures, joined by Disney, filed the copyright lawsuit on November 11, 1976, nine months after Beta recorders were introduced in the United States. Although Lew Wasserman, the president of Universal's parent, MCA, had warned Morita of Universals's intention a month earlier, Morita was appalled. MCA and Sony were engaged in friendly negotiations for the manufacture of MCA's laser disk, and friends do not sue each other in Japan.[5]

The issue had some similarity to the cable system controversy that had just been settled by a revision of the copyright law (see Chapter 7). Television stations pay royalties, based in part on the number of viewers in their coverage areas, for the broadcast of copyrighted material. If other distribution media make programs available to additional viewers, they in effect get a free ride. The importation of distant signals was the medium at issue in the cable controversy. For VCRs, it was the ability to record a program and play it back at a time other than its regular broadcast.

The Universal–Sony copyright case rivaled the Armstrong–RCA patent dispute (see Chapter 3) in complexity, and a detailed discussion of the arguments and counterarguments is beyond the scope of this narrative.[6] Among Sony's defenses were the claims that as a manufacturer it should not be held responsible for the use of its product, that the law had no right to intrude on private actions within one's home, and that the plaintiffs had voluntarily transmitted their copyrighted material over the public airways.

---

[5] Morita was continually shocked by the litigiousness of U.S. business. For example: The procedures for listing a stock on a public exchange are bewilderingly complex. The number of law students in the United States exceeds the number of practicing lawyers in Japan. Business contracts in the United States attempt to foresee everything that could go wrong and specify the remedies; whereas Japanese contracts customarily have a clause stating that the parties will negotiate a settlement in the event of disagreement.

[6] For a good description of the litigation, see Lardner, *Fast Forward*.

After more than two years of "discovery," during which millions of document pages were studied and thousands selected for evidence, the trial began on January 30, 1979. The judge correctly prophesied that the case would be appealed, no matter who won, and he took particular pains to follow approved courtroom procedures. The trial ended in October with a victory for Sony. The judge ruled that "home-use recording from free television is not patent infringement, and even if it were, the corporate defendants are not liable and an injunction is not appropriate." Sony's defenses had been effective.

As the judge had predicted, Universal and Disney appealed the decision to the Ninth Circuit Court of Appeals in San Francisco. There followed another lengthy round of pleadings and counterpleadings, which ended nearly three years later on October 19, 1981, when the appeals court reversed the lower court in a unanimous decision. The decision was not only unanimous but also highly critical of the original decision, stating that the judge had failed to understand the copyright law.

The appeals court's decision was a terrible blow to Sony and indeed to the entire home recorder industry. Conversely, it was hailed as an enormous victory by the motion picture producers, who stated that the courts had "gone back to the basics." Sony won two important victories, however. The appeals court did not assess damages; instead it remanded this issue to the lower court. And it did not issue an injunction forbidding VCR sales; rather it stayed action pending the outcome of an appeal to the Supreme Court. Sony and its competitors were able to stay in business.

Two more years of pleadings and oral arguments followed as the Supreme Court considered the case. Its final verdict came on January 17, 1984, seven years after the suit was originally filed, when it reversed the appeals court in a split 5 to 4 decision. It was uncomfortably close, and it would have been a disaster for the VCR industry if it had gone the other way. VCR manufacturers would have been denied the right to stay in a profitable business, and they would have been subject to enormous damage awards. Morita, of course, was delighted, but his joy was tempered by the fact that the Beta format was losing the competitive battle with VHS.

Beta versus VHS

One has to suspend disbelief to understand the history of the battle between the Beta and VHS formats. Beta started with all the advantages:

- It had a two-year head start in the marketplace.
- It was an excellent product.
- Its picture quality was inherently superior to that of VHS (although not by a significant amount).

- It came from a company that had an outstanding reputation for the quality of its products.
- A substantial library of tapes prerecorded in this format was available for sale or rent.
- Its price was right.
- It was enjoying very healthy sales.

In Morita's view, the most important advantage was that the Beta format was capable of producing marginally better picture quality. To enter the market with an incompatible format with slightly inferior picture quality after Beta had become so well established and had acquired such an excellent reputation seemed suicidal to many in the industry. The shorter playing time of the Beta recorder was a handicap that outweighed its head start and other advantages, however, and it was ultimately fatal in Beta's competition with VHS.

Sony had ample warning that the short playing time of Beta would be a problem. Matsushita's stated reason for refusing to accept this format as a standard was its short playing time, although there were additional competitive considerations. Sony attempted to persuade MITI, the Japanese government's commercial bureau, to decree a standard, but MITI refused, and it was left to the marketplace to decide.

At the time of the first Sony–Matsushita negotiations in 1974, Sony proposed a one-hour playing time and Matsushita a two-hour time. The Sony design was further advanced—and therefore more difficult to change—and Beta recorders had a one-hour playing time when they were placed on the market in Japan in 1975. At that time, Morita strongly believed that one hour was sufficient.

The reaction in the marketplace, particularly in the United States, soon gave indications that one hour was not enough, and Sony engineers began working on a two-hour design to match Matsushita's. The two-hour Beta recorder was announced in early 1977, and U.S. deliveries began in the autumn of that year.

Sony was shooting at a moving target, however, because Matsushita, partly at the urging of RCA, had successfully completed a four-hour design, a major engineering feat. Thus, when Matsushita began delivering VHS recorders to the United States in late 1977, it was a two-hour Beta format versus a four-hour VHS.

The key to penetrating the U.S. market was RCA. Although RCA had not been successful in designing its own VCR, it had a powerful marketing organization. It was the leader in color receiver sales with more than 25 percent of the market, and it had decided to use this strength to sell RCA brand VCRs obtained from a Japanese manufacturer. With RCA's marketing strength, the format it chose would be assured of a major market share.

The choice of format was not an easy one for RCA. Choosing VHS would require it to overcome Sony's established position and encourage the development of a new library of prerecorded tapes in the VHS format, a process that would at best require some time. While the public's preference for longer playing time was predicted by market studies, it had not yet been confirmed by experience. The decision was sufficiently difficult and important that Griffiths, RCA's CEO, became personally involved. He and Roy Pollack, the head of RCA's consumer business, were both strong believers in long playing time, and they decided for Matsushita's VHS. This made it certain that the Beta–VHS competition would be intense.

After RCA's decision, other manufacturers chose sides, and as VHS recorders came on the market in 1977, each format had a coterie of supporters:

Format	Brand	Manufacturer
Beta	Pioneer	Sony
	Sony	Sony
	Sears	Sony
	Toshiba	Sony
	Zenith	Sony
VHS	Hitachi	Hitachi
	JVC	JVC
	Magnavox	Matsushita
	Panasonic	Matsushita
	RCA	Matsushita
	Sharp	JVC
	Sylvania	Matsushita

It did not take the market long to decide. VHS recorders sold well from the beginning, and by 1980 Beta's market share had plunged to 35 percent (Figure 9–4). The VCR market was growing so rapidly that Beta sales actually increased for the next five years in spite of its eroding market share, reaching a peak of about 900,000 units in 1985. But as the VCR market leveled off that year, the market share erosion was not offset by a growth in the total market, and the sales volume dropped to a level that could not sustain a business. Like MacArthur's old soldiers, Beta just faded away. Its death knell was sounded, though not officially, in January 1988 when Sony announced that it was adding VHS recorders to its line.

The VCR Market Burgeons

The VHS format not only won the competitive battle with Beta but it also benefited from an enormous growth in the U.S. market for VCR recorders (see Figure 9–1). The market grew at a moderate rate from 1976 to 1982 to an

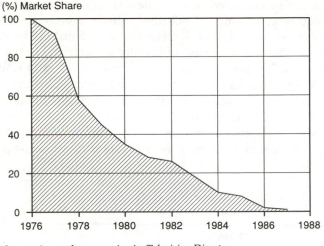

(%) Market Share

Source: Annual summaries in *Television Digest.*

■ **Figure 9–4** Beta's market share, 1976–1987.

annual volume of more than two million. It then accelerated, and by 1985 the annual volume approached twelve million, a level it has since maintained. VCRs have become a major component of the U.S. entertainment industry, and their manufacture and sale has become big business. An equally healthy market has developed for the production of blank tapes and the sale and rental of prerecorded tapes.

Super VHS

Super VHS (see the earlier box on technical parameters) uses an FM carrier range of 5.4 to 7.0 MHz as compared with 3.4 to 4.4 MHz for standard VHS. It also records the monochrome and color components of the signal separately, thus avoiding the spurious effects that result from the use of a subcarrier.

The Super VHS (SVHS) format is made possible by the use of metal particle tape in which the magnetic medium is tiny particles of an alloy rather than iron oxide as is used with standard tape. Shorter wavelengths can be recorded on the particle tape, and this property can be used either to reduce the speed and increase the playing time or increase the frequency range. SVHS takes advantage of the latter option. A disadvantage of metal tape is its higher cost.

The market appeal of SVHS is uncertain. It was given wide publicity at its first major U.S. introduction at the 1987 Summer Consumer Electronics Show. Its ability to record 430 lines of horizontal resolution versus 340 for broadcasting and 240 for VHS created something of a sensation at the show. It also created consternation among broadcasters because for the first time consumers could create a better picture at home than could be received over the air. The market response was not as enthusiastic, and dealers have been skeptical as to whether SVHS is worth the added cost.

The subjective improvement in picture quality is not as great as would be suggested by the increase in resolution from 240 to 430 lines and the elimination of the subcarrier. Further, to achieve full SVHS picture quality, the picture must be displayed on a video monitor or receiver with a video input connection so that the video signal from the VCR bypasses the radio frequency circuitry in both the VCR and the receiver. Receivers with this feature have only recently entered the market.

In spite of these problems, the most likely (but by no means certain) outcome will be that SVHS and other super formats will ultimately be successful in the marketplace. Their first customers will be well-heeled audiovisual hobbyists. As the public's taste for better quality pictures is stimulated by the interest in HDTV (see Chapter 10) and the steady upgrading of the quality of broadcast transmissions and receiver designs, the super formats' market will broaden, and they may become a major factor in the industry.

The 8mm VCR Format

The development of the 8mm format was another major step forward in recorder technology. The original Beta design used less than one-third as much tape as the U-Matic, and 8mm uses less than one-half as much as Beta. The reduction was made possible by the use of metal tape and many incremental improvements in recorders.

No single company was dominant in the design of this product. Nearly every major recorder manufacturer in Japan and Europe engaged in an 8mm design program. Recognizing the difficulties that resulted from the lack of agreement on a standard format for half-inch tape, an all-industry group adopted a set of 8mm standards early in 1983.

Although industry members agreed on a standard 8mm format, they agreed on very little else. The years 1984 and 1985 were marked by mass confusion as the industry attempted to evaluate 8mm's potential. Some manufacturers forecast that it would replace half-inch tape completely, others that it would be used only in camcorders and portable applications, and still others that it would not be used at all. Sony and JVC represented the opposite poles in this controversy.

Having lost the VHS–Beta battle, Sony was looking for a third alternative, and 8mm was an attractive candidate. In April 1985, it told its dealers and distributors that 8mm would be the basis for the complete video system of the future and that the 8mm camcorder was only the first of a full line of 8mm products that would include VCRs.

On the other side, JVC launched an all-out attack on 8mm in January 1986. It wrote a "confidential" letter to its dealers that began with the assertion that "8mm video is not a world standard" and ended with a claim that "8mm video cannot be considered a serious competitor of VHS." Among the alleged problems were a lack of compatibility of recordings made on metal particle and metal evaporated tape, a tendency for pictures to jump because of the slow tape-to-head speed, the use of monaural audio, and the extreme precision required in manufacturing.

The future of 8mm VCRs is not clear. Sales figures to date (1989) are not readily available, but they have been small, partly because of a limited amount of prerecorded software in the 8mm format. In the future, sales should receive a boost from the introduction of portable products such as the Sony Video Walkman, which includes a miniaturized playback unit with a three-inch picture tube display. As of this writing, it appears that the format has a developing potential for VCRs. The most immediate application for the 8mm format has been in camcorders, as described in the next section.

■ CAMCORDERS

The Role of Camcorders

The camcorder, a combined television camera and tape recorder, is the latest success story in consumer electronics. It performs the same function as home movies but eliminates many of the disadvantages of film.

A television recording can be played back immediately (there is no waiting for film processing), which greatly enhances its appeal. The recording medium can be used repeatedly, if desired, at considerable savings in cost. The recording can be played back on a standard television set and can be shown in a well-lit room without the nuisance of pulling a projector and screen out of the closet. Television recordings are probably more durable, since they are not subject to long-term fading, which is a problem with color film. Last but not least, audio can easily be recorded on the same tape.

These advantages have been recognized by the buying public, and the U.S. sales growth has been impressive:

1984	NA*
1985	516,000
1986	1,168,768
1987	1,604,153
1988	2,043,835

* NA = not available

With retail prices exceeding $1,000, camcorders have become a $2 billion industry in only a few years.

The Technology of Camcorders

Camcorders are another of the remarkable recent technical achievements of the recorder industry. They are small and light (typically they weigh from four to six pounds versus ten to fifteen pounds for the professional camcorders used by broadcasters), they are easy to operate, and their picture quality is as good as or better than 8mm film. Their performance has been made possible by rapid technical advances in their two basic components, the camera and the recorder.

Camcorder Cameras The camera tube or functionally equivalent solid-state device is the key component of camcorder cameras. In addition to providing satisfactory picture quality, it must be small, have adequate sensitivity, use very little power (since camcorders are battery operated), and be low priced. Tubes or devices with this combination of specifications were not available until the late 1970s.

Both tubes and solid-state components are used in camcorders. The tubes are half-inch photoconductive types such as the saticon, and the solid-state components are charge-coupled devices (CCDs) (see Chapter 5). Because of their small size and low power drain, CCDs are the most widely used, and this usage will increase as their designs are improved in the years ahead.

Camcorder Recorders Four recording formats have been used for camcorders: standard VHS (there is also some use of SVHS), a smaller and lighter cassette version of VHS known as VHS-C, 8mm, and a version of Beta known as BetaMovie. BetaMovie has disappeared along with Beta, but the other formats are still in active use.

Standard VHS has the advantage that the cassettes can be played back on regular VHS VCRs without the hassle of attaching the camcorder or installing an adapter. VHS-C cassettes can be played back by the camcorder recorder or in a VCR with a VHS-C cassette adapter. 8mm cassettes can be played back by the camcorder recorder or with an 8mm VCR. The fact that there are very few of the latter is a deterrent to the sales of 8mm camcorders.

Precise market share statistics are not available, but in order-of-magnitude terms, standard VHS has 65 percent of the market, VHS-C 10 percent, and 8mm 25 percent. Each of these formats has significant advantages, and they will probably continue to coexist for the foreseeable future.

■ A LOOK AHEAD

There has been an explosion in the technology of home recorders during the past ten years, and the era of rapid advances has not ended. Worldwide, there are more than a dozen major manufacturers of home recorders, and each has a large and highly talented engineering staff. Each is competing vigorously to maintain and increase its share of a large and growing market. Each will attempt to bring out product enhancements and new products that will be attractive to the buying public. The next few years will be interesting indeed.

10

■ ■ THE NEW TECHNOLOGIES

Advances in the technologies of television are taking place at an accelerating pace. Worldwide, more than a dozen major companies and countless smaller ones are devoting thousands of man-years of engineering effort annually to the development and design of new systems and products. They are supported by the work of an equal number of scientists engaged in the basic and applied research that will be the foundation of future designs. Much of this is supported by technically related industries, such as military electronics and computers. As a consequence, television technology has made more progress in the past decade than in the previous fifty years, and even greater progress can be expected in the decade to come.

This book is a history, not a forecast, but it seems appropriate to close with a status report of the major new technologies that are likely to have a profound effect on television's future. The technologies to be considered are high-definition television (HDTV), digital systems, solid-state devices, lasers, and fiber optics.

New technologies is not a precise term for all these developments. HDTV could perhaps be better defined as a system rather than a technology, and solid-state devices have been on the market for forty years. But HDTV is made possible by a number of new technologies, and the greatest advances in solid-state devices are probably still in the future.

■ HIGH-DEFINITION TELEVISION

Two Points of View

HDTV, television with higher technical quality than standard broadcast television, has received wide publicity in recent years, not only in the trade press but also in the business and general press. It has dominated the technical agenda of broadcast professional societies, and it has been discussed in depth at the meetings of broadcast and cable trade associations. It has attracted the attention of the U.S. Congress and has been the subject of a number of congressional hearings. As with most new developments, there are great differences in the prognosis for its future.

The HDTV enthusiasts propound one point of view. They believe that HDTV will become a major entertainment medium within a few years. They view the future of broadcast television with alarm (or perhaps with anticipation if they represent competing media) because of its difficulty in offering full HDTV in the spectrum available for broadcasting. They predict that HDTV will come at first via VCRs, cable (eventually fiber-optic), and satellites, which do not suffer broadcasting's spectrum limitations.

Some enthusiasts forecast that HDTV will become the linchpin of the United States' commercial electronics industry in future years—not only of consumer electronics but also of computers and solid-state products. In one of his first public statements after his appointment as President' Bush's secretary of commerce, Robert Mosbacher averred that "HDTV is not just another stage in TV—not just another consumer good, but a whole generation of electronics."

The proponents of this view hold that HDTV may become the means by which the Japanese can destroy the U.S. electronics industry because of their current lead (though not a large one in many areas) in HDTV technology. To forestall this disaster, they advocate a cooperative effort by the entire U.S. industry, aided by legislation providing limited immunity from the antitrust laws and financial support from the U.S. Treasury. There has even been a suggestion that part of this support should come from the Department of Defense, with the military justification being the production of high-definition satellite and battlefield images.

An opposing point of view was expressed in the authoritative trade paper, *Television Digest:*

> Wishful thinking and panic seem to be mixed in roughly equal portions where HDTV is concerned. Based on history of TV industry, many of today's arguments and statements about advanced TV systems seem to be grounded firmly in myth. Here are some common conceptions being advanced that are at least partly contradicted by fact, history or common sense.
>
> [Conception] "(1) Americans desperately want better pictures, and public opinion is forcing industry to go to advanced TV systems."
>
> [*TV Digest*'s response] "Recent MIT study, augmented by opinion research of HBO and CBC [Canadian Broadcasting Corporation], begins to put situation in perspective. While Americans may be most demanding about program material, they're notoriously passive in their acceptance of technological picture deficiencies.
>
> "There's no question that Americans—or any viewers—can be educated to advantages of good pictures. Larger screen TV sets particularly will cry out for pictures that can be viewed from closer distances. But all available information indicates American demand for HDTV may arise from national pride, threat of competition, or visions of future markets—but not from any spontaneous demand from viewing public, contentedly watching Knott's Landing on maladjusted sets, worn-out lead-in wires and 150 lines of horizontal resolution."

A number of other popular conceptions and *Television Digest*'s responses followed, ending with:

[Conception] (6) "HDTV will come along in about 5 years."

[*TV Digest*'s response] ". . . 10 to 20 years seems far more likely, allowing time for usual court challenges. . . . Advanced systems short of HDTV not requiring frequency spectrum reallocation could come more quickly."[1]

Television Digest was not alone in this opinion. An article in the *Wall Street Journal* stated:

Consumers are a fickle bunch, and high-definition TV has at least two fundamental flaws. Most people buy a TV set on the basis of price, and the early high-definition sets—expected in the early 1990s—will cost $2,500, or about 10 times as much as conventional sets. Most people watch TV on sets smaller than 20 inches—and high-definition sets of that size don't offer improved pictures.[2]

Only time will tell which of these views is correct.

Early History of HDTV

The Japanese are generally given credit for being the pioneers in HDTV. NHK, the Japanese broadcasting company, began research in high-definition systems in 1968, and this culminated in a demonstration of its system at an SMPTE conference in February 1981. The results were variously described as awesome, extraordinary, and breathtaking. Its spectrum bandwidth, 30 MHz, was excessive for broadcasting, but it was the ancestor of the various MUSE systems (described later in this chapter) introduced later.

In fact, however, the Japanese were not the original pioneers. That honor probably belongs to RCA's Otto Schade, who began research in high-definition systems shortly after the end of World War II. He also was able to produce pictures that were quite extraordinary, considering the limitations of the camera tubes and circuitry then available.

In 1948, he began the publication of a series of papers describing the results of his research.[3] They contained his classic study of the factors that determine the picture quality of film and television systems and introduced new criteria for evaluating quality that were equally applicable to film and television. His work provided the basis for the development of aperture

---

[1] "Fact, Fancy, and Fallacies," *Television Digest*, 9 May 1988.

[2] "Will High-Definition TV Be a Turn-Off?" *Wall Street Journal*, 20 January 1989.

[3] Schade published a series of four papers under the title "Electro-Optical Characteristics of Television Systems." They appeared in successive issues of *The RCA Review*, Vol. 9, in March, June, September, and December 1948.

correction and other signal-processing techniques for obtaining the maximum possible definition in bandwidth-limited systems.

Schade has not received the recognition he deserves, possibly because he was ahead of his time. Twenty years would elapse before television pickup tubes and other components would become available to take full advantage of his research.[4]

The NHK demonstration was the catalyst for widespread participation by the television industry. The SMPTE showing was seen primarily by professional engineers, but CBS hosted a semipublic demonstration in Washington a month later for the press, government officials, and other broadcasters. This aroused wider interest, and at least a dozen companies joined NHK and Sony (which had developed much of NHK's equipment) in the quest for a system that would be accepted as a standard by regulatory authorities and the marketplace. This search continues, as described later in this chapter.

High Definition Defined

Conventional broadcast television, either NTSC or PAL, has picture quality, which, in the layperson's terms, is roughly equivalent to that of 16mm film. It is generally agreed that the goal of most high-definition systems is *to achieve the quality of 35mm film but in the wide-screen format commonly used by the motion picture industry.* To reach this subjective goal, it was first necessary to express it in engineering terms.

In these terms, the definition of a picture is the sharpness with which the edges of objects are reproduced, or as *Webster's North Collegiate Dictionary* defines it, "distinctness of outline or detail (as in a photograph)." This property can be described as the degree to which the picture appears to be in focus. Sharp edges are indeed a basic characteristic of HDTV pictures, but *definition* in this context has come to have a broader meaning and to encompass *all* the properties that viewers identify as contributing to picture quality.

In addition to picture characteristics, the quality of the sound is important. Viewing a television program is a total audiovisual experience, and the perception of picture quality can be enhanced or degraded by the sound quality.

---

[4] The prose style of his papers also might have been a factor. Schade apparently thought in German, his native language, and the style of his original papers was often like that of a schoolboy's literal translation. It was difficult to comprehend, even by those skilled in the art.

The Elements of Picture Quality

The most important measurable properties of a television picture are as follows:

1. Signal-to-noise ratio, or in popular terms the amount of "snow" in the picture
2. Contrast ratio, or the ratio of the brightness of the brightest and darkest areas of the picture
3. Colorimetry, or the faithfulness with which the original colors in the scene are reproduced
4. Aspect ratio, or the ratio of the picture width to its height
5. Definition, or image sharpness as described previously

HDTV performance can be achieved with the first three characteristics without amendment of current transmission standards, but the last two require modifications.

Signal-to-Noise Ratio   Communications systems are always troubled by some level of electrical disturbances. In voice systems, these are heard as noise—hiss, crackle, pop, or hum. Thus *noise* has become a generic term that describes any unwanted electrical disturbance in a transmitted signal, even in television signals where the effect is not audible noise but rather speckles, or "snow," in the picture. The degree of visual impairment is determined by the ratio of the amplitude of the wanted signal to the unwanted noise, and the signal-to-noise ratio is a fundamental criterion of picture quality.

Technical advances spurred by competition have greatly improved the signal-to-noise ratio of modern television systems. The use of photoconductive rather than photoemissive tubes (see Chapter 5), together with low-noise solid-state amplifiers in television cameras, makes it possible to produce video signals that are virtually noise free. Further improvement can be achieved in receivers where digital signal processing can remove much of the noise that may have crept into the signal en route.

As a result of these advances, a noisy picture on a home receiver is usually due to a weak signal from the broadcast station. With an adequate signal delivered to the receiver, the signal-to-noise ratio of modern television systems is high enough to meet the demanding requirements of HDTV.

Contrast Ratio   Nearly everyone has had the experience of observing the effect of turning on the room lights during a showing of 35mm slides. The picture, which appeared crisp and in focus in a darkened room, suddenly seems washed out. Nothing has happened to the projected image, but its appearance is changed because the contrast ratio is reduced by the external illumination of the darker areas of the picture. This effect dramatically illustrates the importance of the contrast ratio in achieving high-definition pictures.

Quite obviously, the contrast ratio of a television picture can be increased by brighter whites and/or darker blacks. Since television pictures (unlike photographic slides) are usually observed in well-lit rooms, it is important that the black areas be as dark as possible even when the tube is illuminated by room lighting.

In recent years, there has been a substantial improvement in the contrast ratio of television receivers (see Chapter 5), and modern picture tubes can produce images that meet the demanding requirements of HDTV, even when viewed with normal room lighting.

Colorimetry   The NTSC standards embodied in the FCC rules specify the colorimetry of the three primary colors—red, green, and blue—which are incorporated in the transmitted signal. Nearly all the colors that appear in nature can be reproduced with the NTSC standard primaries. The colorimetry of receivers is not controlled by the FCC, and many picture tube manufacturers have chosen to use phosphors with nonstandard primaries in order to achieve greater picture brightness. This trade-off is controlled by the desires of the customers. The color distortion in commercial receivers is not great, and most of the public is unaware of it.

Aspect Ratio   When the original NTSC monochrome signal standards were established, the aspect ratio for motion pictures was 4 to 3—that is, the picture width was four-thirds of its height. The NTSC adopted this as the standard for television, and it has remained unchanged. Subsequently, the motion picture industry found that wide-screen pictures were popular with the public, and movies commonly have a larger aspect ratio.[5] There is now fairly unanimous industry agreement that HDTV must have a greater aspect ratio than standard television to give the wide-screen effect, and sixteen by nine is the most common proposal. Changing the aspect ratio would clearly require a change in transmission standards.

Picture Definition   In its narrow meaning, picture definition is measured by the geometric sharpness of the picture's edges. It is determined by the number of scanning lines, the bandwidth, the quality of the camera and

---

[5] There is a problem in showing wide-screen movies on four-by-three television. One solution is to record the movie on tape and rerecord only a portion of the complete image having the desired four-by-three dimensions. The portion to be rerecorded can be selected by the operator.

picture tube, and the processing of the signal. Because the number of scanning lines and the bandwidth (which is directly related to the spectrum requirements) must be standardized (for broadcasting by government regulation), these factors receive the most attention.

Attempts to specify the definition of television systems have been confused by photographic practice. The definition of a photograph is customarily specified by its resolution; that is, the fineness of detail that it reproduces or resolves. This is an appropriate measurement in photography because there is a strong correlation between resolution and edge sharpness. A photograph with high resolution will also have sharp edges.

Schade, in his papers in *The RCA Review,* and others have shown, however, that resolution is not a completely satisfactory criterion of the definition of television systems because resolution and edge sharpness are not necessarily correlated. A better measure, Schade found, was the response of the system to alternate black and white lines of varying widths. (The width of the lines is specified by *television line number,* which is the number of lines that fit into a dimension equal to the picture height.) The response of a system as a function of line number is called its *aperture response,* and the better the aperture response of an imaging system the sharper its pictures.

Figure 10–1 compares the aperture response of an NTSC television system (525 scanning lines, 4 MHz bandwidth, and full aperture correction) with typical projected 16mm and 35mm film images (after Schade). The aperture response of film falls off continuously as the line number increases

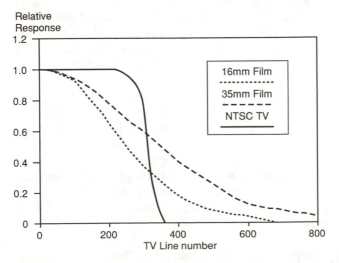

■ **Figure 10–1** The aperture response of film and television systems.

(that is, as the lines become more closely spaced), but measurable response (and hence resolution) continues to a very high line number.

The aperture response of a television system cuts off rather sharply at a point determined by the bandwidth in the horizontal direction and by the number of scanning lines in the vertical. With modern camera pickup tubes and the use of electronic signal processing techniques, such as aperture correction, the response of television systems at line numbers below their limiting resolution can be maintained near unity. As a result, the edge sharpness of television pictures can equal or exceed that of photographs even though the limiting resolution is significantly lower. In the systems shown in Figure 10–1, the perceived definition of the 16mm film images would be about the same as that of NTSC television. The establishment of this equivalence was very important in deriving requirements for bandwidth and scanning lines in HDTV systems.

The Viewing Ratio   The capability of the human eye to perceive fine detail and the sharpness of edges plays a key role in the design of HDTV systems. Schade showed that at a viewing ratio of 4 (the ratio of the viewing distance to the length of the picture diagonal with a 4 to 3 aspect ratio) or at a viewing distance of approximately seven times the picture's height the eye is unable to perceive picture detail that is finer than about 320 lines in the vertical dimension. (The exact ratio depends upon the brightness and contrast ratio of the picture, the signal-to-noise ratio, and the visual acuity of the observer.) Since this degree of detail can be transmitted by a standard NTSC system, *HDTV pictures will appear no better than NTSC pictures when viewed from a distance of seven times picture height or greater.* The converse of this is that NTSC pictures will appear fuzzier at closer viewing distances.

Since many television viewers now sit at a distance from the receiver of more than seven times the height of current picture tubes, larger sets are needed to achieve the benefits of HDTV. The converse is that HDTV is almost a must (or at least highly desirable) for anyone with a large-screen set.

This limitation of the human eye is recognized in some of the performance specifications that have been proposed for HDTV systems. They commonly require that pictures can be viewed without perceptible loss of definition at distances as short as three times picture height.

As noted earlier, the definition of pictures produced by a well-designed NTSC television system is approximately equal to that of 16mm film images. An estimate of the scanning line and bandwidth requirements of HDTV can be made, therefore, by comparing the areas of frames of 16mm and 35mm film expanded to a wide-screen format.

Scanning Line Requirements   The height of a 35mm film frame is approximately twice that of a 16mm frame. This leads to the conclusion that

HDTV should have about twice as many scanning lines as NTSC. Most of the HDTV systems that have been proposed meet this requirement either with exactly twice the NTSC standards—1,050 lines—or with slightly more—1,125 lines as proposed by NHK or 1,250 as proposed in the European Eureka system.

Most of these systems propose *interlaced scanning* in which half the lines are scanned in each field as with NTSC (see Chapter 4). A few proposals specify *progressive scanning* in which all lines are scanned in every field. This eliminates certain minor spurious effects that result from interlacing, but the bandwidth requirements are doubled and it is likely that progressive scanning will be used only in production systems where there is no rigid limitation on available bandwidth.

Bandwidth Requirements   The area of a frame of standard 35mm film is approximately four times as great as that of 16mm film and this is increased further by one-third in the 16:9 format. For HDTV, this would require increasing the 4-MHz NTSC bandwidth by a multiplier of 4×1.3, or to 21 MHz. This was approximately the bandwidth used by NHK in its early demonstrations.

A video bandwidth of 21 MHz would require nearly 30 MHz of the radio frequency spectrum, even with vestigial sideband AM, the most efficient modulation method in terms of spectrum use. With FM as used in satellite transmissions, the spectrum requirements would be even greater. Given the chronic shortage of spectrum space, much of the HDTV research has been directed toward the development of systems that will approximate the performance of a 21-MHz video signal with a substantially lower bandwidth. Efforts to reduce the bandwidth requirements take advantage of two phenomena.

The first phenomenon is the large amount of redundancy in a video signal. Most of the information in a video signal is repeated, frame after frame, and the only new information is that caused by motion in the picture. The result is that only a portion of the video spectrum is used; a fine-grain examination of it shows peaks of energy separated by the frame frequency. There are small gaps between these peaks,[6] and techniques for utilizing these gaps for bandwidth reduction have been developed. A related technique is to portray only the static information in the picture in high definition and to show moving objects in lower definition.

The other phenomenon is the reduced ability of the eye to perceive sharp

---

[6] These gaps in the video spectrum were recognized early in the development of television technology: Pierre Mertz and Frank Gray, "A Theory of Scanning and Its Relationship to the Characteristics of the Transmitted Signal in Telephotography and Television," *Bell System Technical Journal*, Vol 13, July 1934, 464–515; and Pierre Mertz, "Television—the Scanning Process," *Proc. IRE*, October 1942, 529–537.

edges and fine detail on the periphery of vision. Given this characteristic, the picture definition on the sides of a wide screen can be lower than at the center, and with a consequent reduction in bandwidth requirements.

Efforts to Establish HDTV Broadcast Standards

The development of HDTV technology was accompanied by an equally intense effort to establish standards for broadcasting HDTV programs. The FCC, somewhat confused and overwhelmed by the plethora of HDTV systems being proposed, took two steps looking toward the eventual establishment of HDTV broadcast standards. In July 1987, it established the joint FCC–Industry Advanced TV Advisory Committee, and on September 1, 1988, it ruled that any system approved for broadcasting in the United States would have to be compatible with the NTSC system. The latter requirement meant that existing receivers must be able to receive HDTV broadcasts (although of course not with HDTV technical quality).

The FCC–Industry Advanced TV Advisory Committee   The assignment of the joint FCC–Industry Advanced TV Advisory Committee was to study system proposals, sponsor field tests, and make standardization recommendations to the FCC. The candidate systems were not to be limited to full HDTV (35mm quality) but could include those that were superior to NTSC but below HDTV standards. These systems were described as extended definition television (EDTV). The generic term covering both EDTV and HDTV is advanced television (ATV). By December 31, 1988, twenty systems had been submitted to the committee for field-testing.

The Compatibility Requirement   The FCC policy decision requiring HDTV compatibility simplified the selection problem. An important effect of this ruling was to eliminate the Japanese MUSE-E system. The decision was interpreted by some as being anti-Japanese, but the FCC denied this, pointing out that other MUSE variants (MUSE-E and its variants are described later in this chapter) were left in the race.

The FCC further ruled that there would be no reallocation of channels for HDTV; broadcasters would have to use existing VHF and UHF channels. If an HDTV system required more than the 6 MHz of spectrum space included in a single channel, the spectrum requirements would have to be met by using a *primary* channel and part or all of an *augmentation* channel. The augmentation channel would not necessarily be contiguous to the primary channel and would probably be in the UHF spectrum. Anticipating this eventuality, the Commission put a freeze on UHF applications pending a final decision on HDTV.

In an equally important decision, the commission stated that it would not establish standards for nonbroadcast transmission and distribution media such as satellites, cable, and VCRs.

In spite of the simplification implicit in the compatibility requirement, the commission is faced with an awesome task in selecting a broadcast standard for HDTV. Past experience indicates that several years are likely to elapse before a decision becomes final.

ATV Systems Proposed to the TV Advisory Committee The various ATV systems proposed to the FCC advisory committee by December 31, 1988, are shown in the following table. It includes a somewhat arbitrary

■ ATV SYSTEMS PROPOSED TO THE TV ADVISORY COMMITTEE

Proponent	Output Signal
Single-Channel, Enhanced-Definition, NTSC-Compatible Systems	
BTA/Japan	NTSC/EDTV
Del Rey	NTSC/EDTV
Faroudja	NTSC/EDTV
High Resolution Science	Modified NTSC
MIT	NTSC/EDTV
NHK/MUSE-6	NTSC/EDTV
Production Service	NTSC/EDTV
SRI/ACTV-1*	NTSC/EDTV
Single-Channel, Noncompatible Systems	
MIT	—
NHK Narrow MUSE**	EDTV
Zenith**	EDTV
NTSC Channel Plus One-Half Augmentation Channel	
NHK MUSE-9	NTSC/EDTV
Philips	NTSC/EDTV
NTSC Channel Plus Shared Augmentation Channel	
NYIT (Glenn)	NTSC/EDTV
NTSC Channel Plus Full Augmentation Channel	
Osborne	NTSC/HDTV
Philips	NTSC/HDTV
SRI/ACTV-II	NTSC/HDTV
Satellite Transmission Systems	
NHK MUSE-E	HDTV
Philips	NTSC/HDTV
Scientific Atlanta	NTSC/HDTV

* The David Sarnoff Research Laboratories of the Stanford Research Institute (formerly RCA Laboratories). The acronym is advanced compatible television.

** Achieves compatibility by simulcasting on a second, 6-MHz channel.

classification of HDTV systems, which have the potential of meeting the 35mm film criterion, and EDTV systems, whose performance may be somewhat lower. All the compatible systems can be received on NTSC sets, but special receivers are required to achieve enhanced or high definition and the large aspect ratio.

Standards for TV Production

In addition to the need to establish standards for HDTV *broadcasting*, it would be desirable to establish HDTV standards for the *production* of television programs. In this process, program segments are shot and recorded in HDTV, the segments are edited while still in the HDTV format, and a final master tape is prepared. A standards converter is then used to translate this tape to an NTSC, PAL, or HDTV format for broadcasting. The use of HDTV rather than NTSC in the production process results in superior technical quality, even though the recording is later converted to broadcast standards. Further improvement can be achieved by carrying out the production process in the digital format and with the red, green, and blue signals or their derivatives. With a combination of these techniques, it is now possible to produce picture quality that is competitive with 35mm film, which is still the most common production format. The FCC has no jurisdiction over production standards, and agreement must result from an industry consensus.

Strenuous efforts have been made by international standardizing bodies to reach agreement on standards for HDTV production. The limitations on bandwidth are much broader, and the objective has been to adopt a standard that would result in extremely high technical quality and that could be readily converted to the major broadcast standards.

Establishing an international standard has proved to be very difficult, and after several years of effort, it was still not accomplished in 1989. The major issue is the field rate: Should it be fifty or sixty fields per second? The Europeans favor fifty fields because this is the standard for their power sources and television systems. The Japanese and Americans want sixty fields. A more subtle issue is whether to use precisely the power source rate—50 or 60 Hz—or the broadcasting rate, which is a submultiple of the color subcarrier—59.94 Hz for NTSC.

In January 1988, the SMPTE, the Advanced Television Systems Committee, and the American National Standards Institute approved the standards proposed by the Japanese, 60 fields and 1,125 lines, by a less than unanimous vote. This standard had previously been rejected by the CCIR even though many European broadcasters favored it, but because of Japan's leadership in HDTV production, it is a de facto world standard, at least for the present.

The MUSE and ACTV Systems

A description of the many HDTV systems is beyond the scope of this narrative, but the MUSE and ACTV families illustrate many of the technologies that have been used to achieve HDTV performance with a minimum of bandwidth requirements. They are also leading contenders for widespread adoption by the industry.

The MUSE Family   MUSE (Multiple Sub-Nyquist Sampling Encoding) is a bandwidth compression technique that allows HDTV signals to be transmitted in narrower transmission channels. Its underlying principle is to spread the transmission of the stationary portions of a scene over several frames. By slowing the transmission rate, the bandwidth requirements are reduced. The image of the moving objects is transmitted more rapidly with some loss of definition.

MUSE uses dot interlaced sampling, not unlike the original RCA dot sequential color system. The original sampling rate is 48.6 MHz. This is more than adequate for the transmission of HDTV pictures, but it requires a bandwidth of 24.3 MHz. To reduce the bandwidth, this signal is sub-sampled at a submultiple rate.

For MUSE-E, the highest definition member of the MUSE family, the subsampling rates are 24.3 MHz (half the original rate) for the stationary portions of the picture and 16.2 MHz (one-third the original rate) for the moving portions. By slowing the transmission of the stationary signal, both components can be transmitted in an 8.1-MHz band.

By using different subsampling rates, the trade-off between bandwidth and definition can be controlled. Three alternatives have been proposed to the TV Advisory Committee:

- MUSE-9 requires a single NTSC channel plus half an augmentation channel.
- Narrow MUSE requires a single NTSC channel but is incompatible.
- MUSE-6 can be transmitted in a single NTSC channel and is compatible.

The ACTV Family   The complete embodiment of the ACTV system uses a primary channel and an augmentation channel. The primary channel, called ACTV-1, can be viewed with either a standard NTSC receiver or an ATV receiver. The pictures on the ATV receiver have 1,050 lines and a 15:9 aspect ratio, and they are significantly sharper than those on NTSC. The complete ACTV system, ACTV-2, adds an augmentation channel to increase the sharpness and produce a true HDTV picture.

ACTV-1 uses sophisticated signal processing to achieve EDTV performance in a 6-MHz broadcast channel. It can accept as its input a 525-line progressive scan or a 1,050- or 1,125-line interlaced scan signal. This signal is converted to the digital format for processing and is transformed into four components:

1. A standard 525-line NTSC signal containing the information in the central 4:3 portion of the picture plus an additional signal containing the low-frequency information from the side panels. The latter is time compressed so that it appears as narrow bands that are hidden by the side bezels of home receivers.
2. A low-amplitude signal that contains the high-frequency components of the side panels.
3. A low-amplitude signal that contains the high-frequency (5.0- to 6.2-MHz) components of the center portion of the picture.
4. A "helper" signal that contains vertical detail used to reconstruct the missing lines in the 525-line NTSC picture and produce 1,050 lines in the ATV display.

Signals 2, 3, and 4 are added to the transmitted signal by quadrant modulation, so they are virtually invisible on an NTSC receiver. The ATV receiver processes them and delivers a 6-MHz, 1,050-line, 16:9 signal to the picture tube. Frequency information above 6 MHz is transmitted on the augmentation channel and can be added to the primary signal to provide HDTV performance.

The Commercial Status of HDTV

HDTV is currently in an engineering rather than a commercial phase. It has been used to a limited extent for television production, but its progress in this industry has been slow. The high cost of the production equipment and the reluctance of the film industry to adopt television production techniques have been major deterrents. Although it has advantages for the production of programs in the NTSC format, it is possible that wider acceptance will await the introduction of HDTV in broadcasting and/or VCRs and cable. The use of HDTV in the standard broadcast bands cannot come until standards are approved by the FCC or foreign regulatory agencies. This will probably not occur for a number of years.

NHK appears to have the most fully developed plans for the introduction of HDTV outside the standard broadcast bands. Test transmissions of the MUSE-E system began in April 1989, and NHK announced that it will begin broadcasting MUSE-E from a direct broadcast satellite in 1990. There is also a great deal of interest in HDTV VCRs, but to date there are no recorders or receivers available in the commercial marketplace.

■ DIGITAL SYSTEMS

Transmission by digital signals—that is, by a series of coded pulses—was the original mode for long-distance communications. Smoke signals and the Morse code are examples. Alexander Graham Bell's demonstration that the sound vibrations created by the human voice could be converted to an analog, or continuously varying, electrical signal was one of the break-throughs that made telephony possible (see Chapter 1). Analog transmission was then considered to be a more advanced technology than digital.

During the past five decades, enormous progress has been made in digital technology. Much of this has been motivated by the needs of the computer industry. As a result, the perceptions of the analog and digital formats are reversed. Digital technology is now considered advanced, and analog is believed by many to be obsolete.

Digital technology has had a great fascination for the engineering profession. It is technically elegant, and it is can be subjected to sophisticated mathematical analysis. These properties, together with its very real tangible advantages, have made it a favorite of engineers to the extent that it is sometimes oversold. It is variously described as "the wave of the future" and "inevitable," even in situations where it is impractical, and suggestions that there may be a place for the analog format in future communications systems are often dismissed a little contemptuously.

In fact, there is a place for both formats in ratio and television. Our ears and eyes respond to analog stimuli. Hence the output signal from a microphone or television camera is in analog form. At the receiving end, the input to the loudspeaker or picture tube must be an analog signal. The bandwidth requirements for digital transmission are enormous (as described later in this chapter) and often impossible or impractical to achieve. In simple systems, the cost of converting and reconverting the signal from analog to digital cannot be justified.

There are, however, applications in which the digital format plays an important or indispensable role in the transmission, storage, and processing of audio and video signals. The sophisticated signal processing required for HDTV would be difficult or impossible if it remained in analog form. As will be described shortly, the usefulness of the digital format has been greatly enhanced by advances in two other technologies—solid-state devices and fiber optics. The use of the digital format also makes many of the rapid developments of computer technology available to radio and television.

Analog and Digital Signals

The process of converting an analog signal to digital form (A/D conversion) is shown in Figure 10–2. The result of this conversion is a train of pulses or bits at rates of 6.75 to 17.7 MHz.

The Properties of Digital Signals

Digital signals have both advantages and disadvantages as compared with analog. Perhaps their greatest advantage is their relative immunity to degradation as they are subjected to the distortions of transmission and recording systems. When analog signals pass through a series of transmission or recording mediums, they are progressively degraded in signal-to-noise ratio, linearity, and possibly in bandwidth. The rate of degradation can be reduced by improving the performance of the medium, but it cannot be brought to zero. In contrast, digital signals suffer virtually no degradation, particularly if a technique known as parity checking is used, unless the quality of the medium is exceptionally poor. For example, as many as fifty generations of recordings and rerecordings have been made on digital videotape recorders with very little loss of quality. The fifth generation from even the best analog recorder would probably be unusable.

Another extremely useful property of digital signals is their ability to be stored temporarily in a solid-state memory. Once stored, they can easily be retrieved, modified, or erased. This ability is the heart of many signal-processing circuits. The importance of this property has been greatly enhanced by the dramatic reduction in the cost of solid-state memory during recent years.

The major disadvantage of digital signals is their enormous bandwidth requirements. These requirements make digital television impossible for broadcasting and usually impractical for point-to-point microwave or satellite circuits. Given the chronic shortage of radio spectrum space, it seems unlikely that digital transmission of video will become common on any radio medium in the foreseeable future.

Until recently, the recording of digital video signals on magnetic tape recorders was beyond the state of the art. This limitation has now been removed, although somewhat expensively, by improvements in the technology of tape and recording heads (see Chapter 6).

Similarly, fiber optics provides at least a partial solution to the bandwidth problem for digital transmission circuits.

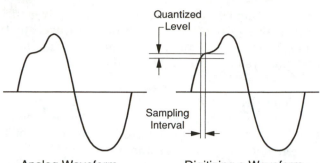

Analog Waveform   Digitizing a Waveform

■ **Figure 10–2** Digitizing an analog waveform. The drawing on the right shows the method for converting an analog signal to digital form (A/D conversion). The signal is quantized by defining discrete levels, typically 256, each of which represents a range of levels in the analog signal. Its amplitude is then sampled at regular intervals, and the quantized level is identified for each sample. Standard sampling rates are as follows:

Component signals
Y	13.5 MHz
R − Y	6.75 MHz
B − Y	6.75 MHz
NTSC signals	14.3 MHz
PAL signals	17.7 MHz

With an 8-bit code, the bit rate is eight times the sampling rate—216 Mbps for all three component signals, 114.4 Mbps for NTSC, and 141.6 Mbps for PAL.

Each quantized level is assigned a unique 8-bit pulse code, which is transmitted or recorded as a digital signal. The digital signal is reconverted to analog form (D/A conversion) by a circuit that recognizes the codes and regenerates the original analog waveform. A standard international sampling rate is 13.5 MHz, giving a bit rate of 108 Mbps and a nominal bandwidth requirement of 54 MHz.

Applications of Digital Technology in Radio and Television

Signal Transmission The intercity transmission of digital television signals requires the use of fiber optics, the only medium that has adequate bandwidth. All the major interstate communications carriers are installing extensive fiber-optic circuits, and they should become available for video transmission as well as for voice and data.

The use of digital transmission by the radio networks for program distribution by satellite was described in Chapter 8. Cable systems may ultimately convert to digital transmission with fiber optics. This is far in the future, however, partly because of the large investment in existing plants and partly because the cost of new all-fiber systems is not yet competitive with coaxial cable. (Fiber is now being used in some systems, but only in the trunks and the signals are analog.) Experience has shown, however, that the costs of new technologies invariably come down, and the time may come when all-fiber, all-digital systems will be the standard.

Video Recording   Digital video recorders are now available for professional use (see Chapter 6). They are more expensive than analog recorders, but their use is almost indispensable for complex production and editing processes in which multigeneration copies are required.

Audio Recording   Compact disks (CDs) record the audio signal in a digital format (see Chapter 3). At the present time, this is the ultimate in audio recording quality.

Digital audiotape recorders (DATs) using magnetic tape have been designed and produced, but their appearance in the marketplace has been blocked by the opposition of the record industry. The quality of rerecordings is so good that widespread piracy is feared. Although Sony ultimately won a court battle involving a somewhat similar situation with VCRs (see Chapter 9), the courts were divided, and the implications of their rulings were by no means clear. At the present time, the manufacturers of recorders and the recording industry are trying to negotiate a mutually satisfactory solution.

Signal Processing   The ability to store digital signals in a solid-state memory where they can be easily retrieved for manipulation and playback at a later time has made a wide variety of signal-processing circuits possible. The cost of some types of this circuitry has now been reduced sufficiently to make it economical for use in receivers. These circuits can improve most of the elements of picture quality, including the sharpness, signal-to-noise ratio, and colorimetry.

Standards Conversion   The ability to store digital signals is the basis for modern standards converters (for example, devices that can convert a 525-line, 60-field NTSC signal to a 625-line, 50-field PAL signal). Previous methods did not work well, and the results were marginal. The results from present-day converters are quite satisfactory, although some loss in quality is inherent.

Signal Security   Digital signals can be encoded by a variety of techniques to prevent unauthorized reception. This technology has been highly developed for national security purposes, and digital transmission is preferable when a very high degree of signal security is desired.

■ SOLID-STATE TECHNOLOGY

No single technology has had as profound an effect on the radio and television industries as solid-state physics. It is difficult today to visualize broadcasting if the designs of its production, storage, transmission, and receiving equipment were based on vacuum tubes. It surely would be a smaller, less dynamic industry, and many of the accomplishments we take for granted would be impossible.

The Transistor

The era of solid-state devices began in 1947 with the research of William Shockley and his colleagues at Bell Laboratories. He was a physicist engaged in a study of the properties of *semiconductors.* Under different electrical or physical conditions, these semiconductors can act either as conductors or insulators, and Shockley discovered a way to use this property to achieve amplification of electrical signals. The resulting device was named the transistor.

Shockley's discovery had a tremendous impact on the electronics industry. The transistor was functionally equivalent to the vacuum tube, but it had many potential advantages. It was much smaller, consumed less power, and, in theory at least, would never wear out. In addition, it was soon found that the transistor had other useful properties as well as its ability to amplify. It could function as a very fast acting electrical switch, and it could act as a storage device, or memory, for the pulses in a digital system. These properties were essential for computers, and they eventually found important applications in radio and television as digital technologies were adopted by these industries.

Transistors first came on the market in the early 1950s. Initially, they were limited in their frequency range and power-handling capability, and their first important commercial application was in portable radios, which were soon given the generic name transistors.

Transistor design made rapid progress during the 1950s, and the introduction of the first all-transistor television tape recorder by RCA in 1961 (see Chapter 6) was something of a milestone. It contained a wide variety of audio, video, radio frequency, servo, pulse, and control circuits, some of which operated at fairly high power and at high frequencies. With the transistors available at the time, the RCA recorder's design was barely possible, and it was a technical tour de force. It marked the beginning of the end of the vacuum tube era for broadcast station equipment.

Technical progress in the design of semiconductor components has continued at a rapid pace, and there are very few functions (for example, very high power transmitter amplifiers) that cannot be performed better with solid-state devices than with vacuum tubes because of their inherent advantages (described previously).

The Integrated Circuit

The designers of equipment that used transistors as switches or memory devices had a special problem. To perform the desired functions, it was necessary to use thousands, and eventually hundreds of thousands, of them. This led to the "tyranny of connections," as each transistor had to be connected to the others with a number of leads. Accomplishing this by hand-soldering, the traditional technique in the electronics industry, was time-consuming, costly, and only marginally reliable. The use of printed circuits for the connections helped somewhat, but an upper limit was reached in the number of transistors that could be used in a single circuit. This limit was far below the number needed to perform the desired functions.

A solution was found in the *monolithic integrated circuit*, which was invented almost simultaneously in 1958 and 1959 by Jack Kilby of Texas Instruments and Roberty Noyce of Fairchild Semiconductor.[7] They found that most of the elements of an electrical circuit, including transistors and conductors, could be produced on a small block of silicon (the chip), by proper treatment of the surface. The circuit layout is made on a series of masks with much larger dimensions (for example, 500 : 1) than those of the actual circuit produced on the chip. The mask images, each corresponding to a particular circuit element, such as the conductors, are transferred to the silicon by a photographic process, and each area is given the treatment required for that type of circuit element.

As techniques for the fabrication of integrated circuits developed, it became possible to increase the number of individual circuits geometrically. It is now possible to include 250,000 or more circuit elements on a single chip. This is called very large scale integration (VLSI). The intricacy of such a chip has been compared to a road map of the United States that shows all the country roads as well as the main highways.

The reduction in circuit costs has been equally dramatic. The manufacturing cost of a VLSI chip is now only a small fraction of a cent per circuit element. This has been of particular benefit to the computer industry with its requirements for large memories, but it also made practical many components of modern television systems that require large memory capacity. Examples are signal processors, graphics generators, and standards converters.

---

[7] For an interesting history of this invention, see T.R. Reid, *The Chip* (New York: Simon & Schuster, 1984).

Image Pickup Devices: CCDs

The television pickup tube was one of the last to succumb to solid-state technology. The performance of photoconductive tubes is so good that it is difficult to improve on them. Nevertheless, CCDs (see Chapter 5) are gradually eroding the market share of vacuum tubes.

CCDs began to be used in television film cameras in the late 1970s, and as their sensitivity has increased, they have been used increasingly in cameras for live pickup. They are widely used in portable television cameras and camcorders for home use. Their small size and low power consumption make them particularly suited for this application. They also are used in professional camcorders, and it is probably only a matter of time before they supersede tubes in studio cameras. Sony made an important step in that direction when it introduced a CCD studio camera at the 1989 NAB convention.

■ LASERS

The technology of lasers (a rather strained acronym meaning light amplification by stimulated emission of radiation) is a postwar development. The first lasers were built in 1960, and scientific papers describing their generation, properties, and applications began appearing soon afterward. Since then their use in commercial, medical, and military applications has proliferated, and it will grow even faster in the future.

The Technology of Lasers

The most important property of laser radiation is its *coherence*. This quality can be visualized by comparing the wake of a small boat proceeding slowly through calm waters to the waves produced by a strong wind. The waves in the wake are smooth and well defined—that is, coherent—while wind-produced waves tend to be jumbled and irregular. The wave motion in ordinary light more nearly resembles wind-produced waves.

The coherence of laser radiation gives it a number of interesting properties. The most important of these in radio and television applications are its abilities to be focused on an extremely tiny spot and to be guided by a thin optical fiber.

Laser radiation can be produced by a variety of sources, including gases, liquids, and solid-state semiconductors. The spectral range of laser types extends from visible to far infrared. Most laser applications in radio and television use solid-state sources operating in the infrared region of the spectrum.

Applications of Lasers
in Radio and Television

The two principal applications of lasers in radio and television are audio-disk and videodisk recording and fiber optics. The use of lasers in audio recorders, or CDs, was described in Chapter 3. Their use in video recording was described in Chapter 9. Their use in fiber optics is described in the next section. Laser technology is relatively new, and important new applications will no doubt be developed in time.

■ FIBER OPTICS

History

Like lasers, fiber optics was developed in the postwar years. The two technologies are closely related because the useful applications of fiber optics depend on the availability of laser radiation sources.

The principle of fiber-optic transmission—the guiding of radiant waves through a thin glass strand by internal reflections from its surface—was first predicted on a theoretical basis in 1966. By 1970, glass with sufficiently low losses became available for experimental systems, and the favorable properties of lasers as a radiation source had been discovered.

During the next decade, there was rapid progress in the technology of both glass and lasers, and commercial applications of fiber optics began in 1979. This progress has continued, and as a result of fiber optics' wide bandwidth and freedom from electrical interference, it has become one of the most important transmission media of the communications industry.

Fiber-Optic Technology

Single Mode and Multimode   There are two classes of fiber-optic systems, single mode and multimode (Figure 10–3). Single-mode fiber has lower attenuation than multimode fiber, as low as 0.3 dB per mile versus 0.6 to 1.5 dB for the wavelengths commonly used. It also has a greater bandwidth. Bandwidths of 25 GHz can be obtained with single mode, while *modal dispersion* resulting from the difference in the time delay of rays following different paths typically limits the bandwidth of multimode systems to 1 GHz (still a very impressive figure).

Single-mode fiber was introduced later than multimode because of the difficulty of manufacturing the incredibly thin core and the need to develop laser generators in the far infrared region. These problems have now been solved, and single-mode fiber is widely used in long-haul wideband cir-

Multimode

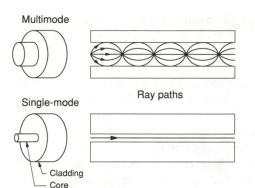

Ray paths

Single-mode

Cladding

Core

■ **Figure 10–3** Single-mode and multimode fiber. In each case, the optical conductor consists of a thin core surrounded by a cladding of another type of glass. The cladding has a lower index of refraction, so radiation striking it at a glancing angle is reflected. Multimoding occurs when the laser signal is transmitted over several different paths, with different lengths and time delays. Multimoding is undesirable and can be prevented by the use of an extremely thin core and laser radiation in the far-infrared region with a wavelength greater than 1.1 micrometers. The core diameter for single-mode fibers is typically 7 to 10 micrometers; for multimode the diameter ranges from 50 to 85 micrometers.

cuits. It is difficult to install and repair, however, and multimode fiber is still used in short-haul circuits with more modest bandwidth requirements.

Digital and Analog Transmission   Digital transmission accomplished by pulsing the radiation source is the natural mode for fiber optics. Its wide bandwidth is ideally suited to digital signals, and it has created the means for a much wider use of the digital format in communications systems, including radio and television. It is used exclusively for long-haul circuits.

The advantages of the digital format are outweighed, however, by the cost and complexity of A/D and D/A circuits in some applications, particularly those involving shorter hauls. In these cases, analog modulation can be used. The analog signal to be transmitted modulates a radio frequency subcarrier, which in turn intensity-modulates the laser generator. It is advantageous to use FM because of the nonlinearity of the modulation process for lasers.

Applications of Fiber Optics in Radio and Television

Radio and television will benefit from the rapid growth in the use of fiber optics by the major communications carriers in their intercity circuits. A properly designed and maintained fiber-optic system is capable of extremely high performance, particularly with digital transmission, and this quality will become available for networking and other broadcast services as well as for the telephone.

The introduction of fiber optics into cable systems will provide an even greater benefit, since it will eliminate many of the problems of wired systems. Many major MSOs such as TCI and ATC have announced their intentions to make a major commitment to fiber optics, and the installation of hybrid fiber–metallic cable systems has begun (see Chapter 7).

Fiber will come to cable on an evolutionary basis, partly because of the industry's large fixed investment in wired systems, partly because of the escalating costs of installing new systems, and partly because of the need for further engineering development, particularly in reducing the costs of fiber systems. This process is well under way, and the next decade should witness a major shift in the cable industry from wire to fiber.

■ A LOOK AHEAD

The progress of radio and television technology has been so rapid during the past three decades that it is difficult to foresee an equally high rate in the future. But the odds are very high that the rate will continue and that progress will continue in directions that are only dimly visible at this time. Television is a dynamic and exciting industry, and it is attracting some of the world's leading scientific and engineering talent. It is not the purpose of this book to forecast where this will lead the industry, but it seems certain that the individuals and organizations that have the responsibility for producing and distributing news, entertainment, and educational programs will be supplied with ever more effective tools.

■■ GLOSSARY

ACTV: Advanced Compatible Television. A high-definition television system developed by the Stanford Research Institute and NBC.

Additive color systems: Imaging systems in which the hue is determined by combining colored light components on a dark background. Color television is an additive system. *See* Subtractive color systems.

Advanced TV Advisory Committee: An ad hoc FCC–industry committee formed in 1987 to study and recommend HDTV standards.

AFCCE: The Association of Federal Communication Consulting Engineers. An association of radio and television engineering consultants who practice before the FCC.

Alpha wrap: A helical scan format in which the tape path goes completely around the drum, giving it a cross section suggesting the Greek letter alpha. *See* Omega wrap and Half-wrap.

Alphanumeric symbols: A generic term specifying characters that may be either numbers or letters.

AM: Amplitude modulation.

Amplitude modulation (AM): The process of altering the amplitude of a radio frequency signal in accordance with the amplitude of the modulating signal.

AMST: Association of Maximum Service Television Telecasters, Inc. A trade association representing the interests of larger television stations.

Analog signal: A representation of a continuously varying quantity, such as a sound pressure wave. *See* Digital signal.

ARRL: The American Radio Relay League. An association of radio amateurs.

Aspect ratio: The width-to-height ratio of a television or film picture.

ATV: Advanced television. A generic term for all types of television systems with better performance than standard broadcast; includes EDTV and HDTV.

Augmentation channel: A second television channel proposed to be assigned to broadcast stations to transmit the additional information required for HDTV.

Basic services: Services provided by cable television systems for a minimum basic monthly fee. *See* Premium services.

Beam tilt: Directing the main beam of a high-gain UHF television transmitting antenna slightly downward to provide a stronger signal along the surface of the earth.

Beta: A videocassette tape format developed by Sony using half-inch tape. *See* VHS.

Bias: In electron tube and transistor circuits, a fixed voltage applied to control elements; in magnetic audio recording, a high-frequency waveform added to the audio signal to improve the quality of the recording.

Blanking pulses: Pulses transmitted as part of the television signal to turn off the picture tube scanning beam during the retrace time for the vertical and horizontal scans.

Broadcast Technology Society of the IEEE: *See* IEEE.

Burst: Eight cycles of the color subcarrier superimposed on the horizontal blanking signal behind the synchronizing pulse to synchronize the hues transmitted from the camera with the hues reproduced on the receiver.

C band: A portion of the electromagnetic frequency spectrum used by microwave and satellite communications systems. C-band satellite uplinks operate in the range 5.925 to 6.425 GHz, and downlinks operate in the range 3.700 to 4.200 GHz.

C-Quam: A method of broadcasting stereo on AM radio developed by Motorola, Inc.

Cable Act of 1984: A congressional act for the regulation of the cable television industry.

Cable-ready receiver: A receiver equipped to receive both cable television and broadcast channels without the use of an external set-top converter.

Camcorder: A combined television camera and recorder, widely used for ENG.

Canadian Association of Broadcasters (CAB): The trade association of Canadian broadcasters. Its address is Box 627, Station B, Ottawa, Ontario K1P5S2, Canada.

Carrier: A radio frequency current, voltage, or wave that is modulated in frequency, amplitude, or phase with the signal voltage.

Cathode ray tube: An electron tube equipped with an electron gun that emits a narrow electron beam. In television application, this beam strikes fluorescent screen on the face of the tube.

CATV: Cable television.

CCD: Charge-coupled device.

CCIR: Comité Consultatif International de Radio Communication. An international standards body for all forms of radio communication.

CD: Compact disk.

Charge-coupled device (CCD): A solid-state photosensitive element used as the pickup device for television cameras.

Clear channels: AM broadcast frequencies on which only one high-power station or one high-power and a limited number of lower power stations are allowed to operate. *See* Regional channels and Local channels.

Coaxial cable: A cable for the transmission of high-frequency signals. It consists of an inner conductor surrounded by an insulating material and a cylindrical outer conductor.

Color subcarrier: A high-frequency signal superimposed on the monochrome signal to transmit the color information. The intensity of the color is indicated by the amplitude of the subcarrier, and its hue is indicated by the phase.

Communications Act of 1934: Superseded the Radio Act of 1927. It gave the government the authority to regulate all forms of electrical communications, wired and wireless. The Federal Communications Commission (FCC) was established to administer its provisions.

Communications Satellite Act of 1962: An act that established the mechanism for U.S. participation in international communications. *See* Intelsat and Comsat.

Compact Disk (CD): A recording medium for audio or video in which the signal (digital for audio, analog for video) is recorded on a small disk with a laser beam. The playback device also uses laser technology.

Component recording (or transmission):   The recording (or transmission) of the brightness and color components of a color signal separately rather than by a composite signal such as NTSC or PAL.

Comsat: A private but quasi-public company established by the Communications Satellite Act of 1962 to be the chosen instrument for U.S. participation in Intelsat, the international satellite organization.

Copyright Royalty Tribunal (CRT):   An organization authorized by Congress to collect blanket copyright fees from cable television systems and distribute the proceeds to copyright owners.

CRT: Copyright Royalty Tribunal; also cathode ray tube.

D layer: The lowest ionized layer in the ionosphere.

DATS: Digital Audio Transmission Service. A system using the digital transmission mode for distributing radio programs by satellite.

DBS: Direct broadcast by satellite. Broadcasting directly to individual homes by high-power satellites. DBS uplinks have been allocated the frequency range 17.3 to 17.8 GHz, while downlinks use the range 12.2 to 12.7 GHz.

Decibel (dB):   A commonly used unit for specifying the ratio between electrical or acoustic power levels. It is equal to ten times the common logarithm of the power ratio. In some cases, a reference power is defined as "zero dB," and actual powers are specified in decibels against this standard.

Definition: Narrowly, a television or film image's sharpness or appearance of being in focus. Broadly, a term describing overall picture quality.

Deintermixture: A policy for the assignment of television station frequencies, advocated by some but not adopted by the FCC, that would have avoided assigning UHF and VHF channels to the same city.

Digital signal: A signal format in which the information is transmitted by a series of coded pulses. *See* Analog signal.

Directional antenna: An array of antennas designed to concentrate the radiation of energy in certain directions and suppress it in others. For AM, directional antennas are used to avoid excessive interference with other stations operating on the same frequency. For FM and television, they are used to concentrate the radiated energy along the earth's surface.

Discriminator: A circuit in an FM receiver that converts frequency variations in the incoming signal to voltage variations, thus demodulating the FM waveform.

Downlink: The radio frequency circuit from a satellite to an earth station.

Drop: The connection from the distribution cable to a cable television subscriber's home.

Duopoly: The ownership of two or more broadcasting stations with overlapping service areas by a single individual or corporation.

E layer: The middle ionized layer in the ionosphere.

Earth station: A terrestrial installation for transmitting and/or receiving signals to and/or from a satellite.

EDTV: Extended-definition television. Television systems with better quality than standard broadcast but not meeting HDTV standards.

Effective radiated power (ERP):   A measurement of the intensity of radiation from a directional transmitting antenna in a single direction. It is equal to the power that would be required from a reference antenna to produce the same radiation intensity. For FM and television stations, the reference antenna is usually a half-wave dipole. For satellites, it is an isotropic radiator.

EIA: Electronic Industries Association.

Electromagnetic radiation: The transmission of energy by a wave consisting of periodic variations in electric and magnetic fields. Its behavior is described by Maxwell's equations.

Electromagnetic spectrum: The range of wavelengths or frequencies of electromagnetic radiation. It includes radio waves, infrared radiation, visible light, ultraviolet light, X rays, and cosmic rays in descending order of wavelength.

Electron gun: A structure mounted in front of an electron-emitting cathode that focuses the electrons into a narrow beam.

Electronic Industries Association (EIA):   A trade association of electronics manufacturers; formerly the Radio Manufacturers Association. Its address is 2001 Eye Street NW, Washington, DC 20006.

ENG: Electronic news gathering: The recording of news events on magnetic tape with a television camera at the scene.

ERP: Effective radiated power.

Exciter: The component of an FM transmitter that generates the original FM signal.

F layer: The highest ionized layer(s) in the ionosphere.

Federal Communications Commission (FCC):   An independent U.S. agency formed by the Communications Act of 1934 to administer its provisions.

Fiber optics: A medium for information transmission consisting of a small glass fiber that conducts visible or infrared light along its length as the result of internal reflections from its inner surface.

Field: An image that includes the lines in one vertical scan, one-half the total in a frame in systems using interlaced scanning.

Flying spot scanner: A device primarily used to generate television signals from film. The film is scanned with a bright spot of light, and the light transmitted through the film is converted to an electrical signal by a phototube.

FM: Frequency modulation.

Footprint: The radiation pattern on the earth resulting from a satellite transmission.

Format: (1) The predominant program content of a radio station, such as Classical or News. (2) The parameters of a recording system.

Frame: A complete image in a television system, which includes all the scanning lines. *See* Field.

Frequency modulation (FM):   The process of altering the frequency of a radio frequency signal in accordance with the amplitude of the modulating signal.

FSS: Fixed service satellites. A class of satellites, defined by FCC Rules, that can be used for the transmission of voice, data, video, and audio signals between fixed locations.

Fukinuki holes: Small gaps in the energy spectrum of a television signal separated by the frame frequency (in the United States, 30 Hz).

Geosynchronous satellites: Communications satellites located 22,300 miles above the equator and rotating with the earth so that they appear to be stationary when observed from the earth.

Ground wave: A vertically polarized medium or long wave that is guided along the earth's surface as the result of its electrical conductivity.

Half-wrap: The helical scan format used for videocassette recorders. It is a form of omega wrap in which the tape passes around one-half the drum.

HDTV: High-definition television.

Headend: The installation that receives the signals for a cable system and processes them for distribution. Typically it includes high-gain, off-air receiving antennas, a satellite receive-only earth station, and microwave receiving antennas together with the frequency translators, amplifiers, and multiplexes required to generate the cable television signal.

Helical scan: The video recording format now almost universally used, both in professional and consumer recorders. The tape is wrapped in a helix around a revolving drum on which the recording heads are mounted. The combined motions of the tape and drum produce recording tracks that are slanted across the tape. Also called slant track.

Hertz (Hz): The basic unit of frequency. One hertz equals one cycle per second.

High-definition television (HDTV): Television systems producing better picture quality and usually a wider aspect ratio than standard broadcast television. It is generally defined as television with the same picture quality as 35mm theatrical film.

Homesat: Small TVRO used for home reception of satellite signals.

Hybrid satellites: Satellites that have both C-band and Ku-band communications facilities.

IEEE: Institute of Electrical and Electronics Engineers.

Institute of Electrical and Electronics Engineers (IEEE): A professional society of electrical and electronics engineers. It was formed after World War II as the result of a merger between the American Institute of Electrical Engineers and the Institute of Radio Engineers. It has a number of specialized societies, including the Broadcast Technology Society. Its address is 345 E. 47th Street, New York, NY 10017.

Integrated circuit (IC): A complete electrical circuit combining the functions of all or most of the types of conventional discrete components—transistors, resistors, capacitors, and inductors—fabricated as a unit on a single substrate. The complexity of ICs has increased enormously as the technology has progressed. The term *VLSI* (very large scale integration) is used to describe the largest ICs.

Intelsat: An international consortium that owns and operates communications satellites used for international communications. *See* Comsat.

Interlaced scanning: A scanning pattern in which alternate horizontal lines, odd and even, are scanned on successive vertical scans.

International Telecommunication Union (ITU): The telecommunications arm of the United Nations.

Ionosphere: A series of ionized layers of the earth's atmosphere varying in height from 50 to 90 kilometers. The ionization results from solar radiation, and the height and density of the layers varies with latitude, the time of day, the season, and the sunspot cycle.

Isotropic radiator: A hypothetical antenna that radiates power equally in all directions. It is used for the mathematical analysis and specification of antenna systems.

ITU: International Telecommunication Union.

Kennelly–Heaviside layer: The ionosphere.

Kinescope: A cathode ray tube (CRT) used for the display of television or other electronic images.

Klystron: A radio frequency power amplifier tube widely used at UHF frequencies and above.

Ku band: A portion of the electromagnetic frequency spectrum used by microwave and satellite communications systems. Ku-band satellite uplinks operate in the range 14.0 to 14.5 GHz, and downlinks operate in the range 11.7 to 16.2 GHz.

Laser: An acronym for light amplification by stimulated emission of radiation. Electromagnetic radiation in the visible, infrared, or near-infrared regions of the spectrum that is coherent—that is, the wave motion is regular rather than jumbled as in ordinary light.

Laservision: A system of video recording on disks using a laser beam.

Launch vehicle: A device for placing a satellite in orbit. It may be an expendable rocket or the reusable space shuttle.

Local channels: AM broadcast frequencies on which a large number of low-power stations are allowed to operate, each serving a single community. *See* Clear channels and Regional channels.

Local spots: Brief advertising messages intended to be broadcast on a single station. *See* National spots.

Long waves: Low-frequency (10 to 300 kHz) electromagnetic radiation with a very long wavelength (30,000 to 1,000 meters).

Longitudinal recording: A system of recording audio or video signals on tape in which the recorded tracks are parallel to the tape. *See* Transverse recording.

Look angle: The elevation and azimuth of the path from the earth to a satellite as seen from the earth.

LP: Long-playing record, generally operating at $33\frac{1}{3}$ rpm.

Master control: The switching system in a broadcast station that selects the program source (studio, network, film, etc.) to go on the air.

Maxwell's equations: The equations expressing the relationship and behavior of the electric and magnetic fields in electromagnetic radiation.

Medium waves: Electromagnetic radiation in the frequency range 300kHz to 3MHz (wavelength 1,000 to 100 meters). Standard AM broadcasting, 550 to 1,600 kHz, is in the middle of the medium-wave band.

Microphone, pressure: A microphone that responds to the variations in air pressure of a sound wave.

Microphone, velocity: A microphone that responds to the variations in air velocity of a sound wave.

MII: A half-inch helical scan videocassette format developed by Panasonic for professional broadcast use.

MITI: The Ministry of International Trade and Industry of the Japanese government.

Modulation: The process of altering the characteristics, usually frequency or amplitude, of a radio frequency signal (the carrier) to convey intelligence. *See* Amplitude modulation and Frequency modulation.

MSO: Multiple system operator.

Multipath distortion: The distortion of an FM broadcast signal when the transmission arrives at a receiving location over multiple paths (for example, a direct and several reflected components).

Multiple system operator (MSO): An organization owning many cable television systems.

Multiplexing: Adding two or more signals for transmission as a single signal.

MUSE: Multiple Sub-Nyquist Sampling Encoding. A high-definition television system developed by NHK and Japanese manufacturers.

Must carry: An FCC rule requiring cable television systems to carry the signals of most or all local stations. It was declared unconstitutional by the courts.

NAB: National Association of Broadcasters.

NASA: National Aeronautics and Space Administration. The federal agency charged with the responsibility for developing the civilian space program.

National Association of Broadcasters (NAB): The major trade association of the U.S. broadcasting industry. Its annual convention has traditionally been the primary vehicle used by manufacturers of broadcast equipment to introduce new products. Its address is 1771 N Street NW, Washington, DC 20036.

National Cable Television Association (NCTA): The leading trade association of the cable television industry. It was founded in 1952, and its annual convention is a focus of the industry's business, regulatory, and technical activities. Its address is 1724 Massachusetts Avenue NW, Washington, DC 20036.

National Radio Systems Committee (NRSC): An ad hoc committee formed by the NAB and EIA to develop methods for improving the quality of reception in the AM radio band.

National spots: Brief advertising messages intended to be broadcast on a regional or national basis over many stations. *See* Local spots.

National Television System Committee (NTSC): An ad hoc committee convened from time to time by the NAB and EIA to study and recommend television standards and specifications.

NCTA: National Cable Television Association.

NRSC: National Radio Systems Committee.

NTSC: National Television System Committee. Also the term used to denote the color system developed by the Committee in 1952.

Offset carriers: A technique for reducing interference between nearby stations operating on the same channel by intentionally separating their carrier frequencies by one-half the line frequency (7,875 Hz).

Omega wrap: A helical scan format in which the tape path is reversed before and after it passes around the drum, giving it a cross section resembling the Greek letter omega. *See* Alpha wrap and Half-wrap.

Open-skies policy: The policy adopted by the FCC in 1972 providing that any financially qualified U.S. citizen or company could apply for authorization to operate a communications satellite.

Orbital arc: The circle or a portion of the circle 22,300 miles above the earth in the equatorial plane. *See* Prime orbital arc.

PAL: Phase Alternating Lines. A modification of the NTSC color television system that is widely used in Europe. The phase of the color subcarrier is reversed on alternate lines to minimize the visual effects of distortions of the signal in transmission.

Pay-per-view: A service provided by cable systems in which the customer is charged a separate fee for each program viewed.

Plumbicon: A storage tube that uses photoconductivity—the change in conductivity of a substance (lead oxide for the Plumbicon) when exposed to light.

Polarization: The direction of the electric field in an electromagnetic wave (for example, vertical, horizontal, or circular). With circular polarization, the direction of the electric field rotates as the wave proceeds.

Preemphasis: Intentionally increased amplification of a band of frequencies, usually the higher frequencies, in an audio or video signal before transmission. The signal is then subjected to complementary deemphasis at the receiving end, thus reducing the noise in the output signal.

Premium services: Services provided by cable systems for an additional monthly fee. *See* Basic services.

Primary colors: In additive color systems, red, green, and blue. In subtractive color systems, magenta, yellow, and cyan (often described as red, yellow, and blue).

Prime orbital arc: The portion of the orbital arc at a given location for which the look angle is greater than 5 degrees above the horizon for C band and 20 degrees for Ku band.

RARC: Regional Administrative Radio Conference. An ad hoc regional conference convened periodically by the ITU to establish frequency allocation and usage policies. *See* WARC.

Raster: The pattern formed by the scanning beam of a television pickup or display device.

Refraction: The bending of the path of an electromagnetic wave as it passes obliquely across the boundary between two media of different densities.

Regional channels: AM broadcast frequencies on which a number of medium-power stations are allowed to operate. *See* Clear channels and Local channels.

Resolution: The ability of a television or film image to distinguish fine detail. It is related but not identical to definition.

Saticon: A recently developed storage tube for live pickup that uses photoconductivity (the change in conductivity of a substance when exposed to light).

Scanning: The technique for converting a two-dimensional image into an electrical signal by simultaneously scanning the image from top to bottom and left to right with a small spot that detects the image brightness at each point.

SECAM: Sequential Colour Avec Memoire. A color television system developed by French engineers to minimize the visual effects of distortions of the signal in transmission.

Semiconductors: A class of materials whose conductivity varies widely under different physical and electrical conditions. They are the basis for the operation of transistors and integrated circuits, known generically as solid-state devices.

Shadow mask: A technique used by most tricolor kinescopes. Color phosphors are deposited on the inside of the kinescope faceplate, either in dot or line patterns. A perforated mask, which allows the electron beam or beams to strike only the phosphors corresponding to the signal modulating the beam, is placed immediately behind the faceplate.

Shortwaves: Electromagnetic radiation in the frequency range 3 to 30 MHz (wavelength 100 to 10 meters).

Sidebands: The frequencies produced on either side of the carrier as the result of modulation.

Signal-to-noise ratio: In a television picture, the ratio of the signal amplitude to unwanted disturbances, which usually result in snow in the picture.

Sky wave: Electromagnetic radiation that is transmitted by means of reflection from the ionosphere.

Slant track: *See* Helical scan.

SMPTE: Society of Motion Picture and Television Engineers.

Society of Motion Picture and Television Engineers (SMPTE): This professional society evolved from the Society of Motion Picture Engineers (SMPE) after World War II to encompass the technical aspects of both visual media. The *Journal of the SMPTE* is a leading publication of television technology. The SMPTE has been a leader in establishing technical standards for television. Its address is 595 W. Hartsdale Avenue, White Plains, NY 10607.

Space shuttle: A device for launching satellites. After reaching an altitude of approximately two hundred miles, the satellite is ejected from a cargo bay and the shuttle returns to earth as a fixed-wing aircraft.

Standards of Good Engineering Practice: Documents issued by the FCC, first for AM and later for FM and television, describing the technical policies and standards required for broadcasting stations.

Stereophonic effect: The perception of the sources of sound in three dimensions. It requires binaural hearing.

Storage tubes: A class of tubes used for converting light images into television signals. Their photosensitive surface is continuously exposed to an image of the scene, and the effect of the exposure is stored until it is removed once each frame by a scanning process.

Subtractive color systems: Imaging systems in which the hue is determined by selectively absorbing color components from white light. Painting and color photography are subtractive systems. *See* Additive color systems.

Super VHS: A modified VHS format using component recording and a greater bandwidth.

Superheterodyne receiver: A receiver that converts the frequency of all incoming signals to a single intermediate frequency for amplification.

Synchronizing signals: Pulses transmitted as part of the television signal to synchronize the scanning of the picture tube with that of the camera.

Syndex: Syndicated exclusivity.

Syndicated exclusivity: An FCC rule that permits the owner of a program or his or her agent (syndicator) to grant a local station the exclusive right to broadcast or distribute it in the station's market.

Syndication: The sale of the right to broadcast or distribute radio and television programming. *See* Syndicated exclusivity.

TelePrompTer: A prompting device for actors and speakers consisting of a continuous roll of paper with the script printed on it that is advanced as the speaker proceeds.

Transistor: A device based on semiconductors that is capable of amplification and rectification of electrical signals. Also, popularly, a portable transistorized radio.

Transponder: The active element in a communications satellite. It receives an uplink signal from the earth, amplifies it, shifts its frequency, and retransmits it to the earth via a downlink.

Transverse recording: (1) A form of audiodisk recording in which the recorded signal is a series of horizontal undulations of the groove. (2) A system of recording audio or video signals on tape in which the recorded tracks are at an angle to the tape.

Tropospheric propagation: Propagation of radio waves around the earth's curvature and past the geometric horizon by refraction resulting from variations in the density of the atmosphere.

TT&C: Tracking, telemetry, and control. The system for monitoring and controlling the location and status of an in-orbit satellite.

TVRO: Television receive only. An earth station equipped to receive satellite signals with television programming.

Type C helical scan: A helical scan recording format standardized by the SMPTE for one-inch tape.

U-Matic: A videocassette format developed by Sony and using three-quarter-inch tape. It has been designated Type E by the SMPTE.

UHF: Ultrahigh frequency. Electromagnetic radiation in the frequency range 300 to 3,000 MHz. Television channels 14 and up are in the UHF band.

Uplink: The radio frequency circuit from an earth station to a satellite.

VCR: Videocassette recorder.

Very large scale integration (VLSI):   Integrated circuits of great complexity.

Vestigial sideband: A method of transmitting an amplitude modulated television signal in which the lower sidebands produced by the higher frequency components are removed to reduce the bandwidth requirements.

Vestigial sideband FM: An FM signal in which the higher frequency sidebands resulting from modulation have been removed. It is a key technology in videotape recorders.

VHF: Very high frequency. Electromagnetic radiation in the frequency range 30 to 300 MHz. FM broadcasting and television channels 2 to 13 are in the VHF band.

VHS: A videocassette tape format developed by Matsushita in competition with the Beta format. It has been more successful in the marketplace.

Vidicon: A storage tube that uses photoconductivity—the change in conductivity of a substance (antimony trisulfide for the vidicon) when exposed to light. It is widely used for film pickup and also for live pickup where high sensitivity is not required.

Viewing ratio: The ratio of the viewing distance to picture height or diagonal.

VLSI: Very large scale integration.

WARC: World Administrative Radio Conference. An ad hoc worldwide conference convened periodically by the ITU to establish frequency allocation and usage policies. *See* RARC.

Weber's law: The principle that the perception of the relative magnitudes of physical sensations is proportional to their ratios.

■■ BIBLIOGRAPHY

■ BIOGRAPHIES AND AUTOBIOGRAPHIES

Bilby, Kenneth. *The General: David Sarnoff.* New York: Harper & Row, 1986.

Brown, George H. *and part of which I was: Recollections of a Research Engineer.* Princeton, N.J.: Angus Cupar Publishers, 1982.

Carneal, Georgette. *Conqueror of Space: The Life of Lee DeForest.* New York: Horace Liveright, 1930.

De Forest, Lee. *Father of Radio.* Chicago: Wilcox & Follett, 1950.

Dreher, Carl. *Sarnoff: An American Success.* New York: Quadrangle/New York Times Book Company, 1977.

Everson, George. *The Story of Television: The Life of Philo T. Farnsworth.* New York: W.W. Norton, 1949. Reprint. New York: Arno Press, 1974.

Fessenden, Helen. *Fessenden: Builder of Tomorrow.* New York: Coward-McCann, 1940. Reprint. New York: Arno Press, 1974.

Goldmark, Peter C. *Maverick Inventor: My Turbulent Years at CBS.* New York: Saturday Review Press, 1973.

Hofer, Stephen F. "Philo Farnsworth: Television's Pioneer," *Journal of Broadcasting* 23 (Spring 1979): 153–165.

Lessing, Lawrence. *Man of High Fidelity: Edwin Howard Armstrong.* New York: J.P. Lippincott, 1956. Reprint. New York: Bantam Books, 1969.

Lyons, Eugene. *David Sarnoff: A Biography.* New York: Harper & Row, 1966.

Morita, Akio. *Made in Japan: Akio Morita and Sony.* New York: E.P. Dutton, 1986.

Paley, William S. *As It Happened: A Memoir.* Garden City, N.Y.: Doubleday, 1979.

Wallen, Alberta I. *Genius at Riverhead: A Profile of Harold H. Beverage.* North Haven, Maine: North Haven Historical Society, 1988.

■ HISTORIES

Aitken, Hugh G.J. *Syntony and Spark: The Origins of Radio.* New York: John Wiley, 1976.

———— *The Continuous Wave: Technology and American Radio, 1900–1932.* Princeton, N.J.: Princeton University Press, 1983.

Archer, Gleason L. *History of Radio to 1926.* New York: American Historical Company, 1938. Reprint. New York: Arno Press, 1971.

Baker, W.J. *A History of the Marconi Company.* London: Metheren & Company, 1970.

Barnouw, Erik. *A Tower in Babel: A History of Broadcasting in the United States to 1933.* New York: Oxford University Press, 1966.

———— *The Golden Web: A History of Broadcasting in the United States, 1933–1953.* New York: Oxford University Press, 1968.

―――― *The Image Empire: A History of Broadcasting in the United States Since 1953*. New York: Oxford University Press, 1970.

―――― *Tube of Plenty: The Development of American Television*. New York: Oxford University Press, 1975.

Blake, George. *History of Radio Telegraphy and Telephony*. London: Chapman and Hall, 1928. Reprint. New York: Arno Press, 1974.

Briggs, Asa A. *The Birth of Broadcasting: The History of Broadcasting in the United Kingdom*. Vol. 1. London: Oxford University Press, 1961.

―――― *The Golden Age of Wireless: The History of Broadcasting in the United Kingdom*. Vol. 2. London: Oxford University Press, 1965.

―――― *The War of Words: The History of Broadcasting in the United Kingdom*. Vol. 3. London: Oxford University Press, 1970.

―――― *Sound and Vision: The History of Broadcasting in the United Kingdom*. London: Oxford University Press, 1979.

Broadcasting magazine editors. *The First 50 Years of Broadcasting*. Washington, D.C.: Broadcasting Publications, Inc., 1982.

Douglas, Susan J. *Inventing American Broadcasting, 1899–1922*. Baltimore: Johns Hopkins University Press, 1987.

Gelett, Roland. *The Fabulous Phonograph: 1877–1977*. 2d ed. New York: Macmillan, 1977.

Graham, Margaret B.W. *RCA & the Videodisc: The Business of Research*. New York: Cambridge University Press, 1986.

Greenfield, Jeff. *Television: The First Fifty Years*. New York: Abrams, 1977.

Hess, Gary Newton. *An Historical Study of the DuMont Television Network*. New York: Arno Press, 1971.

Lardner, James. *Fast Forward: Hollywood, the Japanese, and the VCR Wars*. New York: W.W. Norton, 1987.

Leinwoll, Stanley. *From Spark to Satellite: A History of Radio Communication*. New York: Charles Scribner's Sons, 1979.

MacLaurin, W. Rupert. *Invention and Innovation in the Radio Industry*. New York: Macmillan, 1949. Reprint. New York: Arno Press, 1971.

Reid, T.R. *The Chip: How Two Americans Invented the Microchip & Launched a Revolution*. New York: Simon & Schuster, 1984.

Sobel, Robert. *RCA*. New York: Stein & Day, 1986.

Sterling, Christopher H., and John M. Kittross. *Stay Tuned: A Concise History of American Broadcasting*. Belmont, Calif.: Wadsworth, 1949.

Udelson, Joseh H. *The Great Television Race: A History of the American Television Industry, 1925–1942*. University of Alabama Press, 1982.

Warren Publishing, Inc., editors. *Enter DBS: The Satellite Story*. Washington, D.C.: Warren Publishing, Inc., 1985.

■ GENERAL BOOKS

Codel, Martin. *Radio and Its Future*. New York: Harper, 1930. Reprint. New York: Arno Press, 1971.

Head, Sydney W., and Christopher H. Sterling. *Broadcasting in America: A Survey of Television, Radio, and New Technologies*. 5th ed. Boston: Houghton Mifflin, 1988.

MacLaurin, W. Rupert. *Invention and Innovation in the Radio Industry.* New York: Macmillan, 1949. Reprint. New York: Arno Press, 1971.

Magnant, Robert S. *Domestic Satellites: An FCC Giant Step Toward Competitive Communications Policy.* Boulder, Colo.: Westview Press, 1977.

Mirabito, Michael, and Barbara Morgenstern. *The New Communications Technologies.* Boston and London: Focal Press,1990.

Sarnoff, David. *Looking Ahead: The Papers of David Sarnoff.* New York: McGraw-Hill, 1968.

■ REFERENCE BOOKS

American Radio Relay League. *The Radio Amateur's Handbook.* Newington, Conn.: ARRL, annual.

Bartlett, George. *NAB Engineering Handbook.* 6th ed. Washington, D.C.: National Association of Broadcasters, 1975.

Benson, K. Blair. *Television Engineering Handbook.* New York: McGraw-Hill, 1986.

British Broadcasting Corporation. *Handbook.* London: BBC, annual.

Broadcasting/Cable Yearbook. Washington, D.C.: Broadcasting Publications, Inc., annual.

Cable & Station Coverage Atlas. Washington, D.C.: Warren Publishing, Inc., annual.

Consumer Electronics Video Data Book. Washington, D.C.: Warren Publishing, Inc., annual.

Electronic Industries Association. *Consumer Electronics.* Washington, D.C.: EIA, annual.

Fink, Donald G., and Donald Christiansen. *Electronic Engineers Handbook.* 2d ed. New York: McGraw-Hill, 1982.

Inglis, Andrew F. *Electronic Communications Handbook.* New York: McGraw-Hill, 1988.

National Telecommunications and Information Administration. *The Radio Frequency Spectrum: United States Use and Management.* Washington, D.C.: NTIA, annual.

Television & Cable Factbook: Cable & Services Volume. Washington, D.C.: Warren Publishing, Inc., annual.

Television & Cable Factbook: Stations Volume. Washington, D.C.: Warren Publishing, Inc., annual.

■ TECHNICAL BOOKS

Ennes, Harold E. *Principles and Practices of Television Operations.* Indianapolis: Howard W. Sams, 1953.

Fink, Donald G. *Television Engineering.* New York: McGraw-Hill, 1952.

Fink, Donald G., and David M. Luytens. *The Physics of Television.* Garden City, N.Y.: Anchor Books, 1960.

Miya, K. *Satellite Communications Technology.* Tokyo: KDD Engineering and Consulting, Inc., 1981.

Schade, Otto H., Sr. *Image Quality: A Comparison of Photographic and Television Systems.* Princeton, N.J.: RCA Laboratories, 1975.

Shannon, Claude E., and W. Weaver. *The Mathematical Theory of Communication.* Urbana: University of Illinois Press, 1949.

Wentworth, John. *Color Television Engineering.* New York: McGraw-Hill, 1954.

Zworykin, V.K., and G.A. Morton, *Television,* New York: John Wiley, 1940.

Zworykin, V.K., G.A. Morton, and L.E. Flory. *Television.* 2d ed. New York: John Wiley, 1954.

■ PERIODICALS

Broadcast Engineering
Broadcasting
Cablevision
Journal of the Audio Engineering Society
Journal of the SMPTE
Satellite Week
Spectrum (IEEE)
Television Digest
Transactions IEEE, Broadcast
Transactions IEEE, Consumer Electronics
Transactions IEEE, Magnetics

■ ■ INDEX

ABC, 322, 383, 436
  effect of postfreeze TV assignment
    policies, 200
  founding, 79
ABC radio, 143, 150, 437
Aberdeen, University of, 6
Academy of Television Arts and
    Sciences, 302
Acoustical Society of America, 305
Acoustics, 15
ACTV HDTV system, 485–486
Additive color systems, 238
Advanced TV Advisory Committee,
    482
Aepinus, 3
AIEE, merger with IRE, 130
Alaska, State of, 418
Alexander, Ernst, 166
Alexanderson alternator, 38, 40
All-channel receiver law, 209
Alpha wrap, helical scan, 346
Alphanumeric characters, 29
AM radio,
  adjacent channel interference, 83
  antennas, 66
  antennas, directional, 82, 87, 104
  channel spacing, 82
  clear channel breakdown, 102
  clear channels, 64
  control equipment, 96
  coverage calculation, 69
  decline, 110
  early RCA tubes, 69
  economic basis, 65
  first advertising, 65
  first stations, 63
  forecast, 111
  frequency assignment rules, 83
  ground-wave service, 85
  local channels, 64
  local radio era, 100
  program formats, 101

receiver sales, early, 73
receiver sales, 1955–1970, 109
receivers, 66, 96
regional channels, 64
service areas, 84
service quality, improvement of,
    103
sky-wave service, 85
station classes, 80, 81
station coverage, 85
station population, 57, 99
stereo, 104
studios, 93
superheterodyne receivers, 67
technical regulation, 101
time sales, 58
transistors, effect of, 108
transmitter modulation, 88–91
transmitters, 66, 88
versus FM, 60, 126
Amateur radio, 52
America's Cup, 381
American Marconi Company, 14, 52,
    53, 54
American National Standards
    Institute, 484
American Physical Society, 305
American Telephone & Telegraph
    Company. See AT&T
Ampere, Andre-Marie, 2, 4
Ampex Corporation, 299–300, 302,
    337, 344, 348, 351, 353, 356
  financial success, 326, 340, 341
  financial troubles, 340, 341
  quadruplex recorder breakthrough,
    310–324
  stock prices, 327
  RCA competition, 329–336
  RCA patent agreement, 330
  RCA TR-22 recorder, effect of, 335
Ampliphase modulation, 91
Amplitude modulation, 41

Analog transmission, 30

Anderson, Charles E., 313, 314, 319, 327

Andover, Maine, 413

Anik satellite, 399

Antennas
  antenna farms, 226
  candelabra, 226, 227
  directional, 82, 87, 104
  FM, 146, 147
  gain, 222
  stacked, 226
  superturnstile, 225
  television, 221–230

AP Radio, 437

Aperture response, 479

Apogee kick motor (AKM), 397

Apollo mission, 394

Areas of Dominant Influence (ADIs), 210

Ariane rocket, 399

Armstrong, Edwin Howard, 14, 42, 126, 140, 141, 176
  begins FM research, 121
  biography, 113
  outrage at RCA, 123
  RCA patent battle, 150–154
  Sarnoff friendship, 121
  suicide, 113, 117, 153

ASC satellites, 422

Astoria, Oregon, 360

AT&T, 88, 115, 121, 165, 166, 188, 212–214, 408, 409, 414, 415, 417, 418
  enters broadcasting, 70
  leaves broadcasting, 72

ATC, 376, 377

Atkinson, L.B., 169

Atlas/Centaur rocket (launch vehicle), 396

ATV (advanced television), 483
  proposed systems (table), 483

Audible sound, frequency range, 18

Audio Digital Distribution Service (ADDS), 437

Audio Engineering Society, 305

Audio tape recording,
  cassette, 107
  development of, 105
  electrical bias, 107
  principle of, 106
  reel-to-reel, 106

Audion, 13

Audubon Electronics, 373

Automatic Scan Tracking (AST), 348

Autovon network, 419

Baird, James Logie, 162–164

Baird Television Company, 163

Baker, W.R.G., 263, 267

Bakker, Jim, 383

Bandwidth requirements, HDTV, 481

Barraud, Francis, 23

Baruch, Ralph
  biography, 379

BASF, 344

BASF recording format, 440, 443

BBC (British Broadcasting Company), 163, 175

BCN format, 351

Bedford, Alda V., 193, 244, 268

Beers, George L., 152

Bel, 17

Belknap, J.E. & Associates, 366, 374

Bell, Alexander Graham, 487
  biography, 30

Bell Atlantic, 389

Bell Laboratories, 1, 39, 294

Bellingham, Washington, 360

Bellows, Henry N., 80

Berlin Conference of 1903, 44

Berlin Conference of 1906, 45, 51

Berlin, Irving, 372

Berliner, Emil, 20
  joins Eldridge Johnson, 21

Berliner gramaphone, 20

Betacam, 352

Betacam SP, 352

Beta VCR (video cassette recorder)
  initial enthusiasm, 463
  introduction to marketplace, 462
  manufacturers, 467
  market share, 468
  recording format, 443, 444, 449

Beta versus VHS, 465–467

Beverage, Harold, 123

BEVR, 303

Bishop's College, 49

Block equipment, World War II, 186

Bloom, Edgar, 72

Bologna, University of, 33

Booz, Allen, 273

Bosch, Robert, 175, 351

Bousel, Charles, 30

Boyden, Ronald W., 71

"Bracket" standards, 265

Bradshaw, Thorton, 354, 439, 451
Braun, Karl F., 174
Brewster, Washington, 413
Brigham Young University, 171
British Broadcasting Corporation
    (BBC), 66
British Post Office, 33, 415
Broadcast Electronic Video
    Recording. *See* BEVR
Broadcasting magazine, 432, 433
Brown, George H., 87, 91, 242, 257,
    259, 263, 277, 280, 310, 331
  biography, 252
  and blue banana, 271
Bullard, Rear Admiral, 54
Burns, John, 250, 273, 274
"Burst," 263

C-band satellites
  assignments, 1988, 410
  frequency allocations, 404
  frequency reuse, 406
  homesats (home satellite receivers),
    427
  orbital locations, 406
  versus Ku-band and DBS satellites,
    405
C-Quam stereo system, 104
C-SPAN, 383
Cable Act of 1984, 368
Cable News Network (CNN), 382
Cable Television Association (CTA),
    366
Cable television. *See* CATV
Cable, transatlantic, 29
Cable-ready receivers, 364
Caldwell, Orestes H., 80
Cambridge University, 6, 15, 16
Camcorders, broadcast, 358
Camcorders, consumer
  cameras, 471
  recorders, 471
  sales, 470
  technology, 471
Cameras, color
  CCD, 293–294
  ENG, 291
  field sequential, 283
  General Electric, 285, 287
  Ikegami, 296
  image orthicon, 284–286
  Philips, 286–289
  Plumbicon, 286–291

RCA, 284–286, 289–292
  Sony, 296–298
Cameras, Marius, 317, 318
Cameras, monochrome
  studio, 216–219
  film, 219–221
Cambell-Swinton, Alan, 168, 174
"Can't-carry" rules, CATV, 370
Carey, G.B., 158
Cartridge recorder, broadcast video,
    342–343
Cartrivision, 440, 443
Caruso, Enrico, 27, 49
"Cat's whisker," 44
Cathode ray tube, 9
CATV (cable television)
  A/B switch, 369
  adversarial issues, current, 389
  basic and pay services, 375
  broadcaster relations, 390
  can't-carry rules, 370
  channel assignments, 363, 364
  copyrights, 370, 390
  DBS competition, 390
  distant signal importation, 360,
    366, 374
  FCC regulations, 363
  growth opportunities, future, 391
  local telephone company (telco)
    relations, 389
  market value, systems, 385
  mom-and-pop systems, 360, 365
  must-carry rules, 369
  revenue growth, 385
  satellite program distribution, 361,
    375
  satellite utilization, 1988, 426
  subscriber drops, 363
  subscriber growth, 384, 385
  supertrunk, 362
  system design, 362, 363
  trunks, 362
  U.S. subscribers, 361
Cavendish, Henry, 3
Cavendish Laboratory, 6, 16
CBC (Canadian Broadcasting
    Corporation), 322
CBS, 237, 322, 379, 436, 437, 441
  attracts NBC's stars, 78
  post-war television strategy, 191
  experimental VHF station, 177
CBS field sequential color
  color hearing. *See* Color TV hearing

commitment to manufacture color receivers, 264
demise, 267
FCC petition, 1940, 241
FCC petition, 1946, 241
FCC petition, 1949, 243
first gala broadcast, 266
monochrome sets in use, 266
NPA Order M-90, 267
CBS radio, 126, 143
CCA, 88
CCDs (charged coupled devices), 158, 161, 291, 293, 493
CCIR (Comité Consultatif International de Radio Communication), 275
considers NTSC 275
CD (compact disk, audio), 149
CDV (compact disk video), 447
Challenger space shuttle, 399
Chapin, E.W., 247, 262, 319, 320
Chapin converter, 262, 263
Charge-coupled devices. *See* CCD
Charyk, Joseph V., 413
"Cherry picking," 416
Chicago Federation of Labor, 168
Chicago, University of, 10
Christian Broadcasting Network, 383
Chromacoder, 284
Chrysler, 386
Circular polarization, FM, 138
Clarke, Arthur C., 395
biography, 392
Clear channel breakdown, 102
Clear channel stations, 37
Clifford, Clark, 263, 264
CNN (Cable News Network), 436
Coaxial cable, intercity, 188
Coherer, 33
Colledge, Charles, 287, 288, 334
Color correction, 291
Color TV hearing (1949–1953), 243–266, 271–272
approval, CBS system, 263–265
approval, NTSC system, 272
"bracket standards," 265
comparative demonstrations, 261
court appeals, 264
CTI, 261
early CBS testimony, 258
early industry testimony, 258
early RCA testimony, 258

first CBS demonstration, 261
first RCA demonstration, 259
initial NTSC specifications, 271
final NTSC specifications, 271
NTSC formation, 262
Color television
breakthrough, 274
cameras. *See* Cameras, color
colorimetry, 238, 477
film equipment, 291
historical summary, 237
RCA CT100 receiver, 281, 283
receiver development, 281–283
receiver imports (table), 296
receiver sales, 274
signal transmission, 239
Color systems, international, 275–276
PAL, 276
SECAM, 276
Columbia University, 113, 114, 121, 124
Comité Consultatif International de Radio Communication. *See* CCIR
Commerce Department, 272
Communications Act of 1934, 82, 368
Communications Satellite Act of 1962, 408
Community antenna systems, 360
Compact disks (CDs), 149
Component recording, 356
Composite recording, 357
Comsat, 409, 420, 436
business growth, 413
role and organization, 413
Comsat General, 420, 436
Comsat Laboratories, 414
Conrad Hilton Hotel, 299
Constant luminance color, 270
Contel/ASC (Continental Telephone/American Satellite Company), 422
Continental Electronics, 88
Continuous waves (CW), 49
Contrast radio, 477
CONUS (satellite company), 435
Coolidge, Calvin, 80
Copyright Act of 1976, 370, 390
Copyright Royalty Tribunal, 370
Copyrights, CATV, 370
Coulomb, Charles-Augustin de, 2, 3
Courier satellite, 392, 395
Coy, Wayne, 243, 246, 262–266

Cravath, Swaine & Moore, 152
Cronkite, Walter, 394
Crookes, Sir William, 9
Crookes tube, 9, 174
Crosby, Murray G., 152
Crosley, Powell, 91
Crosley Radio Corporation, 91
CRT. *See* Cathode ray tube;
    Copyright Royalty Tribunal
Crystal detectors, 44
CTI (Color Television Incorporated),
    247, 261
Cushing, Barbara, 77
Cushing, Dr. Harvey, 77
CW. *See* Continuous waves

D-A (digital-to-analog) conversion,
    356, 488–489
Dalmo Victor Company, 312
Daniels, Bill, 374
Daniels, Josephus, 53
David Sarnoff Research Center,
    named, 304
Davy, Sir Humphry, 4
Daytime-only stations, 103
DBS, 388, 404, 431, 432
    applications for initial
        authorizations, 432
    power-antenna size trade-off
        (table), 427
De Forest, Lee, 42, 44, 62, 65, 115
    biography, 13
    broadcasting attempts, 48
    Radio-Telephone Company, 49
    Wireless Telegraph Company, 47
Decibel, 17
Deintermixture, 210
Delta rocket (launch vehicle), 396–398
Dempsey-Carpentier fight, 52, 63
Denny, Charles, 242, 243
Department of Commerce, 63, 64
Department of Justice, 430
Descramblers, 429
Detectors, 12
    coherer, 33, 43
    crystal, 44
    electrolytic, 43
    rectifier, 12
    regenerative, 13
Dieckman, Max, 174
Digital Audio Transmission Service
    (DATS), 437

Digital recording
    advantages, 354–355
    bandwidth requirements, 488–489
    technical challenge, 356
Digital signal applications
    audio recording, 490
    signal processing, 490
    signal security, 490
    signal transmission, 489
    standards conversion, 490
    video recording, 490
Digital systems, 487–490
    bandwidth requirements, 356,
        488–489
    signal processing, 487
    storage, 488
    technology, 487
Digital time base correction, 348
Diode, 11
Direct broadcast by satellite. *See* DBS
Direct-to-home satellite market, 431
Directional antennas, AM, 82, 87, 104
Directional antennas, TV, 222, 231
Discovision Association (DVA), 453,
    454
Disney pay-TV service, 379
Doherty amplifier, 89
Doherty, William, 89, 91
Dolan, Charles, 376
Dolby, Ray M., 313, 314
Dot sequential color system, 244
Dumont, Allen B. Co., 177, 181, 267,
    291

EAILI recorder, 440, 443
Early Bird satellite, 399
Earth stations
    homesats (home satellite receivers),
        427
    receive-only (TVRO), 403
    TT&C (tracking, telemetry &
        control), 403
Echo satellite, 395
EDTV. *See* Extended definition TV
Edinburgh, University of, 6
Edison effect, 12
Edison, Thomas Alva, 12
Edwards, Douglas, 322
EIA (Electronic Industries
    Association), 104, 272
EIRP (effective isotropic radiated
    power), 403

Eisenhower, Dwight D., 190, 395
Electricity
    current, 3
    static, 2
Electromagnetic radiation, 6
Electromagnetic spectrum, 7, 8
Electromagnetic waves, 1, 4
Electromagnetism, 4
Electron
    discovery, 9
    measurement of electrical
        charge, 10
Electron tube
    amplifiers, 32
    diode, 11, 12
    invention, 11
    triode, 11
Electronic Industries Association. *See*
    EIA
Electronic news gathering. *See* ENG
Electronic video recording. *See* EVR
Electronics magazine, 263
Electron tube amplifiers, introduced,
    32
Electrostatics, 3
Embossed plastic tape recording, 440,
    441
Emerson, Faye, 266
EMI (Electric & Musical Industries
    Ltd.), 163
EMI-Marconi, 175
Empire State Building, 176
ENG (Electronic news gathering),
    291, 349, 352, 356, 358, 435
Engstrom, Elmer, 245, 274, 278
    biography, 250
    color TV hearing, 258, 259
ESPN, 382
EVR (electronic video recording), 255,
    302–304, 441
Extended definition television
    (EDTV), 482

Fairchild Semiconductor, 492
Faraday effect, 6
Faraday, Michael, 1, 2, 3, 4, 5
    biography, 4
Fargo, North Dakota, 210
Farnsworth, Philo T., 169, 176, 177,
    180
    biography, 172–173
    patent problem with RCA, 173

Farnsworth Television and Radio
    Corporation, 173, 176
Fasnet yacht race, 381
FCC (Federal Communications
    Commission), 58, 111, 122, 430
    approves NTSC color, 272
    authorizes FM, 127
    CBS color petition, 1949, 243
    color TV hearing, 1949, 237–272
    concern for cochannel TV
        interference, 193
    denies CBS 1947 color petition, 192
    denies CBS 1946 color petition, 242
    establishes NTSC for monochrome
        television, 184
    establishes TV assignment policies,
        200
    FM stereo authorization, 149
    founding, 82
    initial television allocations, 176
    initiates "freeze," 194
    location of FM radio spectrum,
        130–136
    reinstates prewar NTSC standards,
        188
    Standards of Good Engineering
        Practice, 82, 138, 185
FCC laboratories, 261, 262, 319
Federal Aviation Administration
    (FAA), 226
Federal Communications
    Commission. *See* FCC
Federal Radio, 173
Federal Radio Commission. *See* FRC
Federal Telegraph Company, 49
Federal Trade Commission (FTC)
    opposes patent pool, 72–73
Fernseh, 175, 289, 342, 343
Fessenden, Reginald, 14, 46, 62, 65
    biography, 49
    NESCO, 50
Fiber optics
    analog transmission, 495
    analog versus digital transmission,
        388
    digital transmission, 495
    history, 494
    multimode, 494–495
    single-mode, 494–495
Fiber optics, applications
    CATV, 388, 389, 496
    intercity circuits, 495
Field, Cyrus, 29

Field sequential color
    cameras, 283
    system, 237, 239–241, 243
Film cameras, monochrome, 220
Film projectors, 220
Film reproduction systems,
        monochrome, 219–220
Financial News Network, 382
Fink, Donald, 263
Fixed satellite service (FSS), 404
Flechsig, W., 278
Fleming, Sir John A., 12, 14, 44
Fleming valve, 13
Fly, James Lawrence, 178, 183, 184
Flying spot scanner, 158, 220, 291
FM, 118
    1940 frequency allocation, 127
    antennas, 146, 147
    commercial broadcasting begins,
        128
    definition, 117
    effect of stereo, 142
    first public demonstration, 121
    future prospects, 154
    industry establishment forms, 124
    initial post-war enthusiasm, 140
    initial receiver problems, 140
    invention, 113
    narrow-band, 117
    receivers, 147
    revenues, 145
    rule-making hearing, 125
    satellite program distribution, 150
    share of audience, 143, 144
    skywave interference, 128
    spectrum location battle, 130–136
    stagnation, 1948–1957, 140
    stereo, 148
    stations on-air, 143, 144
    technical development, 124–125
    transmitters, 145
    wide-band, 113, 117
    wide-band versus narrow-band,
        119
FM radio regulation
    antenna height, 137
    assignment principles, 138–139
    channel width, 137
    circular polarization, 138
    number of channels, 137
    power, 137
    preemphasis/deemphasis, 138
FM radio versus AM, 126

Fogarty, Joseph, 410, 411, 412
Folsom, Frank, 279
Footprints, satellite, 401
Ford Motor Company, 386
Fort Pierce, Florida, 376
Fowler, Mark, 411
Franchising authority, municipalites,
        369
Franklin, Benjamin, 2
FRC (Federal Radio Commission), 58,
        69
    authority of, 79
    founding, 79
    frequency assignments, initial, 81
    station classes, 80
    television regulation, 166
Fredendall, Gordon, 193
Frederick, Robert, 354
Freeze, television, 197–202
Frequency allocations and
        assignments, 195
Frequency allocations, satellites, 404
Frequency modulation. See FM
Fyler, Norman, 279

Galaxy satellites, 412, 422, 426
Galvani, Luigi, 3
Galvanic pile, 3
Gates Radio, 88, 89
Gauss, Carl Friedrich, 2
Geissler, Heinrich, 9
General Dynamics, 396
General Electric, 14, 55, 70, 72, 91,
        98, 99, 115, 165, 166, 168, 188,
        223, 238, 267, 291
General Instrument, 428, 429
General Motors, 386
General Precision Laboratories, 301
Geosynchronous satellites, 374, 392,
        393
Gifford, Walter S., 71
Gilbert, William, 2
Ginsburg, Charles, 318, 322, 327, 348
    biography, 312–314
Godfrey, Arthur, 266
Goldmark Communications
        Company, 257
Goldmark, Peter, 77, 97, 178, 242,
        243, 258, 260, 278, 280, 283, 284,
        441
    biography, 255
    post-war CBS television, 190

Goldsmith, Alfred N., 167, 278
Goldsmith, T.T., 263
Gramaphone Company, Ltd., 23
Gray, Elisha, 31
Griffiths, Edgar, 449, 450
Grimes, William, 382
Griswold, A.W., 70
Ground controlled approach (GCA),
   392
Ground wave service, 85
Ground waves, 36
GStar satellites, 422
GTE, 422
Gulf & Western Corporation 384
Gundy, Philip L., 320
Gunter, Kenneth, 376

Hansell, Charles, 245
Harris (Intertype) Corporation, 88
Haverford College, 376
Hawkins, Howard, 418, 419
Hazeltine, 270
HBO, 368, 376, 391, 412, 419, 424,
   429, 432, 433
HDTV, 359, 431, 473–487
  aspect ratio, 478
  bandwidth requirements, 481
  colorimetry, 478
  commercial status, 486
  compatibility requirement, 482
  contrast ratio, 477
  definition of, 476
  early history, 475–476
  enthusiasts, 473–474
  picture definition, 478
  production standards, 484
  scanning line requirements, 480
  signal-to-noise ratio, 477
  skeptics, 474–475
  viewing ratio, 480
Hearing
  characteristics, 16
  definition, 16
  frequency ranges, 18
  stereophonic effect, 18
Heaviside, Oliver, 36
Helical scan, broadcast, 313
  analog recording, 349
  formats, 344–346, 357
  markets, 347
Helical scan, consumer. See VCR
Helicon Cablevision, 389
Helmholtz, Hermann von, 16

Henderson, Shelby, 313, 314
Hennock, Frieda, 246, 258, 260, 265
Henry, Joseph, 2
Herold, Edward W., 278
Hertz, Heinrich, 2, 6
High-definition television. See HDTV
High-band recording, 338
Hill-and-dale recording, 20
Himmelfarb, Fred, 288
Hirsch, Charles V., 270
Hitachi, 461
Hobson, Norman, 288
Holography, 441, 442
HOMAG, 53
Home Box Office. See HBO
Home recording formats (table), 440
Homesats (home satellite receivers),
   427
Hooper, Commander, 54
Hoover, Herbert, 64, 66
Hopkins, A.R., 328
Hubbard Broadcasting Company, 435
Hubbard, Gardner, 31
Hughes Communications, 412, 420,
   421, 432
Hughes satellites, 402
Hughes Television Network, 435
Hyde, Rosel, 246, 271
Hytron Division, CBS, 279
Hytron Radio and Electronics
   Corporation, 256

Ibuka, Masaru, 457
Iconoscope, 169–172
IEEE, 305
Ikegami, 296
Image dissector, 169, 172
Image iconoscope, 160, 169, 172
Image isocon, 287
Image orthicon, 160, 172, 186–188
Inglis, Andrew F., 377
Institute of Radio Engineers, 53
Integrated circuits, 15, 492
Intelsat, 414
  competition, 415–416
  growth, 415
  satellites (table), 415
Intercity video circuits, 211
  coaxial cable versus microwave, 212
Interdepartmental Wireless Board, 51
Interlaced scanning, 159
International Telecommunications
   Union (ITU), 408

International Video Corporation (IVC), 348
Ionosphere, 37
IRE
  merger with AIEEE, 130
ITT, 412

Japan
  competition with RCA, 354
  HDTV leadership, 475
  invades monochrome TV market, 233
  product engineering, 455
  VCR dominance, 455
Japan Measuring Instrument Company, 457
Japanese
  cameras, 296, 297
  color receiver imports, 294, 295, 296
  color receivers, 294, 295
  helical scan recorders, 347
  U.S. radio and television exports (table), 295
  videotape recorder, 296
Japanese Victor. See JVD
Jenkins, Charles Francis, 164, 165, 172
Jenkins Television Corporation, 165, 172
Jerrold Electronics, 366
John Hancock Building antenna, 228
Johnson, Eldridge R., 21, 98
  biography, 22–28
Jones, Robert, 199, 247, 258
JVC, 443, 461

Kahn, Irving B., 371–374
  Audobon Electronics, 373
  convicted of bribery, 373
  founds TelePrompTer, 372
Kaltenborn, H.V., 256
Kappel, Frederick, 414
Karlsruhe Polytechnic, 6
KAST, 360
KCBS, 63
KDKA, 61, 62, 63
Kelvin, Lord, 291
Kennelly, A.E., 36
Kennelly-Heaviside layer, 36
Kennett, Missouri, 367
Kesten, Paul, 190, 254, 256
Kilby, Jack, 492

Kinescope, name derivation, 174
Kinescope recording, 300–302
  lenticular film, 301
  ultraviolet, 302
King's College, 6, 394
KING-TV, 360
Klystron tubes, 229
KNX, 63
Koplovitz, Kay, 384
Kozanowski, H.N., 285
KQV, 63
Kreuzer, Barton, 288
KRLD-TV, 226
Ku band
  direct-to-home broadcasting, 430, 432
  satellites, 401, 405
KYW, 63

L-1 coaxial cable intercity network, 188, 213, 264
La Guardia, Mayor, 183
Lamb, Brian, 383
Langmuir, Irving, 14, 115
Lasers (light amplification by stimulated emission of radiation), 441, 493
  applications, 494
  audio recording, 149
  technology, 494
Laservision (Discovision), 443, 447, 452
  Atlanta introduction, 453
  format, 453
  Philips/MCA format, 452
  renaissance, 455
  sales (table), 454
  technology, 452
Law, Harold B., 278, 279
Law, Russell, 279
Lawrence color tube, 281
Lehman Brothers, 337
Lenticular film, 302
Levin, Gerald M.
  biography, 376–379
  founding of HBO, 376
Lind, A.H., 329, 335
Line loading, 291
Local telephone companies. See Telcos
Lodge, Sir Oliver, 40
Lodge, William, 322
London Conference of 1912, 45, 51

Long, George I., 337
Long waves, 36
Loudspeakers, 32
Loughlin, Bernard D., 270
Loughren, Arthur V., 270
Luther, Arch, 335
LVR recorders, 344
Lyons, Eugene, 75

M/A-COM, 428
MacArthur, General Douglas, 458
MacInnes (Armstrong), Marion, 116,
    153
Maeda, Tamon, 458
Magnatape, RCA, 443
Magnavision, 454
Magnavox, 96
Magness, Bob John, 386
Magnetic tape recording (audio). *See*
    Audio tape recording
Magnetism, 4
Mahoney City, Pennsylvania, 360
Malarkey & Taylor, 366
Malarkey, Marvin F. Jr., 366, 367
Malone, John C., 386
Manhatten Cable, 378
Marconi, Annie Jameson, 33
Marconi color cameras, 287
Marconi, Marchese G., 44, 47, 51, 52
    biography, 33
Marconi Wireless Telegraph and
    Signal Company, Ltd., 12, 34,
    46, 156, 304
Master control, 93
Matsushita, 449, 461
Maxey, Alex, 313, 314, 327, 344
Maxwell, James Clerk
    biography, 6
Maxwell's equations, 6, 7
May, Joseph, 158
MCA, 453, 464
McDonald, Eugene, 73
McDonnell-Douglas, 396
MCI, 421
MDS, 387
Medium waves, 36
Meissner, 115
Metropolitan Opera, 27, 49
Microphones, 31, 94
Microwave intercity network, 213
MII recorder, 352
Millikan, Robert, 10
Milwaukee Journal, 177, 182

MIT, 392
Mixed highs (color television), 244
MMDS, 387
Modulation, amplitude
    ampliphase, 91
    Doherty amplifier, 89
    high-level plate, 90
    low-level grid, 89
    pre-electron tube, 42
    pulse width, 90
Monolithic integrated circuits, 492
Morecroft, John, 53, 115
Morita, Akio, 109, 458, 463
    biography, 456
Morse code, 29
Morse, Samuel S.B., 29
Morsey, Chase, 253, 441, 442
Mosbacher, Robert, 474
Motorola, 104
Mount Sutro Antenna, 228
MSOs, 371
    five largest, 386
Multiple system operators. *See* MSOs
MUSE (Multiple Sub-Nyquist
    Sampling Encoding) HDTV
    systems, 474, 475, 485
MUSE HDTV systems
    MUSE-6, 485
    MUSE-9, 485
    MUSE-E, 482
    narrow MUSE, 485
Must-carry rules, 369
Mutual Broadcasting System, 79, 143,
    184

NAB, 104
NAB conventions
    1956, 322
    1958, 332, 333
    1961, 335
    1964, 338
    1965, 287, 338
    1967, 287
    1968, 288
    1979, 296
Nally, Edward J., 54, 61, 62
NASA, 394, 395, 396, 397, 419
    abandons expendable launch
        vehicles, 422
National Aeronautics and Space
    Administration. *See* NASA
National Association of Broadcasters.
    See NAB

National Black Network, 143
National Broadcasting Company. *See* NBC
National Bureau of Standards, 69
National Cable Television Association. *See* NCTA
National Community Television Association, 366
National Public Radio, 143
National Radio Conference, 64
National Radio Systems Committee. *See* NRSC
National Rural Telecommunications Co-op, 430
National Stereophonic Radio Committee Standards, 149
National Telecommunications and Information Administration (NTIA), 412
NATPE (National Association of Television Production Executives), 208
Naval Wireless Communications, 50
NBC, 66, 151, 155, 177, 188, 273, 288–289, 322, 352
  defection of stars to CBS, 78
  founding, 72
  initial television programming, 182
  Red and Blue networks, 78
NBC radio network, 143
NBC Symphony Orchestra, 93
NCTA
  founding, 366
Networks, radio
  decline of, 100
  founding, 72
  heyday, 74
  history, 78
  renaissance, 143
  role, 74
New York Port Authority, 226
New York Theological Seminary, 383
New York Times, 184, 373
New York Yacht Club, 381
NHK, 458
  HDTV activity, 475, 487
Nipkow disk, 158, 162, 237, 255
Nipkow, Paul, 161
Nipper, 23, 25
Nixon, Richard, 412
Norman Bridge Laboratory (Cal Tech), 10
Norton, Kenneth, 130–136

Noyce, Robert, 492
NPA Order M-90, 267
NRSC (National Radio Systems Committee), 104, 111
  AM radio improvement recommendations, 105
NTSC (National Television Systems Committee)
  color standards, 271, 272
  color sytem, 237
  European demonstrations, 275
  formation, 263
  membership, 268
  monochrome standards, 184, 185
  NTSC and RCA color systems, 268

Objectionable interference, AM, 84
Oersted, Hans Christian, 2
Offset carrier technique, 193, 194
Ohm, George, 2, 3
Ohm's law, 3
Olson, Harry F., 93, 305, 306, 313
"One fluid" electrical theory, 3
"Open-skies" policy, 409
Orbital arc
  Canada, 406
  Mexico, 406
  prime (map), 407
  prime (table), 406
  satellite spacing, 407
Orbital slots
  assignment, 409
  authorized, 1988 (map), 423
  authorized, 1988 (table), 410
  spacing, 407
Orion Satellite Corp., 416
Orthophonic records, 27
Owens College, 10

Painter, Patty, 242
PAL color system, 237, 275–276
Palapa satellite, 399
Paley, William S., 74, 155, 249, 267, 284, 303, 441
  biography, 76–78
  CBS postwar TV policy, 190–191
PanAmSat, 416
Patent agreement, RCA-Ampex, 330–332
Parsons, E.L., 360
Paumalu, Hawaii, 413
Pay-per-view services, 379
Pay-TV services, 379

People That Love (PTL), 383
Peoria, Illinois, 210
Percy, Charles, 337
Peterson, Peter G., 337
Petrucci, Steven, 432
Pfost, Fred, 313, 327
Phase alternating lines. *See* PAL
Philadelphia Inquirer, 184
Philco, 70, 176, 181, 184, 366
Philips laser disk, 449
Philips-Bosch, 348
Philosophical magazine, 6
Phonographs, early
  Edison, 20
  mechanical, 18, 21
Photoconductive tubes, 160
Photoemissive tubes, 160
Phototape (RCA), 440
Picture quality, elements, 477
  aperture response, 478
  aspect ratio, 478
  bandwidth, 481
  colorimetry, 478
  contrast ratio, 477
  picture definition, 478
  scanning lines, 480
  signal-to-noise ratio, 477
  viewing ratio, 480
Pierce, G.W., 53
Pittsburgh, University of, 49
Plotkin, Harry, 247
Plumbicon, 160, 186
Plumbicon color cameras, 286–289
Poisson, 3
Polarization, radio wave, 7, 8
Pollack, Roy, 439, 449, 450, 451
Poniatoff, Alexander M., 312, 317, 318
  biography, 311
Poniatoff award, 328
Poplar Bluff, Missouri, 366
Pottsville, Pennsylvania, 366
Poulsen, V., 105
Praise the Lord. *See* PTL
Pratt, Dana, 334
Preemphasis/deemphasis, 138
Primary colors, 238
Product design, 214
Production practices, television, 215
Propagation, radio wave, 35
PTL (Praise the Lord or People That
  Love), 383
Pupin, Michael, 53, 115, 151
Pye, 255

Quadruplex recorders, 460
  Ampex breakthrough, 315
  Ampex development program, 317
  Ampex product design, 323
  critical technologies, 316
  demise, 343
  high-band recorders, 338–339
  initial Ampex in-house
    demonstration, 320
  initial RCA competition, 332–334
  transistorized recorders, 334–335
Quello, James, 369

Radio Act of 1912, 45, 63
Radio Act of 1927, 65, 79
Radio Corporation of America. *See*
  RCA
Radio networks
  decline, 100
  founding, 72
  heyday, 74
  history, 78
  renaissance, 143
  role, 74
Radio trust
  breakup, 98, 99
  conflicts, 69–70
  formed, 54–55
  patent disputes, 70
Radiotelegraphy, 1
Radiotelephony, 1
RARC (Regional Administrative
  Radio Conference), 408
  DBS meeting, 431–432
Rate regulation, satellites, 410
Rayleigh, Lord, 5, 10, 15
RCA, 88, 89, 98, 113, 116, 121–124,
  125, 126, 132, 136, 165, 166, 168,
  178, 188, 267, 301, 366, 411, 412,
  475
  1932 consent decree, 99
  acquires Victor, 99
  Ampex competition, 329
  Ampex patent agreement, 330–332
  antitrust charges, 73
  Armstrong patent battle, 150–154
  begins commercial TV
    broadcasting, 184
  Broadcast Equiment Division, 194,
    272, 287–288, 310, 328, 329, 334,
    353
  early television development, 176
  founded, 54

halcyon years, 275
initial rejection of FM, 122
longitudinal video recorder,
    307–309
patent pool, formed, 14
patent pool, opened to
    licensees, 73
quadruplex video recorders, 329,
    338, 339
Receiver Division, 273
share of equipment market, 203
withdraws from videodisk player
    market, 439, 451
World's Fair 1939 television
    broadcast, 182, 183
RCA Alascom, 418, 419
RCA Americom, 418, 419, 421
satellites, 402
Satcom III loss, 425
versus Western Union, 423
RCA Communications, 245, 413
RCA Laboratories, 193, 323, 441
RCA videodisk, 447–451
design and production, 450
marketing, 450
marketplace response, 451
technology (drawing), 449
technology (table), 448
withdrawal from market, 451
Reagan, Ronald, 412
Receivers
AM, 66, 96
FM, 147
regenerative, 42, 43
superheterodyne, 43
TV, color, 281–284
TV, monochrome, 231–236
Recorders, audio disk
16-inch, 95
33 1/3 rpm LP, 96
45 rpm, 96
78 rpm, 96
compact disks (CDs), 149
Red Seal records, 27
Reflection, radio wave, 7
Refraction, radio wave, 7
Regional Administrative Radio
    Conference. See RARC
Reis, Philip, 301
Reiss, Jeffrey, 379–380
Repeater, relay, 29
Righi, Auguste, 33
Ring, Andrew D., 82

RMA, television standards, 180–181
Roberts, William, 337, 340
Robertson, A. Willis, 383
Robertson, Pat, 383
Rockets (expendable), 422
Roosevelt, Franklin D., 53, 178
Roosevelt, Theodore, 51
Rose, Albert, 186, 278
Rosencrans, Robert, 376
Rosing, Boris, 174
Rowe, W.E., 279
Royal Air Force, 392, 394
Royal Institution of Great Britain, 4,
    5, 10
Royal Military Academy, 4
Royal Society of Great Britain, 169
RTMA, 263
Rumford, Count, 5

Sacramento, California, 369
San Jose State, 313
Sarnoff, David, 52, 62, 74, 78, 98,
    113, 116, 122, 172, 173, 176, 177,
    180, 182, 184, 249, 251, 261, 264,
    265, 273, 279
1947 speech to NBC affiliates, 155
Armstrong patent battle, 150–154
biography, 75
birthday present stolen, 310
birthday presents, 304
founds NBC, 72
"Radio Music Box" memo, 54, 61
RCA founded, 54
"Titanic" loss, 47
Transfers to RCA, 54
Sarnoff, Robert, 251, 253, 442
Satcom III loss, 411
Satcom satellites. See Satellites,
    launches
Satellite Business Systems. See SBS
Satellites
components, 402
frequency allocations, 404
geosynchronous, 393
launches, 1974–1979 (table), 417
launches, 1980–1983 (table), 420
launches, 1984–1988 (table), 422
regulation, 407
signal transmission modes, 401
Satellites, television applications
ABC, 437
ad hoc networks, 435
CATV, 423–426

CBS, 437
ENG, 434, 435
NBC, 436
PBS, 436
program syndication, 434
table, 434
Satellites, radio applications
ABC, 437
AP radio, 437
CBS, 437
FM program distribution, 150
Mutual Broadcasting System, 437
NBC, 437
NPR, 437
RKO, 437
Saticon, 160
Sauter, Jack, 439
SBS, 420, 421, 436
Scanners, 158
Schade, Otto, 475, 476, 479
Schlafly, Hubert (Hub), 372
Schoenberg, Isaac, 175
Schumann-Heink, Ernestine, 27
Scientific-Atlanta, 376, 428, 429
SCORE satellite, 395
Scrambling, 428, 429
Scull & Johnson, 22
Scull Machine Shop, 22
SECAM, 275–276
Seely, Stuart L., 152
Selectavision, 441–443
Seligman, J & W, 27
Sequential Colour Avec Memoire. See
    SECAM
Service Electric Cable, 378
Set-top converters, 364
Shadow mask color kinescopes,
    277–281
Shapp, Milton, 366
Sharer, Kevin, 433
Sheraton Park Hotel, 335
Sherlock, Michael, 352
Shockley, William B., 15
Shortwaves, 36
"Showtime," 379, 380
Silver City, New Mexico, 373
Simpson Thacher & Bartlett, 376
Skouras, Spyro, 372
Sky waves, 36, 85
Slaby-Arco, 46
SMARTS, 434
Smith, David B., 263, 267
Smith, Theodore A., 194, 331

SMPTE, 305, 330, 484
D-1 format, 357
D-2 format, 357
helical recording formats, 350
Type A format, 351, 357
Type B format, 351, 357
Type C format, 350, 351, 356, 357
Type E format, 352, 357
Society of Motion Picture and
    Television Engineers. See SMPTE
Sony, 348, 352, 353, 461, 476, 493
8mm format introduction, 469
Beta VCR format introduction, 463
Beta VCR format withdrawn, 467
Beta-VHS format competition,
    465–467
color cameras, 296–298
digital video recorders, 356
founding, 458
Morita, Akio, biography, 456–460
Type C recorder format
    compromise, 351
MCA copyright battle, 464
U-Matic video recorder, 352, 443
Sony Corporation of America, 459
Sotheby Parke Bernet, 411
Sound
science of, 15
sound sources, frequency
    ranges, 19
Southern Baptist Convention, 383
Space shuttle, 396–399
Spark transmitters, 38
Spectrum, electromagnetic, 7, 8
Speyer & Co., 27
Sponable, Earl, 372
Sputnik satellite, 394–395
St. Paul, Minnesota, 369
Standards of Good Engineering
    Practice
AM radio, 82
FM radio, 138
monochrome TV, 185
Stanford University, 313
Stanton, Frank, 190, 242, 264, 266,
    267, 303
biography, 254
Stereo, AM, 104
Stereo, effect on FM radio, 143
Stereo, FM, 148
Stereophonic effect, 18
Sterling, George, 246
Stolaroff, Myron, 317

Storage tubes, television, 158
Stromberg Carlson, 70
Studio cameras, monochrome,
    216–219
Studios, AM radio, 93, 105
Subcarrier, color, 268–269
Subtractive color systems, 238
Sullivan, Ed, 266
Super VHS format, 468
Superheterodyne circuit, 114–115
Superstation concept, 381
Sykes, Eugene O., 80
Sylvania, 267
SYNCOM satellites, 399
Syndex (syndicated exclusivity), 370

Tainter, Charles, 20
Taylor, Archer, 366
Taylor, Edward, 382
Taylor, John, 382
Telcos versus CATV systems,
    389–340
Teldec (Telefunken, Decca), 440, 442
Telefunken, 14, 115, 172, 175, 442
Telegraph circuits
    first transatlantic, 29
    first transcontinental, 29
Telegraphy, 1, 28
Telephone receivers, 32
Telephony, 1
TelePrompTer Corporation, 372, 373
Television
    all-electronic systems, 168–172
    bandwidth requirements, 159, 178
    blanking pulses, 179
    broadcasting channel
        specifications, 180
    cameras, monochrome, 158
    cochannel interference, 193
    commercial television begins, 188
    costs, operating and capital, 189
    early experimental stations, 168
    early programming, 189
    frame rate, 178
    industry statistics, 1948, 156
    industry statistics, 1946–1987
        (graphs), 205–207
    mechanical systems, 161, 167
    modulation method, 179
    postwar on-air TV stations (table),
        193
    postwar receiver manufacture, 194

scanning lines, number, 178
scanning principle, 158, 159
slow postwar start, 189
standards, 178–185
synchronizing signals, 179
time sales, 1947–1949, 194
transmitter systems, 221
video waveform, 179
Television allocations
    postfreeze, 195
    pre-freeze, 181
Television assignment policies
    effect on broadcasting, 210
    postfreeze policies, 197–202
    pre-freeze policies, 197
    priorities, 198
Television homes, 203
Television station design
    studio, 215
    system elements, 214
Tharp, James, 289
Thomson, Sir J.J., 5, 174
    biography, 9
Time Inc., 376
Tokyo Institute of Technology, 458
Tokyo Telecommunications Research
    Laboratories, 458
Topol, Sidney, 376
Toscanini, Arturo, 93
Toshiba, 344, 440, 443
Tracy, Edwin, 334
Trans-Video Corporation, 366
Transistors, 491
    effect on AM radio, 108
    impact on electronics, 15
    invention, 15
Transmitter systems, television, 221
Transmitters
    AM, 88
    electron tube, 40
    FM, 145
    spark, 38
    UHF, 228–230
    VHF, 221–230
Trevarthen, William, 288
Tricolor picture tubes, 277–281
    development, 279–281
    flat tension mask, 281
    production, 280
    prototypes, 279
    trinitron, 281
Trinity Broadcasting Network, 383
Trinity College, 10

Triode
  invention, 13
  patent litigation, 14
Tsushin Kogyo (Sony), 458
TT&C (tracking, telemetry, and
    command earth stations), 403
Turner, Ted, 382
  biography, 380
TV receivers, monochrome
  factory price trends, 236
  Japanese market invasion, 233
  "Mad Man Muntz," 233
  market, growth and decline, 233,
    236
  RCA market share, 232
  Zenith market share, 233
TVRO (television receive-only earth
    stations), 428, 429, 430
Twentieth Century Fox Film
    Corporation, 372
"Two fluid" electrical theory, 3

U-Matic video recorders, 352, 443,
    461
UA-Columbia, 376, 384
UHF antennas
  beam tilt, 231
  helical, 230
  null fill, 231
  pylon, 230
  V-Z, 230
UHF television
  all-channel law, 209
  CATV, effect of, 210
  FCC support, 209
  future of, 211
  profits, 207–210
  problems, 206
  revenues, 208
  stations on-air, 206
  programs, 208
  technical 210
UHF transmitters
  gridded tubes, 228
  klystron, 228–230
  power levels, 228, 229
Ultrahigh frequencies. See UHF
    entries
United Fruit, 55
United Independent Broadcasters
    (UIB), 78
United States Network, 143

Universal Picture–Sony copyright
    battle, 464
  appeal, 465
  initial decision, 465
  Supreme Court ruling, 465
Universal Pioneer, 453, 454
University of California, 313
University of Minnesota, 250
University of Pennsylvania, 376
USA Network, 384

Van Buren, Martin, 29
Vander Dussen, Neil, 353, 354
VCRs
  ancestry, 460
  competition to broadcasters, 204
  formats (table), 462
  market growth, 467
  technical parameters (table), 463
  technology, 460–466
  versus disks (graph), 444, 446
Venetian blind effect, 193
Vernon Valley, New York, 417
Vero Beach, Florida, 376
Very high frequencies. See VHF
Vestigial sideband FM (TV
    recording), 319–320
Vestigial sideband modulation, 179
VHD format (JVC), 443, 452
VHF signals
  diffraction, 120
  ionospheric transmission, 120
  location in spectrum, 119
  multipath distortion, 120
  reflection, 120
  refraction, 120
  propagation, 120
VHF antennas
  superturnstile, 225
  travelling wave, 226
VHF television
  growth, 202–204
  license demand, 202
  profits, 204–205
  revenues, 204–205
  stations on-air, 203
VHF transmitters, 221–230
VHS, 443
  manufacturers (table), 467
VHS versus Beta, 465–468
Viacom, 379

Victor Talking Machine Co., 23–28, 98, 156–157, 172
  acquired by RCA, 27, 99
Video cassette recorder. *See* VCRs
Video recorder. *See* Quadruplex recorders; Helical scan; VCRs
Video waveform, 179
VideoCipher, 428
Videodisk, 441, 442, 467
  failure in marketplace, 445–446
  manufacturers (table), 444
  RCA program, 447
  versus VCRs, 444
Vidicon, 160, 172, 221
Vienna Technical College, 255
Viewer's Choice, 380
Visual Electronics, 289
VLSI (very large scale integration), 283, 492
Volta, Alessandro, 2, 3
Von Braun, Werner, 394
VPR format, 350

WABC, 63
Waldorf-Astoria Hotel, 439
Walker, Paul, 246
Walson, John, 360
War Production Board, 130
WARC (World Administrative Radio Conference), 112, 431, 432
Wardman Park Hotel, 335
Warner Amex, 380
Washburn, Abbott, 431
Wasserman, Lew, 464
Watson, Thomas, A., 31
Watt, James, 2
Watts, Walter W., 194
WBAL-TV, 226
WBBM-TV, 185, 323
WBZ, 63
WCAU, 78
WCBS-TV, 185
WEAF, 66
Weather Bureau, 49
Weather Channel, 382
Weber, Ernst, 17
Weber's law, 17
Webster, E.M., 247
Welch, Leo, 413, 414
Westar satellites. *See* Satellite, launches
Western Electric, 88, 89, 459

Western Union, 376, 405, 417, 419, 421
Western Union International, 156, 413
Western Union versus RCA, 423
Westinghouse, 49, 55, 61, 62, 70, 91, 98, 115, 116
WFAA-TV, 226
WGN-TV, 382
Wheeler, Senator Burton K., 184
White, Abraham, 47
"White areas," radio coverage, 102
Whitehead, Thomas Clay, 412
Wilkes-Barre, Pennsylvania, 378
Wilson, Woodrow, 53
Wireless communications, 32
  transatlantic circuit, 33
  technical history, 35
Wireless World, 392
WJAS, 63
WJAZ, 64
WJZ-TV, 226
WLW, 500-kilowatt transmitter, 91, 93
WMAQ-TV, 299
WMAR-TV, 226
WNBC-TV, 185
WNEW-TV, 185
WOR-TV, 226, 382
World Administrative Radio Conference. *See* WARC
World Trade Center antenna, 226
World War I, effect on radio, 51
World War II
  effect on FM broadcasting, 128–130
  effect on TV technology, 186–188
  impact on AM radio, 87
World's Fair, 1939, 182
WPIX-TV, 382
WRGB-TV, 185
Wright, Paul, 52
WWJ, 61, 63

Yale Law School, 383
Yankee Network, 125
Young, Owen D., 54

Zenith Radio, 64, 70, 73, 124, 129, 135, 136, 151, 177, 178, 281
Zworykin, Vladimir, 155, 169, 170–172, 174–175, 257